PHYSICS AT THE HIGHEST ENERGY AND LUMINOSITY
To Understand the Origin of Mass

THE SUBNUCLEAR SERIES

Series Editor: **ANTONINO ZICHICHI**, *European Physical Society, Geneva, Switzerland*

1. 1963 STRONG, ELECTROMAGNETIC, AND WEAK INTERACTIONS
2. 1964 SYMMETRIES IN ELEMENTARY PARTICLE PHYSICS
3. 1965 RECENT DEVELOPMENTS IN PARTICLE SYMMETRIES
4. 1966 STRONG AND WEAK INTERACTIONS
5. 1967 HADRONS AND THEIR INTERACTIONS
6. 1968 THEORY AND PHENOMENOLOGY IN PARTICLE PHYSICS
7. 1969 SUBNUCLEAR PHENOMENA
8. 1970 ELEMENTARY PROCESSES AT HIGH ENERGY
9. 1971 PROPERTIES OF THE FUNDAMENTAL INTERACTIONS
10. 1972 HIGHLIGHTS IN PARTICLE PHYSICS
11. 1973 LAWS OF HADRONIC MATTER
12. 1974 LEPTON AND HADRON STRUCTURE
13. 1975 NEW PHENOMENA IN SUBNUCLEAR PHYSICS
14. 1976 UNDERSTANDING THE FUNDAMENTAL CONSTITUENTS OF MATTER
15. 1977 THE WHYS OF SUBNUCLEAR PHYSICS
16. 1978 THE NEW ASPECTS OF SUBNUCLEAR PHYSICS
17. 1979 POINTLIKE STRUCTURES INSIDE AND OUTSIDE HADRONS
18. 1980 THE HIGH-ENERGY LIMIT
19. 1981 THE UNITY OF THE FUNDAMENTAL INTERACTIONS
20. 1982 GAUGE INTERACTIONS: Theory and Experiment
21. 1983 HOW FAR ARE WE FROM THE GAUGE FORCES
22. 1984 QUARKS, LEPTONS, AND THEIR CONSTITUENTS
23. 1985 OLD AND NEW FORCES OF NATURE
24. 1986 THE SUPERWORLD I
25. 1987 THE SUPERWORLD II
26. 1988 THE SUPERWORLD III
27. 1989 THE CHALLENGING QUESTIONS
28. 1990 PHYSICS UP TO 200 TeV
29. 1991 PHYSICS AT THE HIGHEST ENERGY AND LUMINOSITY: To Understand the Origin of Mass

Volume 1 was published by W. A. Benjamin, Inc., New York; 2-8 and 11-12 by Academic Press, New York and London; 9-10, by Editrice Compositori, Bologna; 13-29 by Plenum Press, New York and London.

PHYSICS AT THE HIGHEST ENERGY AND LUMINOSITY
To Understand the Origin of Mass

Edited by
Antonino Zichichi
European Physical Society
Geneva, Switzerland

PLENUM PRESS • NEW YORK AND LONDON

Library of Congress Cataloging-in-Publication Data

Physics at the highest energy and luminosity : to understand the
 origin of mass / edited by Antonino Zichichi.
       p.   cm. -- (Subnuclear series ; v. 29)
    "Proceedings of the Twenty-Ninth Course of the International
 School of Subnuclear Physics ... held July 14-22, 1991, in Erice,
 Sicily, Italy"--T.p. verso.
    Includes bibliographical references and index.
    ISBN 0-306-44301-5
    1. Particles (Nuclear physics)--Congresses. 2. Solar neutrinos-
 -Congresses. 3. Supercolliders--Congresses. I. Zichichi,
 Antonino. II. International School of Subnuclear Physics (29th :
 1991 : Erice, Italy) III. Series.
 QC793.P49 1992
 539.7'2--dc20   92-26658
                                                                CIP

Proceedings of the Twenty-Ninth Course of the International School of Subnuclear Physics
on Physics at the Highest Energy and Luminosity: To Understand the Origin of Mass,
held July 14–22, 1991, in Erice, Sicily, Italy

ISBN 0-306-44301-5

© 1992 Plenum Press, New York
A Division of Plenum Publishing Corporation
233 Spring Street, New York, N.Y. 10013

All rights reserved

No part of this book may be reproduced, stored in a retrieval system, or transmitted
in any form or by any means, electronic, mechanical, photocopying, microfilming,
recording, or otherwise, without written permission from the Publisher

Printed in the United States of America

PREFACE

During July 1991, a group of 99 physicists from 57 laboratories in 27 countries met in Erice for the 29th Course of the International School of Subnuclear Physics. The countries represented were: Algeria, Argentina, Austria, Brazil, Canada, China, France, Germany, Greece, India, Ireland, Israel, Italy, New Zealand, Norway, Pakistan, Poland, Portugal, Rumania, Spain, Sweden, Switzerland, Thailand, Turkey, The Union of Soviet Socialist Republics, the United Kingdom, and the United States of America. The School was sponsored by the European Physical Society (EPS), the Italian Ministry of Education (MPI), the Italian Ministry of University and Scientific Research, the Sicilian Regional Government (ERS), and the Weizmann Institute of Science.

The opening lecture of the School was given by Professor Lev Okun. A few remarks are in order. In the pre-Gorbachev era, Professor Okun's case was the most difficult: in spite of many invitations he was not allowed to come and lecture at Erice. Nowadays the hard times have nearly been forgotten. It is with pleasure that I recall here a discussion I had with my friend Lev Okun in my house during one of his rare visits to CERN. The after-dinner topic was Galileo Galilei and his great discovery; i.e., the equality between gravitational and inertial masses - a discovery that we are celebrating now, four hundred years later. Here is a synthesis of Professor Okun's position: given a massive particle, the only quantity we should consider when talking about masses, is the square of the quadrimomentum, q^2, a relativistic scalar. The so-called gravitational and inertial masses confuse the basic issue. For "the most strenuous defender" of Galileo Galilei - as the great John Bell used to call his friend Nino - this needed a full after-dinner debate. In fact, the first man on this planet to discover that gravitational and inertial masses moving with $v \ll c$ are the same, was Galileo Galilei. Galilei was also the first man who formulated the principle of relativity in such a way to include "all possible experiments". Now it should be clear to all of you why the opening lecture on "The Problem of Mass, from Galilei to Higgs" could only be given by Lev Okun.

As usual, the Course dealt with a set of hot topics in Subnuclear Physics. For example, Jet Rates and Massive Hadron Production in QCD, discussed by Yuri Dokshitzer. The new e^+e^- data from LEP were presented and discussed in detail by Guido Altarelli. Starting from LEP data, John Ellis lectured on the fascinating topic of what could be the Physics beyond the Standard Model. The 200 TeV frontier was visited by Roberto Peccei from the theoretical viewpoint and by William Barletta for the machine properties: maximising energy and luminosity. Supersymmetry with Strings and Fivebranes were the topics of Sergio Ferrara and Michael Duff. The highest Energy available in this Century for experimental physics was the content of John Peoples' lecture, while the SSC project and experimental programme were presented by Fred Gilman.

Without detectors it is impossible to do physics: this is why the LAA Project was implemented in 1987. The LAA Project has been fully reported in previous courses (1989-1990). Furthermore the presentation of technological advances is, necessarily, boring. But the discussion is not, and therefore it is reported.

The 17 keV neutrino problem is now over, but it is interesting to read the lecture by Donald Perkins and the discussion. The time variation of the solar neutrino (Ira Rothstein) and the possible production of a Higgs boson in a photon-photon collider (Doug Borden), were two special seminars added to the School programme.

A distinctive feature of the Erice School is the informal and lively discussion sessions. Thanks to the work of the Scientific Secretaries these have been reproduced with high fidelity. I hope the reader will enjoy the discussions as much as those who were physically present.

Finally my deep gratitude to all my friends and collaborators, in Erice and Geneva, whose efforts and commitment to their work have played a vital role in making the Subnuclear Physics School what it is.

Antonino Zichichi

CONTENTS

OPENING LECTURE

The Problem of Mass: From Galilei to Higgs
  L. Okun   1

QUANTUM CHROMO DYNAMICS

QCD Phenomenology: Jet Rates and Truncated Parton Cascades for Massive Hadron Production
  Yu. L. Dokshitzer   25

THEORETICAL LECTURES FROM 10 to 200 TeV

The Standard Model and Beyond
  J. Ellis   49

Do Weak Interactions become Strong at High Energy?
  R. D. Peccei   89

Geometry and Quantum Symmetries of Superstring Vacua
  S. Ferrara   131

A Duality Between Strings and Fivebranes
  M. J. Duff   169

REVIEW LECTURES

Theoretical Implications of Precision Electroweak Data
  G. Altarelli   209

Novel Neutrino Physics
  D. H. Perkins   251

A Solution to the Time Varying Solar Neutrino Problem
  I. Z. Rothstein   279

Searching for the Higgs Boson at a Photon-Photon Collider
  D. L. Borden   303

Experimental Physics at the Highest Energy (in this Century!)
  J. Peoples   323

THE FUTURE OF HIGH ENERGY PHYSICS

The SSC Project and Experimental Program
  F. J. Gilman   351

Maximizing the Luminosity of Eloisatron, a Hadron Supercollider
 at 100 TeV per Beam
  W. A. Barletta   367

New Detectors for Supercolliders : LAA
  A. Zichichi   387

CLOSING CEREMONY

Prizes and Scholarships, etc   399

Participants   403

Index   415

THE PROBLEM OF MASS: FROM GALILEI TO HIGGS

L.Okun

ITEP, 117259, Moscow, USSR

The subject of my talk was chosen by Professor Zichichi and the title was formulated by him. I have never given such historical talks before: about four hundred years of historical development in some field. Therefore I will stick to physics. My talk will consist of ten parts:
1. Introduction
2. The mass of a particle
3. The mass in XVI-XVII centuries
4. The mass in XVIII-XIX centuries
5. The mass of a system of free particles
6. Interactions and binding energy
7. Quarks and gluons confined in hadrons
8. The pattern of masses of elementary particles
9. Scalars and the nature of mass
10. Conclusions.

As you can see, there will be only two historical parts.

1. INTRODUCTION

The great Galilei started our science, physics, by fighting the prejudices and by bringing to this science its main constituents: observation, experiment, mathematics — and last but not least — common sense.

All these elements could be traced back in the works of his predecessors, but I will not do it. Books about Galilei start usually by castigating Aristotle as a silly person. I will break with this tradition. On the other hand, I will not praise Lucretius Carus whose ideas were often rather close to those ideas of Galilei.

What I would like to do first of all is to define what we are speaking about: to make clear what we mean when we say the word "mass", to make clear the relations between mass and energy, between mass and gravity. It seems to me that only by defining notations, terminology, language would it be possible to go in fifty minutes from Galilei to Higgs.

In the booklet about Erice, which all of us were given by hospitable organizers of the school, you can read that "Erice is a singular synthesis of myth and history ... of fantasy and reality". This applies also to the subject of my talk.

The main message of my talk is that there is only one mass. This message is connected with the poster of our school, according to which $m_i = m_g$. I will insist that there is no m_i, no m_g, no other "indexed" masses, there is only one mass m.

2. THE MASS OF A PARTICLE

In special relativity the mass is defined as a Lorentz invariant quantity defined by the square of the energy-momentum four-vector of a body:

$$m^2 = E^2 - \vec{p}^{\,2}$$

(the velocity of light is used as the unit of velocity).

The velocity of a body is defined by its momentum and energy:

$$\vec{v} = \vec{p}/E.$$

If $m = 0$ (the massless particle) then $|\vec{p}| = E$ and hence $v \equiv |\vec{v}| = 1$.

If $m \neq 0$, then for a body at rest we have

$$v = 0, \quad \vec{p} = 0, \quad m = E_0,$$

where E_0 is the rest energy. By using our starting equations it is easy to express energy E and momentum $\vec{p}$ of a massive particle as functions of its velocity $\vec{v}$:.

$$E = m\gamma,$$

$$\vec{p} = m\vec{v}\gamma,$$

where

$$\gamma = 1/\sqrt{1 - v^2}.$$

It is important to stress that the difference between energy E and the rest energy E_0 is the kinetic energy T:

$$T = E - E_0.$$

When $v \ll 1$, then from

$$(E - m)(E + m) = \vec{p}^{\,2}$$

it immediately follows that

$$T = E - m \simeq \vec{p}^{\,2}/2m$$

and

$$\vec{p} \simeq m\vec{v}.$$

Now, there exists relation between force and the time derivative of momentum:

$$\vec{F} = \frac{d\vec{p}}{dt}$$

which is valid at arbitrary values of v.

In the case of $v \ll 1$,
$$d\vec{p}/dt = md\vec{v}/dt = m\vec{a}.$$

In the non-relativistic case, therefore,
$$\vec{F} = m\vec{a},$$

and, hence, the mass is the measure of inertia: the smaller the mass of a body the easier to accelerate it with a given force.

In the general case, which has to be considered when v is of the order of unity, the differentiation of $\vec{p} = m\vec{v}\gamma$ gives a more complex equation:
$$\vec{F} = \frac{d\vec{p}}{dt} = m\gamma[\vec{a} + \vec{v}(\vec{v}\vec{a})\gamma^2],$$

or, taking into account that $m\gamma = E$,
$$\vec{a} = [\vec{F} - (\vec{F}\vec{v})\vec{v}]/E.$$

You see that in the general case $\vec{a}$ is not directed along $\vec{F}$, and neither m nor E play the role of the measure of inertia.

Of course, one can define velocity and acceleration of a particle in a four-dimensionally covariant way by using instead of t the proper time τ ($\tau = t/\gamma$):
$$u^i = dx^i/d\tau , \quad \dot{u}^i = du^i/d\tau .$$

Then the energy-momentum four-vector p^i is proportional to the velocity:
$$p^i = mu^i ,$$

and the four-vector of fource f^i is proportional to the acceleration:
$$f^i = dp^i/d\tau = m\dot{u}^i .$$

However the clocks usually do not accompany particles in their flights, they are at rest in the laboratory. Therefore we have to express $u^i, \dot{u}^i$ and f^i through $\vec{v}, \vec{a}$ and $\vec{F}$ — the quantities that one actually measures:
$$\vec{u} = \vec{v}\gamma , \; u^0 = \gamma ;$$
$$\dot{\vec{u}} = \gamma^2[\vec{a} + \vec{v}(\vec{v}\vec{a})\gamma^2] , \dot{u}^0 = (\vec{v}\vec{a})\gamma^4 ;$$
$$\vec{f} = \gamma\vec{F} , \; f^0 = m(\vec{v}\vec{a})\gamma^4 .$$

This brings us back to the equations we started with.

In many books you may read about relativistic mass $m = E/c^2$ which increases with velocity, about rest mass $m_0 = E_0/c^2$, about equivalence between mass and energy. Many authors even use the term mass-energy. This terminology is extremely wide-spread and extremely misleading. It stems from the tradition of keeping the non-relativistic relation for relativistic particles: $\vec{p} = m\vec{v}$ instead of $\vec{p} = m\gamma\vec{v}$. Please, don't use this archaic jargon. Use the professional language, which is adequate and precise.

Now let us consider the gravitational interaction of a particle. The key idea of general relativity — the classical theory of gravitation — is to postulate the universality (mass independence) of gravitational acceleration, as proved with increasing accuracy by the experiments of Galileo, Newton and Eötvös. For a non-relativistic particle the coupling to gravitational field is proportional to its mass. However, for a relativistic particle this coupling is proportional to energy-momentum tensor T^{ik}. This tensor is symmetric and therefore has ten components. Gravitational properties of a particle are thus described in a general case not by one quantity — mass, but by ten quantities — components of energy-momentum tensor.

I will not go into full discussion of general relativity, instead I will consider a simple case of a relativistic particle in a weak gravitational field. In this case the energy-momentum tensor T^{ik} may be presented in the form

$$T^{ik} = mu^i u^k \frac{d\tau}{dt} = \frac{p^i p^k}{E},$$

where, as before,

$$p^i = mu^i, \quad u^i = \frac{dx^i}{d\tau},$$

τ - the proper time of a particle, $dt/d\tau = \gamma$. I have omitted a factor which is non-essential to us, namely, $\delta(\vec{r} - \vec{R})$, (see "Classical Field Theory" by L.Landau and E.Lifshitz, sections 87 and 7).

When $v = 0$, the only non-vanishing component of T^{ik} is $T^{00} = m$, where m is just the same mass we have considered before. $T^{00} = m$ remains the dominant component also for $v \ll 1$. In this nonrelativistic limit the gravitational interaction between two bodies of masses M and m is described by the Newtonian potential

$$V = -G_N \frac{Mm}{r}$$

where G_N is the Newtonian gravitational constant:

$$G_N \simeq 6.7 \cdot 10^{-39} GeV^{-2}$$

(we use units in which $\hbar, c = 1$).

In modern physics, there is only one mass m which at low velocities determines both the inertial and gravitational properties of a body ($m_i \equiv m_g \equiv m$). As for high velocities, the mass does not determine either inertia or gravity. Therefore, the notions of m_i and m_g are redundant for slow particles and are irrelevant and even false for fast particle.

3. THE MASS IN XVI–XVII CENTURIES

Now, after giving a short review of modern "mass terminology", I am going to state that Galileo Galilei, in full agreement with modern science, has never written the equation $m_i = m_g$. We have no $m_i = m_g$ today and there was no $m_i = m_g$ four hundred years ago. (It was in between).

The present Galilei festivities are connected in time with the period (1589–1592) when he was a young professor at Pisa. Before that, in 1586, he wrote "La Bilancetta" ("The Small Balance"), a manuscript in which he considered a scale with weights immersed in various liquids. Due to this work a position of professor was given to him in Pisa (with a very modest salary).

In 1590 he prepared a collection of manuscripts dealing with mechanical motion under the title "De motu antiquiora". As I understand, the word "antiquiora" referred to his own "old" papers (at that time he was 26 years old), not to antique papers of ancient authors.

There was no "publish or perish" rule at that time: many manuscripts were not printed at all. Fragments of "De motu" were incorporated later by Galilei into his books: "Le meccaniche" (written in 1593, published in French in 1634 and in Italian in 1649, seven years after Galileo had passed away), "Dialogo" (1632) and "Discorsi" (1638).

There is no definition or discussion of the notion of mass in any of Galilei's writings. As you can easily guess, I have not studied all of them (they were first published in twenty volumes of "Opere" a century ago). But I did read some of them. And I trust historians of physics, who contend that the term "mass" was unknown to Galilei and to other scientists of his time. No m_i, no m_g, no mass at all. Instead of the word "mass" he used the word "weight".

Historians assert also that dropping balls from the Pisa tower was only a gedanken experiment (such experiment was actually performed much later at St.Paul Cathedral in London and was described by Newton in "Principia" (1687)). What Galilei really did were experiments with balls rolling on an inclined plane. These were very ingenious experiments, with bells ringing when balls were passing certain distances.

What was Galilei's main result? His main result was that he established that g – the gravity acceleration – is independent of weight, size, material of a body. It was only approximately true in the air, but he postulated that this is an exact law for the free fall in vacuum. In modern language this means that there is no "fifth force", which is presumed to depend on the material of a body (on its baryon or lepton numbers, or proton-neutron ratio).

It was Newton who introduced the notion of mass. He defined mass as a quantity of matter. (This definition has repercussions even today. In many popular science books and articles, you may read about transformation of matter into energy, say, in electron-positron annihilation. But the fact is that energy is conserved and what is transformed are particles: electron and positron into photons.)

According to Newton, the mass of a body is proportional to the body's density and volume. This definition was considered by many physicists and philosophers as a kind of vicious circle, because it was not defined what density was. Some of them have proposed other definitions of mass. Others argued that identification of mass and matter by Newton implied in fact the atomic structure of matter. Without following this debate, let us note that this identification definitely implied conservation and additivity of mass.

Newton defined mass in "Principia" (1687) – in the main work of his life. Galilei was mentioned in the book ten times and with great respect. Newton first referred to Galilei in connection with the action at distance. (At that point he was arguing against Descartes). But he referred to Galilei mainly when describing parabolic trajectories in gravitational field. This parabola impressed Newton greatly. You can see it from the text of "Principia".

Looking from a distance of four centuries, one of the greatest achievements of Galilei is certainly the formulation of the relativity principle. This formulation is presented in "Dialogo" as a picture of a gedanken experiment in a cabin of a ship.

There are a lot of processes going on in the cabin: the water is dripping, fishes are swimming in an aquarium etc. Galilei insists that by observing all possible processes in a cabin with the windows covered one cannot tell whether the ship is at rest or in uniform rectilinear motion.

This picture is in fact much broader than what is called Galilean relativity or "non-relativistic relativity". In fact it has in it all of relativity. Indeed, there is light in the cabin, and the velocity of light is finite.

The velocity of light was first measured by Romer in 1676 by observing the motion of the satellites of Jupiter discovered by Galilei. As soon as you include the finiteness of velocity of signals in the picture of the cabin, you are immediately forced to generalize the "non-relativistic relativity" to full relativity in Einstein's sense.

The two great gedankenexperiments of Galilei were like two seeds. The campanile of Pisa was the seed of general relativity, the cabin of a ship was the seed of special relativity. The "Dialogo" and "Discorsi" were the cradle of the modern physics in general, and of the modern concept of mass, in particular.

4. THE MASS IN XVIII AND XIX CENTURIES

In order to trace the further evolution of the concept of mass I will mention here only three milestones.

First, it is the discovery by A.Lavoisier (1789) that mass is conserved and additive in chemical reactions.

Second, it is the discovery of the conservation of energy in the process of its transformation from its mechanical form into heat and vice versa (R.Mayer, 1842; H.Helmholtz, 1847).

Third, it is Maxwell's equations (1872) which have opened the direct road to special relativity.

These three discoveries are beyond mechanics and at first sight have little connection to the problem of mass. But I think that they are much more important for the real understanding of mass (remember: mass is equal to the rest energy) than many philosophical treaties on the definition of mass. Quite a number of them were written during these two centuries, discussing whether to define mass as a ratio of F over a, or consider it as a purely mathematical symbol in equations of mechanics, etc. But let us return from the history of physics to physics.

5. THE MASS OF A SYSTEM OF FREE PARTICLES

Let us consider n free particles. Both energy and momentum are conserved and additivie, therefore the total energy E and the total momentum $\vec{p}$ are

$$E = \sum_{i=1}^{n} E_i, \quad \vec{p} = \sum_{i=1}^{n} \vec{p}_i.$$

The word "additive" is extremely important when you speak about the difference between energy and mass. I am stressing this, because usually people somehow forget about this word which should be kept in mind.

The mass of a system of particles is expressed through E and $\vec{p}$ in the same way as the mass of a single particle:

$$m^2 = E^2 - \vec{p}^{\,2}.$$

As an example, let us consider the mass of a system of two photons:

$$m^2 = (E_1 + E_2)^2 - (\vec{p}_1 + \vec{p}_2)^2 = 2(E_1 E_2 - \vec{p}_1\vec{p}_2) = 2E_1 E_2(1 - cos\vartheta)$$

You see that when photons fly in the same direction, $\vartheta = 0, m = 0$. When photons fly in opposite directions, $\vartheta = \pi, m = 2\sqrt{E_1 E_2}$. In the rest frame of the system ($\vec{p} = 0$), the two photons fly in opposite directions with equal energies: $E_1 = E_2 = E_0/2$, so in that frame $m = E_0$: the mass of a system is equal to its rest energy. Please note that here E_0 is the rest energy of the system and that the rest energy of a photon is equal to zero.

Mass being equal to the rest energy, and energy being conserved, it is absolutely clear that mass is also conserved. That means that the mass of any isolated system before any reactions take place in it and afterwards is the same. But there is no additivity of mass. This is demonstrated by the two photons: each of them is massless, but the mass of the two photons may be zero or may be very large; it depends on the energy of the photons and on the angle between their momenta.

You see that the often claimed "equivalence" between mass and energy does not hold! Saying this, I fully realize that I contradict occasional statements of many a great physicist. But what I'm saying is not my invention. It is known to every professional theorist in particle physics. As for the great physicists, I believe that sometimes they did not care too much about dotting all the "i's" and crossing all the "t's", especially when addressing non-physicists.

It is true: whenever you have mass, you have energy. But you cannot reverse this statement. One may have energy and have no mass: an example is a photon.

For energy: conservation — yes! additivity — yes!

For mass: conservation — yes! additivity — no!

The energy is the time-like component of a four-vector in four-dimensional space-time. The mass is a scalar in the four-dimensional space-time. This is our professional language.

6. INTERACTIONS AND BINDING ENERGY

Let us consider now bound systems of particles. We have two long-range static potentials in Nature — Newton's and Coulomb's. We have Quantum Field Theory which describes all interactions including the Newton and the Coulomb potentials as the exchange of virtual particles. But the last statement should be taken with a grain of salt, because of an essential difference between these two potentials and all other potentials and forces. When you go to the limit $c \to \infty$, the former survive, while the latter (magnetic forces, Yukawa potential, exchanges of W-bosons and of gluons) die out.

Nature displays an impressive spectrum of bound systems: nuclei, atoms, molecules... galaxies... . Each of these systems has a certain mass defect which is equal to the difference between the sum of the masses of its constituents and the mass of a system as a whole. In the case of a nucleus, the constituents are nucleons. In the case of a galaxy, the constituents are stars and interstellar matter. With c as the unit of velocity, the mass defect is equal to the binding energy. The binding energy is defined as energy needed to pull the constituents of a body (at rest) apart.

As an extreme example of the mass defect, I shall consider the closed universe. For the universe as a whole the gravitational attraction of galaxies may compensate ("eat up") the sum of their masses (rest energies) and their kinetic energies. In this case the universe is closed and its total mass is equal to zero. For the observed velocities of Hubble expansion, the critical mean density above which the universe becomes closed is

$$\rho_c = 3H^2/8\pi G_N \simeq 10^{29}\,\mathrm{g/cm}^3.$$

Here G_N - is Newtonian constant and H is the Hubble constant: $H \simeq 100\,\mathrm{km/s} \cdot \mathrm{Mps}$.

The observed mean density of luminous matter is about $10^{-2}\rho_c$. There is strong evidence, however, that the actual density ρ is much larger. The evidence comes from the observation of motion of stars on the periphery of galaxies. These stars are moving too fast, thus indicating that the actual masses of galaxies are substantially larger than their visible masses. The same is true for clusters of galaxies. According to the inflationary cosmological scenario, $\rho \simeq \rho_c$ and the universe is flat.

When speaking about the Universe, one cannot avoid mentioning the problem of the energy density of vacuum, the so called cosmological term. According to all evidence, this energy density is vanishingly small. On the other hand, according to quantum field theory it must be tremendously large because of vacuum quantum fluctuations. This is one of the deepest mysteries of physics!

7. QUARKS AND GLUONS CONFINED IN HADRONS

One of the great pinnacles of physics is the discovery (or creation?) of Quantum Chromodynamics. It turned out that Quantum Field Theory can describe particles which have no mass in the normal sense, i.e. in the sense in which I used it before: mass as rest energy of a free particle. Quarks and gluons are never at rest and never free: they dash around inside hadrons, being jailed for life. Therefore, their masses cannot be defined in the usual sense. Although masses of quarks and gluons cannot be defined at large distances, they can be defined at short distances, deep inside their jail cells where they become asymptotically free. At short distances (large momentum transfers q), the role of the strong interaction becomes less and less important, thus we can see quarks almost naked inside their gluonic "fur coats". As for the mass of the "coat", it is about 300 MeV, being determined by the running of the strong coupling constant, $\alpha_s \sim 1/ln(\Lambda_{QCD}/q)$, and, thus, by the radius of confinement: $r_{conf} \sim 1/\Lambda_{QCD}$.

The Lagrangian of QCD is known, but the calculation of hadron masses is a very difficult problem. I will mention here two approaches to the problem: QCD sum rules and computer calculations on lattices.

The QCD sum rules are based on dispersion relations and asymptotic freedom. In this approach, dispersion relations serve as a bridge between the region of asymptotic freedom and the region of confinement. (I remind you that dispersion relations are based on two general principles: analyticity and unitarity of the S-matrix). With QCD sum rules, masses were calculated for a number of particles in the lower part of the hadron spectrum.

The space-time lattices allow calculations in the strong coupling limit and therefore are very promising.

8. THE PATTERN OF MASSES OF ELEMENTARY PARTICLES

The notion of elementarity is time-dependent. In 1991 the particles we consider as elementary are gauge bosons, quarks and leptons.

The "oldest" of the gauge bosons — photon — is "experimentally" massless. What do I mean by this? We usually consider it to be massless, but what we really know is that it is very light. The upper limits on its mass (in fact, lower limits for the Compton wave length) are given by astronomical observations of the magnetic fields of Jupiter and of the galactic magnetic field. These limits are very impressive, but we have no compelling theoretical reason, no theoretical principle that would require the photon to be massless.

Sometimes people say it is gauge invariance. But gauge invariance in the case of abelian symmetry is not sacred. For electrodynamics it means only that the photon is massless. It is tautology. An intrinsic anomalous magnetic moment of the electron is gauge invariant, but it would destroy the renormalizability of QED and the theory would become a mess. A tiny photon mass is gauge non-invariant, but it does not destroy renormalizability and its presence would not spoil the beautiful agreement between QED and experiment. The abelian gauge invariance is neither sufficient nor necessary.

Contrary to photons, gluons (and also gravitons) have to be massless on theoretical grounds: for the consistency of the theory. A tiny mass-term in QCD lagrangian makes that theory unrenormalizable because of ultra-violet divergences.

The masses of W- and Z-bosons essentially give us the Fermi scale, the scale that determines the masses of all known elementary particles. Another parameter which plays an important role in determining this scale is $\sqrt{\alpha_W}$, where α_W is the weak gauge coupling constant ($\alpha_W \sim 1/30$). As a result, the Fermi scale is in the ballpark of several hundred GeV.

Among all elementary fermions, the top quark seems to have the most natural value of the mass: of the order of Fermi scale. More precisely, on the basis of the data on electroweak radiative corrections and on the transitions $K^0 - \bar{K}^0$ and $B^0 - \bar{B}^0$ we expect, that m_t is somewhere between 100 GeV and 200 GeV. We have not yet discovered the top quark just because its mass is so natural.

The masses of lighter quarks are strongly suppressed with respect to Fermi scale:

$$m_b, m_c \sim \alpha m_t,$$

$$m_s \sim \alpha^{3/2} m_t,$$

$$m_d, m_u \sim \alpha^2 m_t.$$

We have no theory explaining this empirical pattern of suppression.

Charged leptons are lighter than their quark partners (down-components of quark doublets):

$$m_\tau < m_b, \quad m_\mu < m_s, \quad m_e < m_d.$$

Neutral leptons — neutrinos — are the most enigmatic particles: we know only upper limits on their masses. According to the 1990 Particle Data Booklet,

$$m_{\nu_e} < 17 eV \text{ at } 95\% \text{ CL},$$

$$m_{\nu_\mu} < 0.27 MeV \text{ at } 90\% \text{ CL},$$

$$m_{\nu_\tau} < 35 MeV \text{ at } 95\% \text{ CL}.$$

Many years ago it was a common opinion that neutrinos are massless because of γ_5-symmetry. This option is preserved by the Standard Model which deals only with massless left-handed neutrinos and right-handed antineutrinos. But at present we know that all discrete symmetries are violated in nature: C, P, CP, T. (Only gauge symmetries stand). Why then respect the discrete γ_5-symmetry? Also, with the advent of grand- and superunification, nobody is afraid anymore of very big or very small numbers in physics. Therefore, the vogue now is to believe that neutrino masses are small but non-vanishing. Unfortunately, I have no time to discuss these extremely interesting particles. I will mention only that they may have Majorana masses in addition to the usual (Dirac) masses.

In connection with CP and all that, let me remark that CPT is okay. The equality of particle and antiparticle masses was tested in the case of K^0 and $\bar{K}^0$ with fantastic accuracy: $\Delta m/m \simeq 10^{-18}$. For all other particles the accuracy is more than a billion times worse, and special experiments to improve it are underway, in particular, for the equality of masses of proton and antiproton and of their gravitational properties. They will get improved limits, but I personally do not expect that any mass-difference will be discovered.

9. SCALARS AND THE NATURE OF MASS

The nature of mass is the most important unsolved problem of modern particle physics. It is really fascinating that this problem may be solved (at least partially) by experiments on future colliders such as SSC or LHC.

Theoretically, the nature and the origin of mass are connected with the expected existence of scalar particles, among which higgses are the most popular ones. Thus, we finally arrived from Galilei to Higgs.

Professor Higgs, undoubtedly, will discuss scalars in great detail in his closing lecture. I therefore consider that my task is accomplished and I shall confine myself only to a few remarks.

The existence of elementary scalars — higgses — is one of several possibilities. Another possibility are composite scalars, for instance, scalar which is a bound state of $t\bar{t}$. They have been discussed very extensively during the last two years. There is also technicolor, in which mass appears due to a confinement mechanism similar to that in QCD. The technicolor model predicts heavy technihadrons (with masses above 1 TeV) and comparatively light pseudogoldstone bosons (technianalogs of pions). Unfortunately, we have no selfconsistent technicolor theory.

If scalars are elementary, then supersymmetric particles must exist. These superparticles cancel quadratic divergences connected with elementary scalars. Otherwise the Fermi scale would blow up to the Planck scale. The minimal standard model Higgs scalar cannot be very heavy, definitely not heavier than 1 TeV. Otherwise the theory would become strongly inconsistent at the level of logarithmic divergences.

If there are no scalars at all, there must exist a strong WW-interaction at energies between 1 TeV and 2 TeV.

The lightest scalars (elementary or not) and superparticles, if they exist, would be discovered at SSC and LHC. But the full exploration of Scalarland and Superland is the task for a bigger machine.

I would like to make here a comparison with the study of ordinary hadrons. To study production and decay of pions, several hundred MeV was enough. To produce strange particles and antiprotons, energies above 1 GeV were needed. To answer that need, the Cosmotron and the Bevatron were built in the 1950's. But even today our understanding of the strong and weak interactions of quarks is far from being complete, and we need machines of much, much higher energies. Thus, it turned out that Quarkland stretches from subGev- to TeV-energies.

Similarly, it is quite natural to expect that the Scalarland and Superland would stretch from subTeV- to PeV-energies. It is fine that such machine as Eloisatron, with energy in subPeV-range, is under active discussion. But it would be better if its energy were raised to 1 PeV.

The detailed study of physics in the range up to 1 PeV may allow extrapolation to physics at the deepest level, the level of the Planck scale. As you know, this scale is determined by the three most fundamental constants: $\hbar, c$ and G_N. At present we believe that it is the Planck scale, where superunification of all interactions, including gravity, takes place. Clues to macroscopic physics are hidden at atomic and nuclear scales. Clues to atomic and nuclear physics are hidden at TeV and PeV scales. Clues to TeV and PeV physics are hidden at Planck scale.

The graphs of the three running gauge coupling constants published recently in CERN Courier may be considered as a crude example of the extrapolation I'm speaking about. These graphs, which are based on the data of DELPHI Collaboration at LEP bring additional support to the observation made a decade ago, namely, that the existence of superparticles drastically improves the convergence of the three couplings to a common grand unification point. (A critical analysis of the experimental uncertainties is given by F. Anselmo, L. Cifarelli, A. Peterman and A. Zichichi in a paper submitted to "Il Nuovo Cimento"). In a similar way, the extrapolation from the PeV scale to the Planck scale may become possible.

10. CONCLUSIONS

Galileo Galilei was the first who limelighted acceleration. Acceleration is the key word, if we think about his fundamental contribution to physics. His accelerators were gravitational. Our accelerators are electromagnetic. We have to build them with the highest possible energies and luminosities in order to unravel the origin of mass and the most profound properties of matter, including gravity.

Chairman: L. Okun

*Scientific Secretaries: P. Hernandez, Z. Pluciennik,
K. Qureshi, U. Vikas*

DISCUSSION

– *U. Vikas:*

What according to you is the significance of the equivalence principle if you do not consider as any kind of a principle the equality of inertial mass and gravitational mass?

– *Okun:*

The local equivalence of gravitational field and accelerated reference frame does not become less significant if instead of using two letters with indices, m_i and m_g, you say that $m_i \equiv m_g \equiv m$ and use only one mass m.

– *U. Vikas:*

Now you are stating this almost like principle of equivalence. But how do we know that the mass involved in acceleration under a force different from gravitation is the same as that under the force of gravitation?

– *Zichichi:*

How do we know that the identity $m_i \equiv m_g \equiv m$ exists? The first man who discovered experimentally this equation was Galilei.

– *Okun:*

We know it from both experiment and theory. Experiments started by Galilei and continued by Newton have achieved very high accuracy during the XX century. As for theory, I refer here to special and to general relativity and to quantum field theory.

In the nonrelativistic limit, the acceleration of a body in a given gravitational field does not depend on its mass. You may call this mass inertial mass and denote it by m_i, but better call it simply mass and denote it by m. You may introduce gravitational mass m_g. Then you conclude from the above independence that m_g is proportional to m: $m_g = km$, where the coefficient k is arbitrary. You may choose, for instance, $k = 1/2$. Then you have to redefine the Newtonian gravitational constant G. However, it is better not to introduce m_g, not to introduce k, but directly conclude that the gravitational force is proportional to m. This immediately fixes the value of G. The indices i and g are thus redundant.

– Sivaram:

What will be the implications of a particle with a negative mass?

– Okun:

At first sight, the sign of the mass is irrelevant in relativistic quantum theory. In the case of the fermions, the unitary transformation:

$$\psi \to \gamma_5 \psi$$

changes the sign of mass in the Dirac equation. In the bosonic case, it is m^2 that enters the Klein-Gordon equation and hence the sign of m is unimportant. However, we should not forget about the relation between mass and energy:

$$m^2 = E^2 - \vec{p}^{\,2}.$$

The energy of a free particle is positive by definition: $E = +\sqrt{m^2 + \vec{p}^{\,2}}$. (The sign is the same for all particles, otherwise the vacuum would be unstable.) Therefore in order to preserve the equation $E_0 = m$ for the rest energy (in order not to write $E_0 = |m|$), we have to define mass as positive.

Forget for a moment about relativity and consider the non-relativistic relation $E_{kin} = mv^2/2$. With negative mass you have to change this equation into $E_{kin} = |m| v^2/2$. Those who don't want to change would obtain a terrible variant of perpetuum mobile: a particle which would accelerate itself by giving its energy to other, normal particles. But now remember relativity and the instability of vacuum in the case of negative rest energy.

– Sivaram:

Could you comment on the antiparticles behaving differently under gravitational force?

– Okun:

Experimentally, the equality of masses of particles and antiparticles have been checked with a very poor accuracy. Galilei taught us that we have to measure everything which can be measured and to make unmeasurable, measurable. If experimentalists can do the appropriate experiments, they should do them.

The inequality of masses of particles and antiparticles would involve the breaking of CPT-symmetry which in a consistent theory is difficult to imagine. Therefore I don't expect that such inequality would be discovered.

– Ellis:

I would agree with you if you restrict your attention to only a graviton exchange of spin 2, but these experiments would be also sensitive to the exchange of vector particles, which would change the sign of the coupling.

– *Okun:*

Yes, that would imply the fifth force, which is definitely very interesting to observe. What I meant is difference of masses, not attraction for particles and repulsion for antiparticles.

– *Sivaram:*

Would you comment on a massive graviton? I believe there is a result which states that, in the case of a massive graviton, the deflection of light in a gravitational field would be 3/4 of the Einstein value.

– *Okun:*

I do not know this specific result. Anyway, a massive graviton because of its longitudinal components would make the theory badly divergent. There is no continuous limit $m \to 0$ for graviton.

– *Junk:*

How would the gravitational force change when a particle is embedded in some medium, e.g., electron in a piece of silicon or neutrino in the sun?

– *Okun:*

I would say that there is nothing fundamental that happens to the gravitational force in a medium. There is no fundamental difference between an elementary particle embedded in silicon and a macroscopic body immersed in water. Both are not free, owing to the electromagnetic interaction with the medium.

– *Junk:*

It seems it might be more fundamental because of the quantum mechanical wave length.

– *Okun:*

A solar neutrino has different wave lengths inside and outside the sun. This is described by the refraction index in the same way as for a photon inside a piece of glass. But I don't see any fundamental problem here as far as gravity is concerned.

– *Rothstein:*

Why is the abelian QED with a massive photon renormalizable, while the non-abelian QCD with a massive gluon is not?

– *Okun:*

This is connected with the emission of longitudinal components of the gauge bosons. In the abelian case, the corresponding amplitudes are suppressed by a factor m/E, where m is the mass of the boson and E its energy. In the non-abelian case, they are enhanced by a factor E/m.

– *Hasan:*

If the Universe is closed and massless, why do we need massive particles at all? What is then the meaning of mass at all?

– *Okun:*

It is not up to us to choose the content of the Universe. Whether Universe is closed or not, nothing would change locally, here in Erice. The mass of a proton does not vanish when the mass of the Universe vanishes. The mass of the Universe is a global property, while the mass of a particle or a star is a local one. Even if you believe in Mach's principle, you cannot say that the mass of a proton is proportional to the mass of the Universe. The closed Universe has vanishing mass, open Universe has infinite mass.

– *Weselka:*

What does it mean to say that the Universe is massless?

– *Okun:*

It means that the absolute value of the potential energy of gravitational attraction between galaxies is exactly equal to the sum of their rest energies and kinetic energies.

– *Zichichi:*

The notion of the total energy of the Universe is not so obvious. There is no way to measure the total energy of a Universe externally.

– *Okun:*

No, but you can do it internally.

– *Ellis:*

In my opinion mass is just an index to label representations of the Lorentz group, which acquires its physical significance from the existence of an asymptotic Minkowski space. The Universe is not sitting in such a space and the comment you made that a closed Universe is massless seems to be independent of the possible existence of such an asymptotic Minkowski space. So how can we talk about the Universe having a mass?

– *Okun:*

There may be an asymptotic Minkowski space, there may be no asymptotic Minkowski space. We don't know. The result I'm quoting is derived in "The

Classical Theory of Fields" by L.Landau and E.Lifshitz, section 111. The closed Universe has vanishing total electric charge and vanishing total four–momentum, obtained by integrating over all space.

As for the external Minkowski space, there is a drastic difference between a massless Universe and a massless particle, e.g. the photon. The energy of the photon depends on the reference frame and there is no frame in which it is equal to zero. The energy of the massless Universe is equal to zero regardless of the choice of reference frame. A bubble of true vacuum created in false vacuum has a similar property. It is an object which is invariant with respect to Lorentz transformations.

– *Qureshi:*

If there is only one mass m, then is it a conceptual mistake on the part of those who use $m = m_0/\sqrt{1-\beta^2}$?

– *Okun:*

It is not an algebraic mistake, but it is a conceptual mistake. If you use units where $c = 1$, then you have two different notations, E and m, for the same quantity. Instead of one conservation law you have now two, for energy and mass, with only one symmetry behind them. You have to speak about mass of a massless photon. All this is very misleading.

– *Czyzewski:*

I would like to ask a question about two kinds of masses: m_i and m_g. If any other feature of matter, for example, baryon number, would give correction to the energy-momentum tensor and hence to the space-time curvature, would it mean that we have two different masses or would it be included in both of them?

– *Okun:*

It would be included in both of them if you use notations m_i and m_g. If there is a force connected with the baryon number, the acceleration of a body in the field of the Earth would depend on the material of the body, because now the Earth would have two different fields: gravitational and baryonic. But still there will be only one mass and the gravitational attraction will be proportional to the mass. There is no difference in this respect between new barionic force and the old Coulomb force.

– *Pluciennik:*

You mentioned in your lecture three ideas: closed Universe, fifth force and confinement. I would like to combine them in my question. Can it be that the Universe is closed due to a new confining force with confinement radius of astronomic scale?

– *Okun:*

A new confining force between particles of ordinary matter (electrons and nucleons) is impossible. The point is that such a force would imply a new degree of freedom like colour. You need at least two degenerate electrons, or protons, or neutrons. This would drastically change the Periodic Mendeleev Table. A new confining force may exist between heavy not-yet-discovered particles. But even in this case the radius of confinement cannot be larger than centimetres, otherwise the gas of relic "gluons" would be too hot ($\sim 3K$) to form "gluonic" strings.

– *Sanchez:*

You have presented mass in the context of gravity, whether relativistic or non-relativistic, at the classical level. When discussing the elementary particle mass spectrum without gravity, you talked, of course, at the level of Q.F.T.

I would like to comment about the combination of the two problems, namely the mass in the context of gravity at the quantum level. When the gravitational radius of a particle (which is proportional to the mass) becomes of the order of the Compton length (which is inversely proportional to the mass), we reach the Planck mass (or Planck energy) scale. This is the highest mass for point-like particles in Q.F.T., compatible with locality. The existence of this highest scale energy has several conceptual implications: a quantum theory of gravity must be finite and it must be a Theory of Everything (that is, all dimensionless numbers must be compatible in it and there is no hope to include gravity in a consistent way in the context of a renormalizable Q.F.T.). These implications are based on renormalization group arguments.

– *Okun:*

I agree with your comment. I would like also to add that the future theory beyond the Planck mass would not use our conventional idea of space-time. It would not involve a metric and would need new concepts for causality.

– *Gourdin:*

Why do you not define mass as an invariant of the Poincare group, P^2, using space-time translation operator P_μ?

– *Okun:*

I have defined mass as a Lorentz scalar: $m^2 = E^2 - \vec{p}^{\,2}$. I agree with you: it is better to define it as a Poincare scalar. This emphasises that conservation of energy and momentum is connected with the 4-dimensional translational symmetry.

– *Gourdin:*

My second question: why do you make a definite difference between an exact SU(3) in QCD with massless gluons and exact U(1) in QED with massless photon? Is it more important for theory or for experiment, this type of difference?

– *Okun:*

As I said this morning, the photon is massless, so to say, experimentally. We know from astronomical observations that the upper limit on its mass is very small. But a non-vanishing mass within this limit will not spoil the agreement between QED and experiment. As for the gluons, they are massless theoretically. Even a tiny gluon mass would make QCD non-renormalizable.

– *Khoze:*

Could you, please, comment on a scenario of a Higgs boson connected with the $t\bar{t}$ condensate.

– *Okun:*

Electroweak radiative corrections tell us that the top quark is not very heavy: $m_t \leq 2m_W$. For the Yukawa coupling of the top, this gives: $g_t^2 \leq 4\alpha_w \sim 1/7 \ll 1$. This coupling is not large enough to have fast running and to produce $t\bar{t}$ condensate. So special four-fermion interaction at much higher energies has to be introduced. The $t\bar{t}$-scenario would be more attractive in the case the mass of the top is of the order of 0.3 TeV or higher.

– *Neubert:*

I would like to come back to the question of the photon mass. Since gauge invariance is a fundamental concept of field theory would not you agree that the discovery of the photon mass would also be a disaster for theory?

– *Okun:*

My point is that gauge invariance in the case of abelian theory is not so fundamental. The non-abelian gauge invariance is sacred, but the abelian is not.

– *Zichichi:*

We had a special session on Gauge Invariance here in 1982, (P.A.M. Dirac, S. Ferrara, H. Kleinert, A. Martin, E.P. Wigner, C.N. Yang and A. Zichichi, "Special Session on Symmetries and Gauge Invariance", in *"Gauge Interactions"*, Plenum Press, New York, p. 125-140, 1984). It would be interesting, as this is a school, if you illustrate the link between the origin and the present meaning of gauge invariance.

– *Okun*:

It seems to me I have to make a digression into history. The term gauge invariance was coined by Hermann Weil in 1919, when he was (unsuccessfully) trying to create a unified theory of electromagnetism and gravity. In German it was Eichinvarianz and it was used by Weyl in its direct sense of scale invariance (Maßstabinvarianz).

With the advent of quantum mechanics, Vladimir Fock in 1926 invented the Klein-Fock-Gordon equation (after Klein but before Gordon) and discovered that the equation is invariant under the transformation

$$A_\mu \to A_\mu + df/dx_\mu, \quad \varphi \to \varphi e^{if}.$$

Fock called it gradient transformation. Note that if i is dropped, the phase factor becomes a scale factor. This observation was made in 1927 by F.London, who thus related the new gradient transformation to Weyl's old gauge transformation. In 1929 Weyl proclaimed Eichinvarianz (in the sense of Fock's gradient invariance) as a general principle. The old name was given thus to a new concept.

– *Neubert*:

But gauge invariance is a fruitful concept to build theories. And I think it would really change the understanding of field theory, of forces if you would not have this concept.

– *Okun*:

Historically, of course, the idea of electromagnetic gauge invariance played a very important role. When the notion of isospin appeared, the abelian gauge invariance was generalized to non-abelian in the late 1930's by Oscar Klein. Then in the 1950's this generalization was independently introduced by Yang and Mills. Today the non-abelian gauge invariance is the basis of the Standard Model. But the abelian gauge invariance, in spite of its historical importance, is not important physically. You may violate it without any penalty.

– *Zichichi*:

Why is there a catastrophe in non-abelian case? What do you lose when you give a tiny mass to gluons?

– *Okun*:

Photons are truly neutral: they do not emit photons. Gluons are charged in the sense of colour. Because of their colour charges, gluons emit gluons. Here is the drastic difference between photons and gluons. If you give mass to gluons via Higgs mechanism, you lose confinement. If you give mass to gluons by hand, you lose renormalizability, because of the catastrophic emission of longitudinally polarized gluons.

– *Zichichi:*

It would be great to break confinement.

– *Hernandez:*

Is it possible to consider the Standard Model, with gauge symmetry broken by explicit mass terms, as phenomenological model? Is it possible to describe in this way, by effective chiral Lagrangian, the non-unitarity in WW-scattering?

– *Okun:*

It is possible. But just because of this non-unitarity the phenomenological model would become non-self-consistent at high energies and of course loop diagrams would be divergent.

– *Danielsson:*

You mentioned dark matter. It explains the discrepancy between observations of stars rotation and our naive expectations. Could it not be that something is fundamentally wrong with our theories and with the notion of mass on a very large scale? Can you give an example of a really good test of gravity at a very large distance?

– *Okun:*

I do not know any other observations except those I have mentioned. If you do not believe in dark matter then these observations could be interpreted as evidence for anomalously strong gravity at large distances. But I don't know any reasonable theoretical framework for discussion of such anomalous behaviour of gravity.

– *Zichichi:*

Several months ago I have seen a paper in which it was claimed that the amount of luminous matter was grossly underestimated because the interstellar space is in fact less transparent than it is usually assumed.

– *Ellis:*

I think that this claim was not generally accepted and certainly it would be very difficult to explain constant rotational velocity in that way.

– *Okun:*

I have not seen this paper.

– *Duff:*

With three parameters G, $\hbar$, c we can form a fundamental mass, a fundamental length and a fundamental time. This is in accord with what we were taught in high school that there are the three basic dimensions of mass, length taught in high school that there are the three basic dimensions of mass, length and time. Does this mean that the Theory of Everything must have at least three fundamental parameters?

– Okun:

Yes, there must be three dimensional fundamental constants. The arguments I am going to present were first formulated, slightly differently, by Matvei Petrovich Bronshtein in the paper "On the possible theory of the world as a whole" which was published (in Russian) in 1933 (Uspekhi Astronom. Nauk. Sbornik 3, p.p. 3–30, Moscow ONTI). Professor Bronshtein was one of the most brilliant theorists. During Stalin purges he was arrested and executed in 1938 when he was 32 years old.

Place G, $\hbar$ and $1/c$ on three orthogonal axes (Fig. 1).

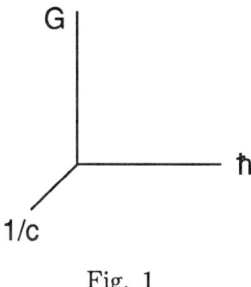

Fig. 1

Now we can construct the cube of theories (Fig. 2).

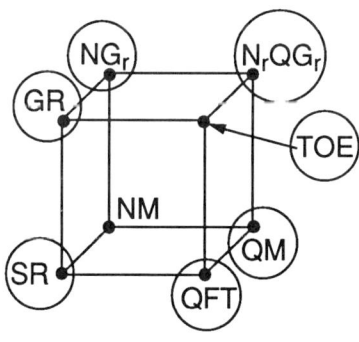

Fig. 2

At the origin (where $G = 0, \hbar = 0, 1/c = 0$), we have Newtonian Mechanics (without gravity, without relativity, without quantum effects). If we take into account G we have Newtonian Gravity. If we take into account finiteness of c, we have Special Relativity. If we take into account $\hbar$, we have Quantum Mechanics.

When we combine NG and SR we obtain General Relativity. When we combine SR and QM we obtain Quantum Field Theory. NG and QM give us Nonrelativistic Quantum Gravity. (This is a science which we don't need at present, because we have not observed objects which are described by this theory. But these objects may exist. These are particles, say, ten times lighter than Planck mass. Gravitationally attracting each other they would form "atoms". If these particles are neutral, it is very difficult to discover them.)

If we combine QFT and GR we obtain Theory of Everything. It is impossible to construct the Theory of Everything staying in one of the planes of the cube of theories. This was stated even earlier, in 1928, by George Gamow, Dimitri Ivanenko and Lev Landau (ZhRFCHO (Journal of Russian Physico-Chemical Society) $\underline{60}$,13(1928)). Let me remark that it was the great Einstein who for many years tried to build a unified theory in Gc-plane, without taking into account $\hbar$.

– *Zichichi:*

We will have a Bronshtein scholarship in Erice.

– *Duff:*

Now that we have set up the straw man, may we knock him down? Namely, now that we all agree that the Theory of Everything must have at least three fundamental parameters, where does that leave the string theory that has less than three?

– *Okun:*

This is a very good question which, I suppose, you will answer in your lecture.

– *Ellis:*

I don't agree with what Lev and Mike have said. I don't think you would need in the Theory of Everything three parameters. I think there will be no parameters at all. I have already told that to Mike at the coffee break this morning.

– *Zichichi:*

A Theory of Everything starting from nothing: fantastic!

– *Okun:*

I fully agree with John. There will be no parameters. G, $\hbar$, c are in fact not what we usually call parameters, they are units in which we will calculate everything. These units will drop out of physics. Physics depends only on dimensionless parameters. The Theory of Everything will give us relations between different dimensionless quantities like α or m_e/m_p, so there will be no free parameters.

– *Ellis:*

Exactly. All we measure is numbers.

– *Okun:*

Yes, all we measure is numbers. Some of them are obviously dimensionless like electron-proton mass ratio. But others like $\alpha = e^2/\hbar c$ involve fundamental units. You have the maximal velocity of signals c, you have the quantum of action $\hbar$ and you have the Planck mass m_P or what you prefer instead. You cannot get rid of them. In a natural theory you measure everything in these natural units. Therefore you must know that such three fundamental units exist.

– *Ellis:*

Yes, historically somehow we grew up with units of mass, length and time, which have no fundamental significance. We re-express them through G, $\hbar$, c.

– *Okun:*

The values of the three fundamental units are of historical origin. When we say that $c = 3 \cdot 10^{10} cm/s$, it depends on our definitions of centimetre and of second. The only fact that we really use is that our unit of velocity is the maximal velocity of signals.

The existence of G, $\hbar$, c is a fundamental property of Nature.

– *Ellis:*

The statement that there is an ultimate velocity is just a statement that Minkowski space has an indefinite metric. It is just group theory.

– *Okun:*

There would be group theory even in the case of infinite velocity — Galilean group.

– *Ellis:*

That would be a different group.

– *Okun:*

Of course, you may speak about Minkowski space, but it is simpler to say that velocity is limited.

– *Levi:*

Is gravitational field the only field that deforms space-time? If yes, why is gravitation so special in comparison to other fields?

– *Okun:*

The space is deformed only by gravitation. This is connected with the fact that the graviton is the only fundamental particle with spin equal to two.

– *Ungkitchanukit:*

Galilei discovered that the motion in gravitational field does not depend on the mass (it depends only on g). But in quantum mechanics mass appears in the phase of wave function. So it seems that in classical mechanics mass drops out while in quantum mechanics it does not.

– *Okun:*

Of course, quantum mechanical phase depends on energy and for massive particles, on mass. But there are lots of effects in classical mechanics as well in which energy and mass are essential. The acceleration of free fall is not the only measurable quantity of classical mechanics.

– *Shabelski:*

Is it possible for different free electrons to have slightly different mass values?

– *Okun:*

This is a very interesting question. There exist about 10^{80} electrons in our Universe. All of them are identical. But maybe some of them are more identical than others? Is it possible that there is a distribution of their masses which is characterized by a quantity δm?

Such a question was asked by Enrico Fermi in 1933. He noticed that a non-vanishing δm would spoil the property of atoms during the long time T of their existence. The non-observation of such effects means that $\delta m \ll 1/T$.

– *Pluciennik:*

What are your sources on Galileo Galilei in addition to the references given in the lecture?

– *Okun:*

- A.Koyré. Etudes Galiléennes. Paris 1939.
- M.Fierz. Vorlesungen zur Entwicklungsgeschichte der Mechanik (Lecture Notes in Physics, Vol. 15) Springer-Verlag. Berlin, 1972.
- Galilée: L'Expérience sensible. Textes de P.Caluzzi, G.Micheli, A.Porta, L.Rosino, G.Taboreli. Editions VILO. Paris, 1990 (French translation of Italian edition: Amilcare Pizzi S.P.A. 1988)
- Antologia Galileana, con traduzioni e note bibliografiche a cura di Carlo Maccagni. Firenze. Barbera Editore. 1964.

QCD PHENOMENOLOGY: JET RATES AND TRUNCATED PARTON CASCADES FOR MASSIVE HADRON PRODUCTION

Yu.L. Dokshitzer

Department of Theoretical Physics, University of Lund,

Sölvegatan 14A, S-22362 Lund, Sweden and

Nuclear Physics Institute, Gatchina, St. Petersburg 188350, Russia

1 Durham Algorithm for Jet Cross Sections

The multijet cross sections has been of primary experimental interest in e^+e^- annihilation studies at (and below) LEP energies [1, 2, 3]. These cross sections are defined in terms of a dimensionless jet resolution parameter y_{cut} and a jet recombination scheme. Final state particles are combined into clusters (**pseudo**particles), which are in turn recombined, according to the prescribed algorithm, until any further recombination would yield clusters that exceed the resolution y_{cut}. The number of remaining clusters is defined as the jet multiplicity. In the original *JADE algorithm* [2] the jet resolution parameter was taken to be of the general form $y_{\text{cut}} = M_j^2/W^2$ with M_j the maximum invariant mass of a pair of final particles (pseudoparticles) and $W = 2E$ the center-of-mass energy, and several different recombination schemes were introduced and studied (for details see [2, 4]).

The multijet cross sections defined in this way prove to be *infrared safe*, so that the experimental data should be directly comparable with perturbative QCD calculations at the parton level. For this reason study of JCS is important both as a test of QCD and as a good means of determining the strong coupling constant α_s.

So far this comparison has been performed with theoretical predictions based on the second order matrix elements in the QCD coupling [4]. Experimentally, it is found that the two-loop theoretical calculations are not able to fit the experimental data[1] for

---

[1] unless very small values of the renormalization scale are used [5]

$y_{\text{cut}} < 0.05$. This is a clear signal of the breakdown of the PT expansion in α_s due to terms enhanced at small y_{cut} values by powers of $\ln y_{\text{cut}}$. In this kinematical region the real expansion parameter, $\alpha_s \ln^2 y_{\text{cut}}$, becomes large and therefore *any* finite-order perturbative calculation cannot give an accurate evaluation of the cross section. The logarithmic terms need to be identified and resummed to all orders in α_s before a reliable prediction can be made.

The appearance of large logarithmic terms is a common feature of any semi-inclusive characteristic of a hard process which is sensitive to soft and/or quasicollinear parton splitting. Successful resummation of such terms depends on the fact that the large logarithmic corrections to the relevant quantity *exponentiate*. By this we mean that terms of the form $\alpha_s^n \ln^m y_{\text{cut}}$ with $n < m \le 2n$ can be combined into an exponential function of less singular terms, in fact with $m \le n + 1$. Once exponentiation has been established, the resummation of leading ($\alpha_s \log^2$) and next-to-leading ($\alpha_s \log$) terms to all orders is reduced to a much simpler, finite-order calculation of the relevant exponent.

The widespread occurrence of exponentiation in perturbative QCD is a consequence of the factorization properties of multiparton amplitudes in the soft and collinear limits, which appear necessary for the cancellation of singularities in inclusive quantities.

Unfortunately, the definition of the jet cross sections (JCS) according to the JADE type of algorithm generates strong *kinematic correlations*, which have been shown [6] to lead to a breakdown of exponentiation even at the leading double-logarithmic level. These kinematic correlations strongly increase with the number of final state partons and therefore even the resummation of the leading terms to all orders appears to be hopeless.

Although these features certainly do not invalidate the continued use of the original JADE jet algorithm, they do suggest that a modified jet-counting procedure that preserves exponentiation of the cross section would be worth finding and investigating. Such an algorithm [7] arose from discussions at the Durham Workshop on Jet Studies at LEP and HERA, December 1990, inspired by the UK Phenomenology Initiative [8]. It uses relative *transverse momentum* in place of *invariant mass* of the original JADE scheme as a jet resolution variable. Such a choice is motivated by the coherence properties [9] of QCD cascades.

For basic elements of the picture of QCD parton cascades, the notion of MLLA (Modified Leading Log Approximation) and LPHD (Local Parton-Hadron Duality hypothesis) the reader is referred to the Erice lectures [10] and the recent reviews [11].

Let us begin by recalling the discussion in previous lectures where the *generating functionals* (GF) which accumulate information about the parton content of QCD jets have been introduced. The notion of GF's is one of the chief supporting pillars of the QCD applications to jet physics. We are familiar with basic *inclusive* jet characteristics, such as mean parton multiplicity, inclusive particle spectra *etc*. (The latter will be under focus in the second Section of this lecture.)

JSC is a typical *exclusive* quantity: we need to find cross sections for production of n pseudoparticles, each of which represents in general a subjet with intrinsic relative parton transverse momenta smaller than the resolution parameter $Q_0 = \sqrt{y_{\text{cut}}} \cdot E$. The GF's prove to be perfectly suited for studying both inclusive and exclusive particle distributions.

1.1 Generating Functionals

The system of two coupled MLLA equations [12] for quark (Z_F) and gluon (Z_G) functionals reads ($A, B, C = F, G$):

$$Z_A(E, \Theta; u(k)) = e^{-w_A(E\Theta)} u_A(E) + \frac{1}{2!} \sum_{B,C} \int^\Theta \frac{d\Theta'}{\Theta'} \int_0^1 dz\, e^{-w_A(E\Theta)+w_A(E\Theta')}$$
$$\cdot \frac{\alpha_s(k_\perp^2)}{2\pi} \Phi_A^{BC}(z)\, Z_B(zE, \Theta'; u)\, Z_C((1-z)E, \Theta'; u)\,. \qquad (1)$$

The first term in the r.h.s. corresponds to the form factor damped situation when the A-jet with energy E and opening angle Θ consists of the parent parton only. The integral term describes the first splitting $A \to B+C$ with angle Θ' between the products. The exponential factor provides this decay being the very first one: it is the probability to have no particles emitted in the angular interval between Θ' and Θ. The two last factors account for the further evolution of the produced subjets B and C having smaller energies and $\Theta' < \Theta$ as the opening angle.

Initial condition for solving this system reads

$$Z_A(E, \Theta; \{u\})|_{E\Theta=Q_0} = u_A(E)\,. \qquad (2)$$

The A-jet with the hardness parameter $E\Theta$ which is set to the boundary value Q_0 where the PT-evolution starts, consists of the only parent parton A. This natural condition is clearly seen from the integral Evolution Equation (1) where the second term disappears at $E\Theta = Q_0$ together with the Born term Form Factor:

$$w_A(Q_0) = 0\,, \qquad \exp(-w_A) = 1\,.$$

MLLA form factors can be found with use of **normalization** property of GF's, namely

$$Z_A(E, \Theta; \{u\})|_{u(k)\equiv 1} = 1\,.$$

Putting $Z \equiv 1$ and $u \equiv 1$ in (1), one obtains integral equations

$$e^{w_A(E\Theta)} = 1 + \frac{1}{2!} \sum_{B,C} \int^\Theta \frac{d\Theta'}{\Theta'} \int_0^1 dz\, \frac{\alpha_s(k_\perp^2)}{2\pi} \Phi_A^{BC}(z)\, e^{w_A(E\Theta')} \qquad (3)$$

which lead directly to the MLLA expressions for total parton decay probabilities

$$w_F = \int^\Theta \frac{d\Theta'}{\Theta'} \int_0^1 dz \frac{\alpha_s(k_\perp^2)}{2\pi} \Phi_F^F(z)\,, \quad (F = q, \bar{q})\,, \qquad (4a)$$

$$w_G = \int^\Theta \frac{d\Theta'}{\Theta'} \int_0^1 dz \frac{\alpha_s(k_\perp^2)}{2\pi} \left[\frac{1}{2}\Phi_G^G(z) + n_f\, \Phi_G^F(z)\right]\,. \qquad (4b)$$

Collinear and soft singularities which are there in Eqs.(1) – (4) are regularized by usual PT transverse momentum restriction

$$k_\perp \approx \min[z, 1-z]\, E\,\Theta' > Q_0\,. \qquad (5)$$

Taking the nth variational derivative of GF Z_A over the probing functions $u(k_i)$ near the "point" $u = 0$ one gets the *exclusive* n-parton cross sections. Expansion of Z_A at $u = 1$ generates *inclusive* parton distributions and correlations.

All the properties [2] of a system of jets produced in some hard interaction can be derived (within the MLLA accuracy) by applying variational derivatives to the proper product of GF's. For example, for e^+e^- annihilation to hadrons at energy $W = 2E$, which is dominated (provided no special event selection is imposed) by two-jet configuration, one has

$$Z_{e^+e^-}(W; \{u\}) = \{ Z_F(E, \Theta \sim \pi; \{u\}) \}^2 . \tag{6}$$

To find the JCS in the "Durham scheme" one has simply to write down an *exclusive* expansion of (6) for

$$Q_0^2 = y_{\text{cut}} \cdot E^2 .$$

Generally speaking there are different "$u_f(k)$" to probe gluons, quarks and antiquarks in the final state. The n-jet rate will be given then by

$$R_n = \sum_{i+j+k=n} \frac{1}{i!\, j!\, k!} \frac{\partial^i}{\partial u_q^i} \frac{\partial^j}{\partial u_{\bar{q}}^j} \frac{\partial^k}{\partial u_g^k} \{ Z_q Z_{\bar{q}} \}|_{u_q=u_{\bar{q}}=u_g=0} \tag{7}$$

It is a matter of simple combinatorics to show that counting the number of jets only one can treat final partons as identical items using a single probing function u for all of them:

$$R_n = \frac{1}{n!} \frac{\partial^n}{\partial u^n} Z_F^2 \Big|_{u=0} .$$

Moreover, since particle momenta are integrated over in the final state, one can treat $u(k)$ as a number, replacing variational derivatives by simple differentiation over u:

$$\int \frac{d^3k}{2k} \frac{\partial}{\partial u} \to \frac{d}{du} .$$

Resulting expressions for JCS which account for all orders of leading and next-to-leading logarithms can be found in Ref.[7]. To illustrate the general method we restrict ourselves for pedagogical reasons to the Double Log (DL) approximation which keeps track of the leading terms $\alpha_s \log^2 y_{\text{cut}}$ only.

1.2 Double Logarithmic Approximation

In this approximation the Master Equation (1) significantly simplifies since it is sufficient to account for soft gluon emissions. This leads to the following DL Master Equation for GF of the *gluon* jet:

$$Z_G(y; u) = u \cdot \exp\left(\int_0^y dy'\, (y-y')\, \gamma_0^2(y')\, [Z_G(y'; u) - 1] \right), \quad y \equiv \ln \frac{k_\perp}{Q_0} . \tag{8}$$

Here we introduced notation γ_0 for the color charge factor depending on $k_\perp$ of the parton[3]

$$\gamma_0^2(y) = 4N_c \frac{\alpha_s(k_\perp^2)}{2\pi} \approx \frac{4N_c}{b(y+\lambda)} ; \quad \lambda \equiv \ln Q_0/\Lambda, \quad b = \frac{11}{3} N_c - \frac{2}{3} n_f . \tag{9}$$

It is convenient for us to "rescale" the functional by replacing

$$Z_G = u \cdot Z .$$

---

[2] angular correlations excluded (sic! see [10, 11])
[3] γ_0 is the *anomalous dimension* determining the rate of the parton multiplicity growth

New GF will generate now additional soft gluons accompanying the parent parton which carries in DLA the whole jet energy (factor $u \equiv u(E)$ in the r.h.s. of (8)). Equation for Z reads

$$Z(y;u) = \exp\left(\int_0^y dy'(y-y')\,\gamma_0^2(y')\,[u \cdot Z(y';u) - 1]\right) \tag{10}$$

It is straightforward to check that the corresponding GF for the quark jet in DLA is simply

$$Z_F = Z^{(C_F/N_c)}. \tag{11}$$

Finally, e^+e^- annihilation is described in DLA by

$$Z_{e^+e^-} = Z^\kappa, \qquad \kappa \equiv \frac{2C_F}{N_c}. \tag{12}$$

Now to calculate JCS one has to expand $Z_{e^+e^-}$ near $u = 0$:

$$Z_{e^+e^-}(u) = \sum_{n=0}^\infty R_{2+n}\, u^n \tag{13a}$$

$$R_{2+n} = \frac{1}{n!}\left(\frac{d}{du}\right)^n Z_{e^+e^-}(u)\bigg|_{u=0} \tag{13b}$$

1.3 Effective Form Factors and JCS

Let us introduce short notation for the integral which appears when expanding $Z_{e^+e^-}(u)$:

$$[f](k) \equiv \int_{Q_0}^k d\Gamma(k')\, f(k') \cdot \exp(-w(k')) \tag{14}$$

with $d\Gamma$ the logarithmic phase space and w the DLA gluon splitting probability determining the gluon form factor F_g

$$d\Gamma(k') = \frac{dk'}{k'}\,(\ln k - \ln k')\,4N_c\,\frac{\alpha_s(k')}{2\pi}, \tag{15}$$

$$w(k) = \int_{Q_0}^k d\Gamma, \qquad F_g(k) = \exp(-w(k)). \tag{16}$$

The two-jet cross section is given by the square of the Sudakov form factor of a quark (cf. Eq.(11))

$$R_2 = Z_{e^+e^-}|_{\{u=0\}} \equiv Z_0 = F_q^2(E) = \exp(-\kappa \cdot w(E)). \tag{17}$$

To present the JCS with $m > 2$ jets it is convenient to introduce Effective integrated gluon Form Factors according to

$$\mathcal{F}_1(k) = [1] \equiv \int d\Gamma(k')\,\exp(-w(k')) \tag{18a}$$

$$\mathcal{F}_{m+1}(k) = [\mathcal{F}_m] \equiv \int d\Gamma(k')\,\mathcal{F}_m(k') \cdot \exp(-w(k')), \quad m \geq 2. \tag{18b}$$

The cross sections then take the form

$$R_3 = Z_0 \cdot \{\kappa \mathcal{F}_1\} \tag{19a}$$

$$R_4 = Z_0 \cdot \left\{\kappa^2 \frac{1}{2!}\mathcal{F}_1^2 + \kappa \mathcal{F}_2\right\} \tag{19b}$$

$$R_5 = Z_0 \cdot \left\{\kappa^3 \frac{1}{3!}\mathcal{F}_1^3 + \kappa^2 \mathcal{F}_1 \mathcal{F}_2 + \kappa\left(\frac{1}{2!}[\mathcal{F}_1^2] + \mathcal{F}_3\right)\right\} \tag{19c}$$

$$R_6 = Z_0 \cdot \left\{\kappa^4 \frac{1}{4!}\mathcal{F}_1^4 + \kappa^3 \frac{1}{2!}\mathcal{F}_1^2\mathcal{F}_2 + \kappa^2\left(\mathcal{F}_1\frac{1}{2!}[\mathcal{F}_1^2] + \mathcal{F}_1\mathcal{F}_3 + \frac{1}{2!}\mathcal{F}_2^2\right)\right.$$
$$\left. + \kappa\left(\frac{1}{3!}[\mathcal{F}_1^3] + [\mathcal{F}_1\mathcal{F}_2] + \frac{1}{2!}[[\mathcal{F}_1^2]] + \mathcal{F}_4\right)\right\} \tag{19d}$$

Implicit Q_0-dependence of R_m is induced by the transverse momentum resolution cut (16). Qualitative behaviour of the DLA JCS is shown in Fig. 1.

Equations (17), (19) express JCS in terms of the Sudakov form factor $\exp(-w)$, its (multiple) logarithmic integrals and their powers. This is how the "exponentiation" looks like for jet cross sections in a simple DL approximation. We leave it for the reader to design a probabilistic scheme explaining the structure of Eqs.(19), which would attribute each term in the $r.h.s.$ to a definite parton generating picture.

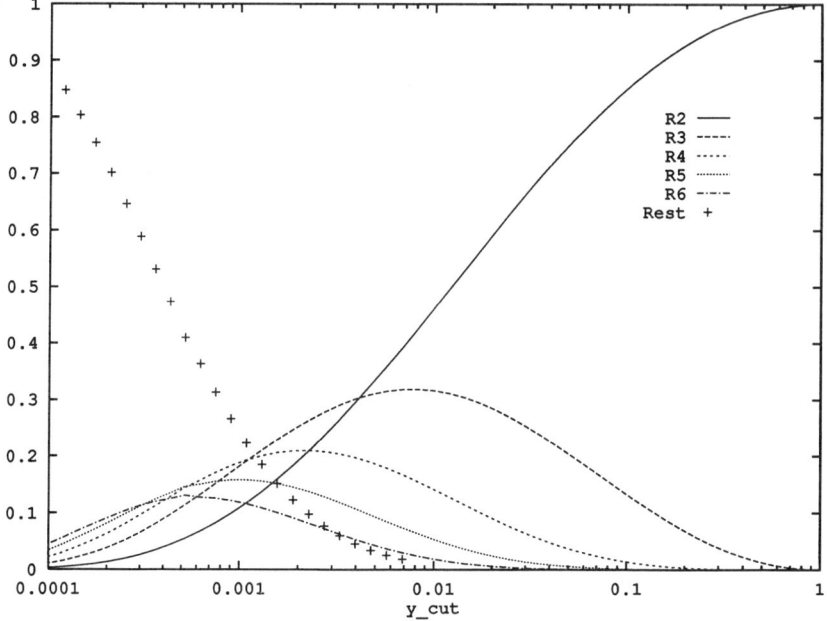

Figure 1. DLA Jet Cross Sections for the LEP energy as given by Eqs.(19). Crosses show the production cross section of $m > 6$ jets.

2 Massive Hadrons from Truncated QCD Cascades

At the last year Erice lecture [10] we have finished our discussion of the MLLA particle spectra by noticing that "for more profound study of LPHD picture conjectures the measurements of identified hadron spectra are of substantial importance". The year 1990/91 has brought a breakthrough in the field marked by the L3 measurement [13] of inclusive π^0s and the OPAL (and coming ALEPH and DELPHI) results on K_s^0 spectra [14]. In accordance with theoretical expectations the value of Λ_{eff} for a pure π^0-sample (which is much closer to the "true" scale parameter determining the value of running α_s) appeared to be essentially smaller than the "mixed" effective quantity extracted from the charged particle data $\Lambda_{eff} \approx 250$MeV. Meantime the *Kaon* spectra peaked at smaller values of $\ell = \ln 1/x_p$. Such *stiffening* of energy distributions of massive hadrons has been first observed for baryons in the PETRA/PEP epoch and has found a natural motivation in the framework of PT picture of QCD cascades [15].

2.1 Selected Properties of Truncated Cascades

Before turning to basic properties of parton spectra we remind the reader of the notation

$$\ell = \ln \frac{1}{x}, \quad Y = \ln \frac{E}{Q_0}, \quad \lambda = \ln \frac{Q_0}{\Lambda}$$

$$z_1 = \sqrt{\frac{16N_c}{b}(Y+\lambda)}, \quad z_2 = \sqrt{\frac{16N_c}{b}\lambda}$$

$$b = \frac{11}{3}N_c - \frac{2}{3}n_f, \quad a = \frac{11}{3}N_c + \frac{2n_f}{3N_c^2}; \quad B = \frac{a}{b}.$$

2.1.1 Mean Parton Multiplicity

can be written in a compact form in terms of the Modified Bessel functions [16] $I_\nu(z)$ and $K_\nu(z)$

$$\mathcal{N}(Y,\lambda) = z_1 \left(\frac{z_2}{z_1}\right)^B [I_{B+1}(z_1)K_B(z_2) + K_{B+1}(z_1)I_B(z_2)] . \quad (20)$$

With z_1 increasing the first term in (20) grows exponentially in $\sqrt{Y+\lambda}$, while the second exponentially decreases and can be therefore practically neglected. It reveals itself only for small values of Y, $Y < \sqrt{\lambda}$, where the parton multiplicity is still small, $\mathcal{N} \sim 1$, and the two terms in (20) act together to provide the initial condition for the jet evolution ($\mathcal{N} = 1$ at $Y = 0$). For well developed cascades, $\mathcal{N} \gg 1$, neglecting the second term can be estimated as producing $O(\mathcal{N}^{-2})$ correction[4].

---

[4]Eq.(20) is written for a *gluon* jet. PT prediction for parton content of a *quark* jet is, roughly speaking, $C_F/N_c = 4/9$ times smaller; $\sqrt{\alpha_s}$ effects in the multiplicity ratio and in the shape of inclusive spectra have been discussed in the last year lecture, see [10]

2.1.2 Mean ℓ

acquires a comparatively simple expression in the same approximation (neglecting terms exponentially decreasing with $\sqrt{Y+\lambda}$), namely [17]

$$\langle \ell \rangle = (Y+\lambda) \cdot \left[\frac{1}{2} + \frac{B}{z_1} \frac{I_{B+2}(z_1)}{I_{B+1}(z_1)} \right] - \lambda \cdot \left[\frac{1}{2} + \frac{B+1}{z_2} \frac{K_{B-1}(z_2)}{K_B(z_2)} \right] \qquad (21)$$

Let me remind you that the Limiting Spectrum corresponds to $z_2 = 0 \, (\lambda = 0)$. Equation (21) shows an influence of finite Q_0 on $\langle \ell \rangle$ and predicts the stiffening of distributions of partons originating from truncated cascades to be energy independent.

How $Q_0 > \Lambda$ affects other spectrum shape characteristics?

2.1.3 Higher Moments

of the distribution can be found for truncated cascades as well. Analytic procedure developed in Ref.[17] was based on explicit solution of MLLA Evolution Equations for the Mellin-transformed spectrum $D(\omega, Y, \lambda)$

$$\ell_m \equiv \langle \ell^m \rangle = \frac{1}{\mathcal{N}} \int_0^\infty d\ell \int_{-i\infty}^{+i\infty} \frac{d\omega}{2\pi i} D(\omega, Y, \lambda) \frac{d^m}{d\omega^m} e^{\omega \ell} = \left(-\frac{\partial}{\partial \omega} \right)^m D(\omega, Y, \lambda) \bigg|_{\omega=0} \qquad (22)$$

After calculating the moments one can construct dispersion and subsequent higher *cumulants* of the distribution among which **skewness** s and **kurtosis** k are the first ones that form the base for the very useful *distorted Gaussian* expression proposed by Fong and Webber in Ref.[18]. The distorted Gaussian representing the spectrum in terms of the scaled variable $\delta \equiv (\ell - \ell_1)/\sigma$ as

$$\overline{D}(\ell, Y) \approx \frac{\mathcal{N}(Y)}{\sigma \sqrt{2\pi}} \exp \left[\frac{1}{8} k - \frac{1}{2} s \delta - \frac{1}{4}(2+k)\delta^2 + \frac{1}{6} s \delta^3 + \frac{1}{24} k \delta^4 \right] \qquad (23)$$

with analytically calculated shape parameters

$$\sigma^2 \equiv \langle (\ell - \langle \ell \rangle)^2 \rangle = \ell_2 - \ell_1^2, \qquad (24a)$$

$$s \equiv \frac{\langle (\ell - \langle \ell \rangle)^3 \rangle}{\sigma^3} = \frac{1}{\sigma^3}(\ell_3 - 3\ell_2 \ell_1 + 2\ell_1^3), \qquad (24b)$$

$$k \equiv \frac{\langle (\ell - \langle \ell \rangle)^4 \rangle}{\sigma^4} - 3 = \frac{1}{\sigma^4}(\ell_4 - 4\ell_3 \ell_1 - 3\ell_2^2 + 12\ell_2 \ell_1^2 - 6\ell_1^4) \qquad (24c)$$

proves to give reasonable approximations of MLLA spectra in the peak region not only for the limiting case but for truncated cascades as well.

For the pure Gaussian $s = \kappa_3$ and $k = \kappa_4$ vanish together with all other higher *reduced cumulants* κ_p which are closely related to the anomalous dimension $\gamma_\omega(\alpha_s)$:

$$\sigma^2 = \int_\lambda^{Y+\lambda} dy \left(\frac{\partial}{\partial \omega} \right)^2 \gamma_\omega(\alpha_s(y)) \bigg|_{\omega=0} \qquad (25a)$$

$$\kappa_p = \frac{1}{\sigma^p} \int_\lambda^{Y+\lambda} dy \left(-\frac{\partial}{\partial \omega} \right)^p \gamma_\omega(\alpha_s(y)) \bigg|_{\omega=0}. \qquad (25b)$$

Knowing the MLLA anomalous dimension (see, e.g., Ref.[10])

$$\gamma(\alpha_s) = -\frac{\omega}{2} + \frac{K}{2} + \gamma_0^2 \left[-\frac{a}{8N_c}\left(1+\omega K^{-1}\right) + \frac{b}{4N_c}\gamma_0^2 K^{-2} \right] + \mathcal{O}(\alpha_s^{3/2}) \quad (26)$$

with $\quad K = \sqrt{\omega^2 + 4\gamma_0^2}; \quad \gamma_0^2 = 4N_c \frac{\alpha_s}{2\pi},$

one can find κ_p explicitly to show that, relative to the lowest order Gaussian spectrum with $\ell_1 = \ell_{max} = \frac{1}{2}Y$, the peak in the ℓ-distribution, according to (23), is shifted up (i.e. to the lower x), narrowed, skewed towards higher x's, and flattened, with tails that fall off more rapidly than a true Gaussian. Referring to the asymptotic behaviour ($n \geq 1$)

$$\sigma^2 \sim Y^{3/2} \quad (27a)$$

$$\kappa_{2n+2} \sim \left(\sqrt{Y}\right)^{-n} \quad (27b)$$

$$\kappa_{2n+1} \sim \left(\sqrt{Y}\right)^{-n-1/2} \quad (27c)$$

one concludes that the higher cumulants ($p > 4$) appear to be less significant for the shape of the spectrum in the hump region $\delta \lesssim 1$.

2.1.4 Dispersion

For example, for the dispersion one obtains from (25a)

$$\frac{d}{dY}\sigma^2 = \left.\frac{\partial^2}{\partial\omega^2}\gamma_\omega\right|_{\omega=0} = \frac{1}{4\gamma_0} - \frac{b}{32N_c} + \mathcal{O}(\gamma_0) \quad (28a)$$

$$\sigma^2 \approx \frac{1}{6}\sqrt{\frac{b}{4N_c}}\left[(\sqrt{Y+\lambda})^3 - (\sqrt{\lambda})^3\right] - \frac{b}{32N_c}\left[(Y+\lambda) - \lambda\right]. \quad (28b)$$

This estimate is valid for large values of λ only, since the asymptotic expansion (26) relied upon $\alpha_s(Q_0) \ll 1$. However the message we get from (28b) is clear: the width of the spectrum becomes smaller with λ increasing, i.e. when one truncates parton branching at finite $Q_0 > \Lambda$.

2.1.5 Skewness and the Peak Position

Without going into detailed technical analysis of particle spectra we can notice one interesting property which concerns the most important characteristic — position of the maximum. From the expression (23) for distorted Gaussian one can see that the nonzero skewness leads to a calculable splitting between ℓ_{max} and ℓ_1, namely

$$\delta_{max} \approx -\frac{s}{2}$$

which means, in a usual x-scale,

$$\ell_{max} = \left(\ln\frac{1}{x}\right)_{peak} = \ell_1 - \frac{1}{2}s \cdot \sigma. \quad (29)$$

Comparing high-Y behaviour of dispersion (27a) with that of skewness given by (27c) ($n=1$) one could expect the difference $\ell_{max} - \ell_1$ to be an energy independent constant.

Let us verify this expectation and show that this splitting happens to be Q_0-independent either.

Skewness vanishes in DLA, as all higher **odd** cumulants do, so that only one subleading term from the MLLA anomalous dimension (26) contributes to $p = 3$ in (25b):

$$\frac{d}{dY}\left(\sigma^3 s\right) = -\frac{a}{2}\left(-\frac{\partial}{\partial \omega}\right)^3 \frac{\omega}{\sqrt{\omega^2 + 4\gamma_0^2}}\bigg|_{\omega=0} = -\frac{3a}{16N_c} \cdot \frac{1}{4\gamma_0}. \tag{30}$$

According to (30) the Y-derivative of the combination $\sigma^3 s$ is basically similar to that for the dispersion squared (cf. Eq.(28a)). Integral over virtuality (25b) preserves this relation so that for any Y and λ an approximate relation holds true, i.e.,

$$\sigma^3 s = -\frac{3a}{16N_c} \cdot \sigma^2. \tag{31}$$

This results in an interesting testable QCD prediction revealing the net *subleading* MLLA effects in parton cascades:

$$\ell_{max} - \ell_1 \approx \frac{1}{2} \cdot \frac{3a}{16N_c} \equiv \frac{11}{32} + \frac{n_f}{16N_c^3} = 0.351 \ (0.355) \quad \text{for } n_f = 3 \ (5). \tag{32}$$

Combining this result with (21) we conclude that QCD predicts the energy-independent shift of the peak position for truncated parton distributions compared to the limiting (*pion?!*) spectrum. This fact can be used to *measure* effective Q_0 values for different hadron species by comparative study of the energy evolution of the peak. Visualization of expected increase of Q_0 with m_h would demonstrate that massive hadrons are produced effectively at *smaller* space-time scales.

Investigation of massive hadron production in jets has scarcely begun and the situation looks rather unclear at present. One definitely needs much more accurate measurements of inclusive spectra in line with reinforced attempts to approach the problem from theoretical side.

To help an interested reader to achieve a more complete understanding of the present status of the problem we conclude the lecture by a short survey of the situation in a form of the letter compiled from the correspondence of the author with his experimental colleagues.

2.2 Letter to a friend-experimentalist

In your letter you asked about the MLLA predictions for inclusive spectra of massive particles in jets.

1. MLLA is a "family" name for perturbative (PT) theoretical predictions of parton spectra from well developed QCD cascades derived with **controllable** accuracy (*i.e.* better than in the leading Double Logarithmic approximation when one deals with gluon branching only in the soft limit making ignorance of energy-momentum conservation). The MLLA provides **asymptotically exact** results, which means that the relative error of the prediction decreases with W as $\alpha_s(W^2)/\pi$.

2. Then comes an additional hypothesis, LPHD, that calculable **parton** distributions can be directly compared with measurable **hadron** spectra. If confinement acts

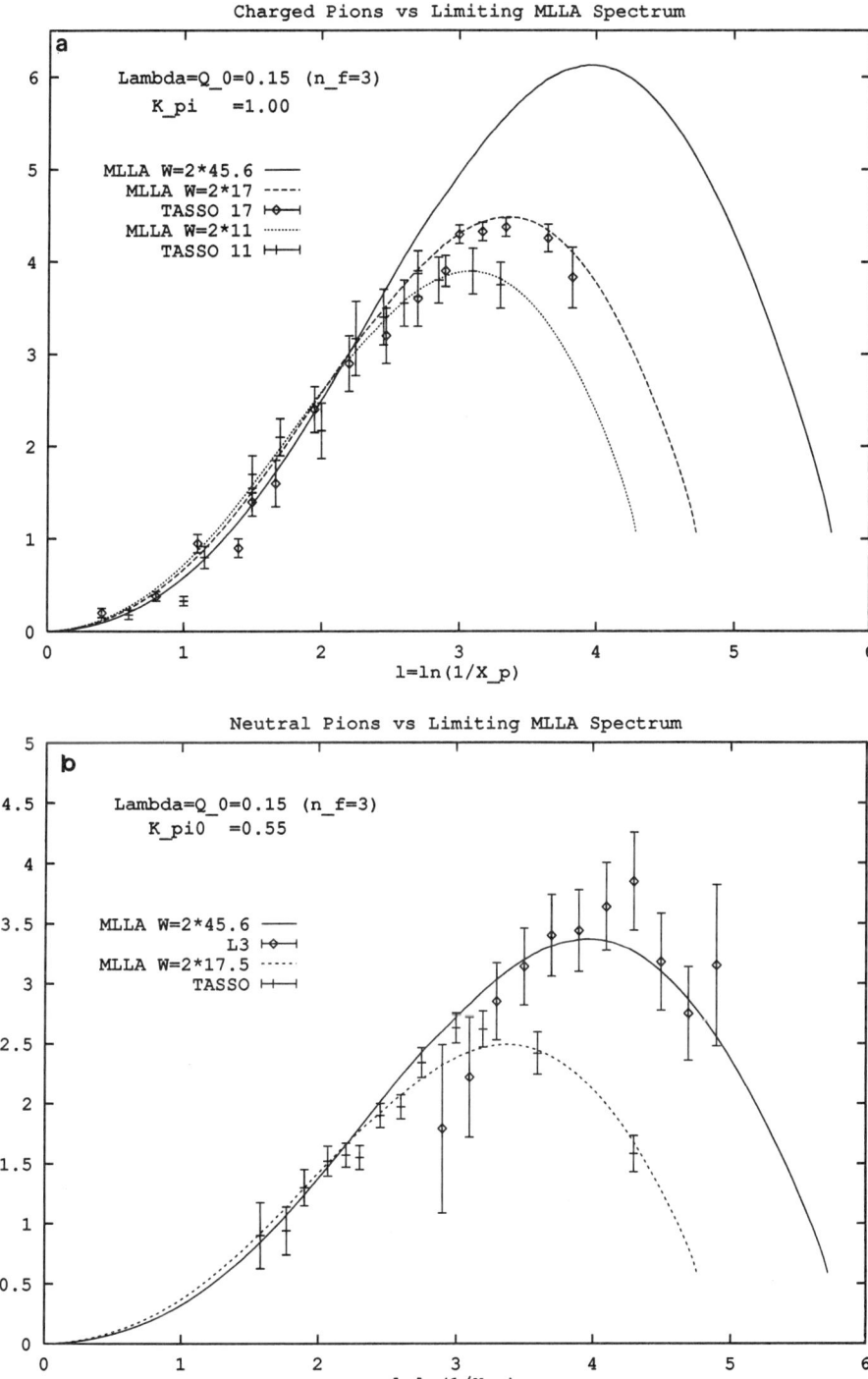

Figure 2. Limiting MLLA description of a) $\pi^{\pm}$ and b) π^0 spectra.

locally in the configuration space (and there are good reasons to believe in "soft confinement") then one could expect no dramatic changes in the shape of particle spectra after hadronization but some smearing of order $\Delta \ell \lesssim 1$ when turning to hadrons from partons. This point must be taken with a pinch of salt: this expectation could work only well far from kinematical boundaries since hadrons are known to have masses while partons within the PT game were treated as massless. So, one has *either* to wait for "asymptotically high" energies (which, by the way, is a usual excuse for theoreticians) *or* attempt to apply this hypothesis to **light** hadrons, *i.e.* pions, for which no large mass effects are expected. Worthwhile to mention that such an attempt clearly amounts to the assertion that there exists a close duality between inclusive resonance decays and pure parton description *continued* down to very small (dangerously small from the perturbative point of view!) relative transverse momenta of partons in the QCD cascade.

3. This is exactly the logics which leads to a notion of "limiting MLLA spectrum": the one-parametric ($\Lambda_{eff} = Q_0$) prediction of parton spectra to be compared directly with yield of pions in jets (with an overall normalization as an additional (and only) phenomenological parameter which is experimentally close to 1 and should be W- and x-independent). The limiting spectrum is rather easy to calculate and people usually take it to compare directly with "ALL CHARGED" data[5] to extract $\Lambda_{eff} \approx 250$MeV. So, this could be the first (most naive but already traditional) thing to do.

4. When looking at pure **pion** samples one finds naturally (twice) smaller value $\Lambda \approx 150$MeV (or even 115MeV obtained by L3 for π^0's), see Fig. 2. This is natural because "ALL CHARGED" is a mixture including massive $K^\pm$ and p and the spectra of **massive** hadrons are known to be **stiffened**. Then comes the next point: how one could account for stiffening?

5. The so called ζ-*scaling* was used as a first trial for this phenomenon [15]. The following procedure has been used

a) present limiting parton spectrum as a function of scaled variable

$$\zeta_{parton} = \ln\frac{xE}{\Lambda} \Big/ \ln\frac{E}{\Lambda}, \qquad 0 \leq \zeta \leq 1;$$

b) construct similar variable for a massive hadron

$$\zeta_{hadron} = y_h / y_h^{Max}$$

with

$$y_h = \ln\frac{p_h + \sqrt{m_h^2 + p_h^2}}{m_h}, \qquad y_h^{Max} = \ln\frac{E_{jet} + \sqrt{m_h^2 + E_{jet}^2}}{m_h};$$

c) ascribe $\zeta_{hadron} = \zeta_{parton}$ to get the hadron distribution in terms of the limiting parton spectrum (Fig. 3).

6. Phenomenologically distributions of different hadrons look similar indeed in the ζ scale. However it is worth looking for better grounded approach to the problem. First thing to be said is that there is NO theory to predict from the first principles

---

[5]Glen Cowan (ALEPH) pioneered such an analysis at LEP; OPAL results were first published [19]

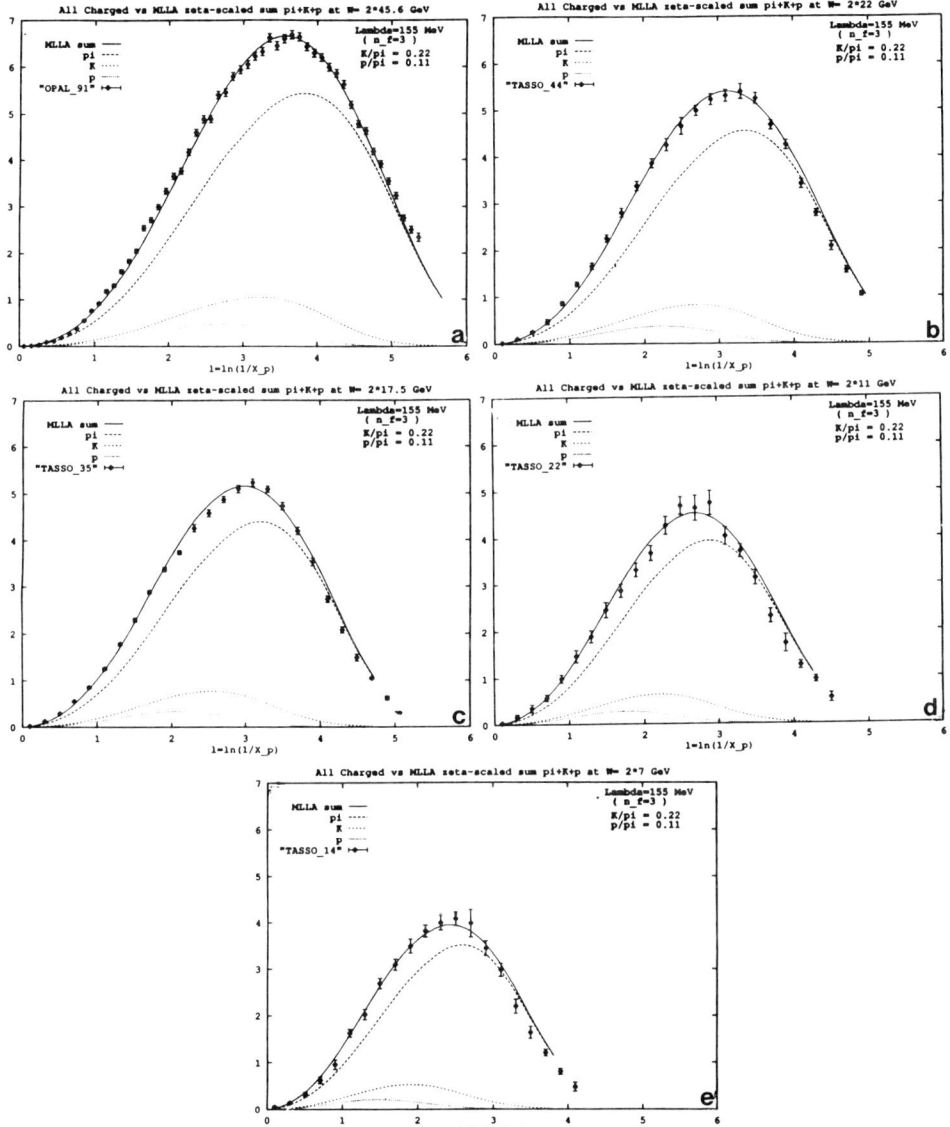

Figure 3. Spectra of charged hadrons versus ζ-scaled sum of $\pi^\pm$, $K^\pm$ and p for different energies.

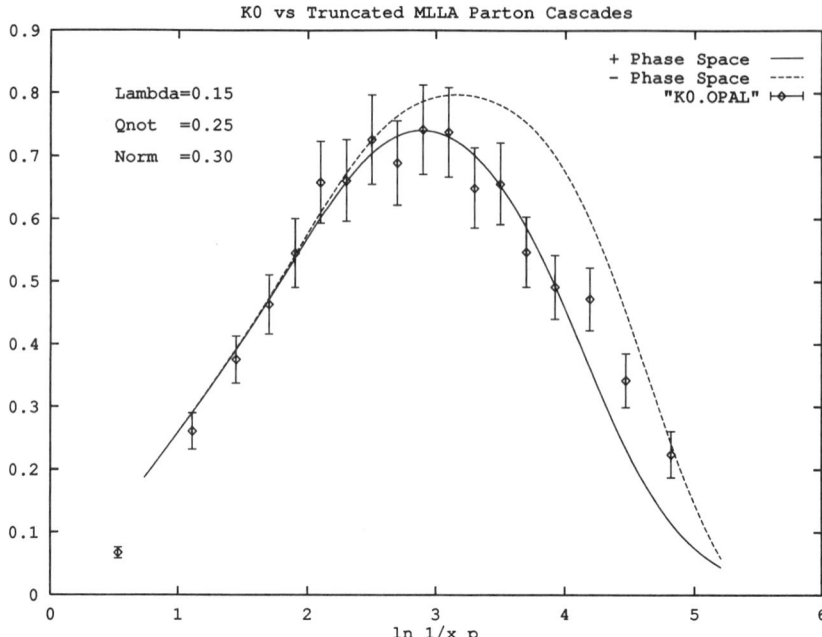

Figure 4. Truncated parton cascades for inclusive spectrum of K_s^0 [14].

distributions of massive hadrons. It looks natural to begin with comparing the hadron yield with calculable parton spectra stemming from parton cascades *truncated* from below at $k_\perp \geq Q_0 > \Lambda_{eff}$ with Q_0 (minimal allowed relative transverse momentum inside PT cascade) increasing with hadron mass. Such a procedure provides stiffening (and natural damping of the multiplicity as well). These "truncated" parton spectra are much more difficult but still possible to calculate numerically [17].

7. Additional parameter Q_0 is not the only new uncertainty however: the asymptotic PT analysis is insensitive to the difference between x_E and x_p of partons involved which makes a big difference for real massive hadrons meantime! There are good reasons to think of PT formula as being written in terms of the **energy** fraction x_E. Then one has to

- recalculate variables when turning from parton to hadron distribution and

- take care of the difference between theoretical $x_E \dfrac{d\sigma}{dx_E}$ and observable $x_p \dfrac{d\sigma}{dx_p}$.

This results in two curves marked "Phase Space" and "NO Phase Space" in Fig.4 for the K_s^0 spectra. Including Ph.Sp. effect leads to decrease of Q_0 since a good deal of stiffening is provided by simple kinematics.

You see how many *MLLA's* are there for massive hadrons at present!

References

[1] ALEPH Collaboration, D. Decamp et al., *Phys. Lett.* 234B (1990) 399; *ibid.* 255B (1991) 623; CERN preprint PPE/90-196;
DELPHI Collaboration, P. Aarnio et al., *Phys. Lett.* 240B (1990) 271; *ibid.* 247B (1990) 167;
L3 Collaboration, B. Adeva et al., *Phys. Lett.* 237B (1990) 136; *ibid.* 248B (1990) 464;
OPAL Collaboration, M.Z. Akrawy et al., *Phys. Lett.* 235B (1990) 389; *ibid.* 252B (1990) 159; *Zeit. Phys.* C47 (1990) 505; *ibid.* C49 (1991) 375.

[2] JADE Collaboration, W. Bartel et al., *Zeit. Phys.* C33 (1986) 23; S. Bethke et al., *Phys. Lett.* 213B (1988) 235.

[3] TASSO Collaboration, W. Braunschweig et al., *Zeit. Phys.* C33 (1990) 187;
Mark II Collaboration, S. Komamiya et al., *Phys. Rev. Lett.* 64 (1990) 987;
AMY Collaboration, I. Park et al., *Phys. Rev. Lett.* 62 (1989) 1713.

[4] G. Kramer and B. Lampe, J. Math. Phys. 28 (1987) 945; *Zeit. Phys.* C34 (1987) 497; *ibid.* C39 (1988) 101; *ibid.* C42 (1989) 504 (E); *Fortschr. Phys.* 37 (1989) 161;
Z. Kunszt, P. Nason, G. Marchesini and B.R. Webber, in 'Z Physics at LEP 1', CERN 89-08, vol.1, p. 373;
N. Magnoli, P. Nason and R. Rattazzi, *Phys. Lett.* 252B (1990) 271.

[5] G. Kramer and B. Lampe, *Zeit. Phys.* C39 (1988) 101;
S. Bethke, *Zeit. Phys.* C43 (1989) 331.

[6] N. Brown and W.J. Stirling, *Phys. Lett.* 252B (1990) 657.

[7] S. Catani, Yu.L. Dokshitzer, M. Olsson, G. Turnock and B.R. Webber, *Phys. Lett.* 269B (1991) 432.

[8] Workshop on Jet Studies at LEP and HERA, *Journ. of Physics,* G17 (1991) # 10 (special issue)

[9] A.H. Mueller, *Phys. Lett.* 104B (1981) 161;
B.I. Ermolaev and V.S. Fadin, *JETP Lett.* 33 (1981) 285.

[10] Yu.L. Dokshitzer, Lectures given at Int. Schools for Subnuclear Physics, Erice, 1989, 1990.

[11] Yu.L. Dokshitzer, V.A. Khoze and S.I. Troyan, in *Perturbative Quantum Chromodynamics*, ed. A.H. Mueller (World Scientific, Singapore, 1989);
Yu.L. Dokshitzer, V.A. Khoze, A.H. Mueller and S.I. Troyan, *Basics of Perturbative QCD*, Editions Frontieres, Paris, 1991.

[12] Yu.L. Dokshitzer and S.I. Troyan, *Asymptotic Freedom and Local Parton-Hadron Duality*, in: Proc. of the XIX LNPI Winter School, volume 1, p.144, Leningrad, 1984; preprint LNPI–922 (1984);
A.H. Mueller, *Nucl. Phys.* B213 (1983) 85; *ibid.* B241 (1984) 141.

[13] L3 Collaboration, B. Adeva et al., *Phys. Lett.* 259B (1991) 199.

[14] OPAL Collaboration, G. Alexander et al., *A Study of K_s^0 in Z^0 Decays*, CERN-PPE/91-86;
ALEPH Collaboration, D. Decamp et al., DELPHI Collaboration, P. Abreu et al., Contributions to the Geneva EPS-LP 91 Conference, July 1991.

[15] Ya.I. Azimov, Yu.L. Dokshitzer, V.A. Khoze, and S.I. Troyan, *Zeit. Phys.* C27 (1985) 65; *ibid.* C31 (1986) 213.

[16] A. Erdélyi et al., *Higher Transcendental Functions*, volume 2, McGRAW-HILL, 1953.

[17] Yu.L. Dokshitzer, V.A. Khoze, and S.I. Troyan, *Int. Journ. of Mod. Phys.* A7 (1992) 1875.

[18] C.P.Fong and B.R.Webber, *Phys. Lett.* 241B (1990) 255.

[19] OPAL Collaboration, M.Z. Akrawy et al., *Phys. Lett.* 247B (1990) 617.

CHAIRMAN: Yu. Dokshitzer

Scientific Secretaries: J. Czyzewski, P. Hernandez, M. Kaur, R. Malik

DISCUSSION I

– *Sivaram:*

Could you clarify why you choose different values of Λ parameter for neutral pions and for charged particles?

– *Dokshitzer:*

The theoretical formula was derived for massless particles and you can consider pions as particles of that type. At the same time, experimental data are taken for all charged particles including heavier hadrons. If you take the theoretical prediction for pions and try to fit the experimental data, you are forced to choose an effective value of Λ which is larger than that for nearly massless π's.

– *Potters:*

Could you give a theoretical definition of Λ_{eff} and explain how it relates to other phenomenological definitions?

– *Dokshitzer:*

Refering to Λ_{QCD} is a renormalization-scale independent way of parametrizing the QCD coupling α_s. To be able to compare values of Λ extracted from different experiments one has to have corresponding theoretical predictions derived including at least first next-to leading corrections, since a finite rescaling say $\Lambda \to 2\Lambda$, would shift the magnitude of the first subleading term by a constant proportional to $\beta_2 \log 2$.

When studying parton cascades theoretical predictions one gets:

$$e^{C_0 \sqrt{\ln \frac{E}{\Lambda}}} \left(\ln \frac{E}{\Lambda}\right)^{C_1} \left(1 + C_2 \left(\ln \frac{E}{\Lambda}\right)^{-1/2} + \ldots\right)$$

where C_0 and C_1 are known numbers which are controlled by the leading Double Log and the subleading Modified Leading Log approximations respectively. As you can easily see in this case rescaling of Λ effects not C_1 but the next subleading coefficient C_2 only. Therefore the value of $\Lambda_{\text{eff}} \approx 150$ Mev we extracted from fitting the $e^+e^- \to$ hadron spectra cannot be directly compared with other determinations of Λ before next-to-next leading calculations improving the MLLA accuracy will be performed.

– *Syed:*

Can I believe the measurement of α_s using jet rate analysis, knowing that the definition of jets is arbitrary?

– *Dokshitzer:*

Provided that your definition of a jet is infrared safe, and it is the case for both original JADE algorithm and the new "Durham" or $K_\perp$ algorithm, the value of α_s you measure will be independent of the particular definition you choose.

– *Passalacqua:*

Since the JADE Algorithm has been revised, should the α_s measurement of the LEP experiment be revised too?

– *Dokshitzer:*

Revising the JADE algorithm makes it possible to improve theoretical predictions for Jet Rates and should thus reduce theoretical uncertainty of the α_s determination by weakening the factorization scale dependence and hopefully an influence of hadronization effects.

– *Weselka:*

We cannot directly observe the complex evolution of a jet, but we can measure the correlations of the resulting hadrons in the final phase space. Do you think that the self-similarity at the evolution stage of the jet could produce self-similar fluctuations in the phase-space (something which is called "intermittency"), or do you think that such correlations are destroyed in the hadronization phase? Can we learn something new from such intermittency studies?

– *Dokshitzer:*

The very idea of the LPHD is that the observed hadrons really follow the distributions of particles which are there at small distances. It means that the hadronization process is local in the phase-space. From this point of view I would not expect any serious thing to happen, even with the phenomena you mentioned. Nevertheless, when studying small rapidity windows you should not forget that the effective hardness of the process is not determined by the total W but is decreased proportionally to the size of the rapidity window. So, at present energies, studying intermittency, or other speaking fractal structure of a jet, means dealing with soft processes mainly.

– *Neubert:*

In order to fit the energy spectrum of kaons, you had to introduce a cut off Q_0, below which the cascade is truncated. Can you predict the dependence of Q_0

on the mass of the produced hadron, or is Q_0 a fit-parameter not well-determined from theory?

– Dokshitzer:

Q_0 is a pure phenomenological parameter that must be obtained by fitting experimental data. The only thing which is clear is that Q_0 should increase with m_h although there exist no experimental data available yet to obtain any precise dependence.

– Schulze:

How do you bring the finally observed hadrons to their mass-shell when truncating the cascade?

– Dokshitzer:

In the predictions we obtain, there are not hadrons at all. You just produce a distribution of massless partons. The only new point is that you stop cascading when the invariant transverse momentum in the splitting reaches the value of Q_0.

– Syed:

Would the $\ell n(1/x)$ distribution look different for quark and gluon jets?

– Dokshitzer:

The particle spectra in these two cases slightly differ in total energy normalization. It is an interesting problem and it is going to be studied experimentally.

– Shabelski:

Do you take into account the strange quarks separately in your cascade and are you sure that their number is exactly equal to the number of kaons?

– Dokshitzer:

No, in the framework of this analytic game your cascades consist of secondary gluons mainly, and at the last stage one has to introduce phenomenological gluon → hadron conversion coefficients, so there is no direct correspondence between the number of kaons and that of strange sea-quarks.

– Lupia:

Just a comment. You said that the cascade mechanism can be applied only to hard processes. Such mechanism can lead to a negative binominal distribution which however, describes also the soft reactions. But one gets the same distribution at the partonic level as well, e.g. in Lund parton shower. So, the occurance of negative binominal distribution at both hadronic and partonic level can support the LPHD.

– *Dokshitzer:*

I did not say that there is no cascading in soft processes. But only in hard processes we can keep everything under control. The fact that there are so many similarities between hard and soft processes, as e.g. the negative binominal distribution, is a puzzle which may be of very high importance and it is still a question to be studied.

– *Kaur:*

Very high P_T jets will be a dominant feature of hard scattering at LHC. What possible deviations of the measured cross-sections at high P_T from the predicted QCD values would you expect, which might arise due to possible manifestation of quark compositeness visible at such large momentum transfer?

– *Dokshitzer:*

I think that the internal QCD uncertainties will be a couple of orders of magnitude larger than any sign of compositeness you would be able to see even at LHC. The main uncertainty will come from poorly known structure functions.

– *Khalatyan:*

In your comparison of $< n_{ch} >$ vs E_{CM} dependence with the experimental data, all the curves have a crossing point at about 35 GeV. Can we understand this effect?

– *Dokshitzer:*

It has no physics meaning. It is the result of the fact that all the curves have been normalized at that energy. The point is the different rates of multiplicity growth predicted by the cascade model without coherence, DLA and MLLA.

CHAIRMAN: Yu. Dokshitzer

Scientific Secretaries: J. Czyzewski, P. Hernandez, M. Kaur, R. Malik

DISCUSSION II

– *Syed:*

Does the string effect vary with a variation in the angle between the quark and gluon jets?

– *Dokshitzer:*

Yes. The ratio of particle flows depends on the geometry of the 3-jet ensemble in a predictable way.

– *Khoze:*

My question is connected with this mystery of 80's & 90's that you mentioned. Do you know at least one experiment where this famous multiplicity ratio N_g/N_q has been measured directly?

– *Dokshitzer:*

As you know one has to be careful when discussing this ratio. The clear unambiguous test could be obtained by comparing, for example:

$$e^+e^- \to q\bar{q} \quad \text{with}$$
$$\chi \to gg$$

at $s^{e^+e^-} = M_\chi^2$ with C-even heavy quark bound state χ used as a source of two-gluon jet events. What people are trying to do in reality is a comparison of quark and gluon jets from 3-jet events where the very notion of multiplicity of an individual jet is not well defined. However if the jets are cleanly resolved then from this type of comparison we could check the theoretical prediction about the ratio of particles from quark and gluon jets. 9/4 is the leading approximation and there are known two subleading terms $C_1 \alpha_s^{1/2} + C_2 \alpha_s$. This correction is negative and of the order of 10%.

– *Zichichi:*

It has been demonstrated in pp colliders that by correct analysis of subtraction of leading effect, we get same multiplicity in e^+e^- and pp collisions. In e^+e^- collision, the multiplicity results from quark induced jets while in the case of pp, there is a contribution from gluon jets resulting in hadronization. If the factor N_g/N_q is 2, then we could not have found such a spectacular agreement, as within $\pm$ 20%.

– *Dokshitzer:*

I can agree with you but partially. If you speak about minimum bias pp interactions with comparatively small momentum transfer, I cannot insist on having a ratio of 2 for the gluon and quark hadronization density. But extra QCD bremsstrahlung which is controlled perturbatively and comes onto stage with increasing the hardness of the process should be different for q and g jets. The hardness of the two processes should be compared. In pp collisions it is determined by P_T and in e^+e^- by the total annihilation energy. The hardness determines an amount of QCD radiation and thus the extra multiplicity of the final state particles. Therefore the rate of multiplicity growth should be about twice as large in case of gluon jets as in case of quark jets. And this must be true for both e^+e^- and pp at the same hardness.

– *Czyzewski:*

You mentioned that depending on whether a reaction goes into the t-channel or u-channel, the colour strings are formed in different angular regions and one should observe strong fluctuations from event to event in angular distributions. Are such fluctuations observed experimentally?

– *Dokshitzer:*

No. Not yet. But this should be definitely done. The most spectacular is the scattering of gluons. In this case one Lorentz amplitude contains by itself two different colour-topologies (i.e. two ways of making colour connections-drawing "strings").

– *Ellis:*

Are you aware of any QCD predictions for the multiplicity flow in 4-jet events in e^+e^- annihilation? It seems to me likely that the probabilistic Lund string picture would completely break down in this case. In view of the great interest in jets at LEP, this is a very topical question.

– *Dokshitzer:*

To my knowledge there was no special study of multiplicity flow pattern in 4 jet events. However according to our experience, the Lund string picture should reproduce the QCD motivated pattern up to $1/N_c^2$ corrections in this case as well.

– *Hernandez:*

I believe there are some projects of using forward products in pp collisions as high energy heavy quark beams to use them afterwards in fixed target experiments, for example to detect τ neutrino. For those experiments quite precise knowledge of the energy spectrum of charmed particles at low P_T is needed. Could you comment on the feasibility of these types of experiments?

– *Dokshitzer:*

For the heavy quarks, it is commonly believed that one may use perturbative approach because the mass of heavy quark itself sets the small distance boundary. There is an observation made by S. Brodsky et al. that in case of heavy quark, there is some peculiar soft physics. There could be some long living fluctuations due to heavy quarks inside the hadrons, present in the initial state. Since the quark participates in the strong interaction and forms a part of the wave function of the initial proton, this can influence the structure functions. It is not easy to predict this contribution rigorously, so the distributions of heavy quarks inside protons in the region of large x appear to be not so easy to control quantitatively.

– *Khalatyan:*

Do you have any suggestions to the experiments for jet registration?

– *Dokshitzer:*

You can choose some algorithm to separate the events in terms of the number of jets. But I would prefer to study the energy-multiplicity correlations being purely at inclusive level to measure characteristics of QCD jets without involving any jet-finding algorithm.

– *Wadhwa:*

For the calculation of azimuthal asymmetry you always consider 3-jet events. What happens in case of 4-jet events?

– *Dokshitzer:*

Azimuth jet asymmetry for pp and p$\bar{\text{p}}$ interactions we discussed gives an example of predictions based on calculations for 4-jet event ensembles.

– *Etzion:*

What is the physical meaning of the negative cross-section you have mentioned in connection with the asymmetry in jets?

– *Dokshitzer:*

It has no physical meaning but is a matter of interpretation only. The cross-section is definitely positive. However, when trying to express the predicted particle flow in terms of a sum of contributions of two-parton antennas in a probabilistic manner, one finds the colour-suppressed $1/N_C^2$ antenna in e.g. $e^+e^- \to q\bar{q}g$ events to give a negative contribution. Negative pieces are problematic to incorporate into Monte Carlo event generators. Therefore up to now string-motivated hadronization models neglected $1/N_C^2$ effects.

– *Sivaram:*

1) Could you clarify how the azimuthal asymmetry depends on energy and on the other parameters?

2) In the string picture is there any estimate of the tension and its relation to the Λ parameter?

– *Dokshitzer:*

1) Azimuthal asymmetry of QCD jets depends on the angle between the jets and the number of colours in your theory, at least asymptotically, and very weakly depends on energy via the factor $(\alpha_s(E\theta_0))^{1/2}$ with θ_0 the size of the jet cone inside

which asymmetry is studied. This is a very slowly decreasing function of the jet energy E.

2) String picture is a phenomenological picture, giving a pictorical representation of hadron production. The "string tension" can be related to the mean value of $< K_\perp^2 >$ in the hadronization process. Formally speaking the string tension can be said to be determined by the same characteristic QCD scale Λ.

– *Gallo:*

To study string effect at LEP, what happens if one uses energy flow instead of particle-flow?

– *Dokshitzer:*

The string effect by definition is a property of soft particles which determine the main multiplicity. When you study the energy flow you select multijet events. The energy weight prepares antenna for you and counting particle for multiplicity flow is a study of soft emission by this antenna.

– *Shabelski:*

We can select 4-quark jet events by selecting events with 2 charmed hadrons and 2 beauty hadrons, all comparatively fast. Can you calculate interference picture for such events?

– *Dokshitzer:*

I am afraid this would be a rather costly way of getting rid of gluon jets. Moreover, since one of the two quark pairs originates from the gluon splitting the coherent large angle gluon radiation by a quasicollinear $c\bar{c}$ or $b\bar{b}$ pair will be identical to bremsstrahlung off a gluon jet.

– *Lupia:*

Recently DELPHI Collaboration has performed a study of multiplicity distribution in 2-jet, 3-jet and 4-jet samples separately. Do you think that the comparison between the 2-jet and the 3-jet sample could give some information on the gluon-jet?

– *Dokshitzer:*

This is not enough. I would like the analysis to be done separately for quark and gluon identified jets otherwise all the effects will be smeared. Let me mention a very promising new technology called neural networks in which the computer network is trained to recognize the pattern necessary for separating the jets. This gives presently about 85% identification of the nature of the jet.

THE STANDARD MODEL AND BEYOND

John Ellis

CERN, Geneva, Switzerland

0 - Beyond the Standard Model

The outstanding problems of the Standard Model of particle physics can conveniently be classified into 3 main categories. The most immediate is that of *Mass*: why are the quark, lepton and electroweak gauge boson masses non-zero? And why are they so small: $m_W/m_P \sim 10^{-10}$? The answer to the first question is presumably some variant of the Higgs mechanism, entailing the existence of at least one physical Higgs boson. The answer to the second question probably involves some additional symmetry, such as technicolour [1] or supersymmetry [2], entailing the existence of many more particles. The answer to both these questions should be found at energies $\leq$ TeV, putting the solution to the problem of Mass within reach of prospective experiments. Then there is the problem of *Unification*: is there a unifying group comprising all the particle interactions, some Grand Unified Theory (GUT) group [3] containing SU(3) × SU(2) × U(1)? This would entail interesting new phenomena such as proton decay and neutrino masses, but the characteristic energy scale would be at least 10^{15} GeV. Nevertheless, some indirect signatures may be found in detailed measurements at present energies, as we shall see later. Finally, there is the problem of *Flavour*: why are there so many types of matter particles, and what fixes the ratios of their masses and the charged current mixing angles? It is often suggested that quarks and leptons are composite, but the scale of such compositeness is most unclear, and there are no very attractive models, so

I will not discuss this possibility in these lectures. All the above problems should find their resolution in a Theory of Everything (TOE), for which the only viable candidate is some variant of string theory. In the second lecture I will discuss ideas for the structure of this TOE, based in particular on the flipped SU(5) GUT [4].

1 - GUTs AND SUPERSYMMETRIC GUTs

1.1 - Introduction to Grand Unification

The Standard Model has many free parameters, namely 3 gauge couplings g_3, g_2 and g_1 and 2 non-perturbative vacuum angle parameters θ_3, θ_2; 6 quark and 3 charged lepton masses as well as 4 Cabibbo-Kobayashi-Maskawa mixing angle parameters, and 2 boson mass parameters which may be taken as m_W, m_H; a total of 20 parameters. In addition, there are the apparently arbitrary choices of SU(3) × SU(2) particle quantum numbers:

$$(u,d)_L : (3,2) \; ; \; u_L^c : (\bar{3},1) \; ; \; d_L^c : (\bar{3},1)$$
$$(\nu,e)_L : (1,2) \; ; \; e_L^c : (1,1) \quad (1.1)$$

as well as the U(1) hypercharges $Y = Q_{em} - I_3$. Why are these chosen in just such a way that electric charge is quantized: $Q_p + Q_e = 2Q_u + Q_d + Q_e = 0$?

The basic idea of grand unification is to exploit the logarithmic evolution of coupling strengths, which permits the hope that $g_3 = g_2 = g_1$ at some exponentially high energy $m_X = \Lambda_{QCD} \exp(0(1)/\alpha_{em})$, permitting the incorporation of SU(3), SU(2) and U(1) into some GUT group G. Baryon stability requires $m_X > 10^{14}$ GeV, while the neglect of quantum gravity is consistent only if $m_X < 10^{19}$ GeV, constraints that are met only if [5]:

$$1/120 < \alpha_{em} < 1/170 \quad (1.2)$$

It is encouraging that the experimental value of α_{em} fits comfortably inside this range!

The simplest GUT model is based on the group SU(5) [3], which has an adjoint $\underline{24}$ representation of gauge bosons:

$$\begin{pmatrix} & : & \overline{X}\,\overline{Y} \\ g_{1,\ldots,8} & : & \overline{X}\,\overline{Y} \\ & : & \overline{X}\,\overline{Y} \\ \ldots & \ldots & \ldots \\ XXX & : & \\ YYY & : & W \end{pmatrix} : Y \propto \text{diag}\left(1,1,1,-\frac{3}{2},-\frac{3}{2}\right) \quad (1.3)$$

The quarks and leptons are assigned to $\bar{5}$ and $\underline{10}$ representations for the first generation:

$$\overline{F}: \begin{pmatrix} d_r^c \\ d_y^c \\ d_b^c \\ \ldots \\ -e^- \\ \nu_e \end{pmatrix}_L \quad ; \quad T: \frac{1}{\sqrt{2}} \begin{pmatrix} 0 & u_b^c & -u_y^c & \vdots & -u_r & -d_r \\ -u_b^c & 0 & u_r^c & \vdots & -u_y & -d_y \\ u_y^c & -u_r^c & 0 & \vdots & -u_b & -d_b \\ \ldots & \ldots & \ldots & \vdots & \ldots & \ldots \\ u_r & u_y & u_b & \vdots & 0 & -e^+ \\ d_r & d_y & d_b & \vdots & e^+ & 0 \end{pmatrix}_L \quad (1.4)$$

Charge quantization is automatic, because Q_{em} is a generator of SU(5), so that $\text{Tr} Q_{em} = 0$ in any representation. In particular, in the $\bar{5}$ representation $3Q_{d^c} + Q_{e^-} = 0$, so that $Q_d = -1/3$ and $Q_p = +1$ follows. The breaking of SU(5) to SU(3) × SU(2) × U(1)$_Y$ is most economically achieved via an adjoint $\underline{24}$ of Higgs ϕ, and the breaking of SU(2) × U(1)$_Y$ to U(1)$_{em}$ via a $\underline{5}$ of Higgs H.

The most characteristic prediction of any GUT is for baryon decay, which is mediated by new massive gauge bosons such as the X and the Y of minimal SU(5) (1.3), which have couplings of the form (dropping internal indices):

$$\frac{g}{\sqrt{2}} X_\mu (\bar{d}\gamma^\mu e^+, u\gamma^\mu u) \quad ; \quad \frac{g}{\sqrt{2}} Y_\mu (\bar{d}\gamma^\mu \bar{\nu}, \bar{u}\gamma^\mu e^+, u\gamma^\mu d) \quad (1.5)$$

yielding an effective d=6 $\Delta B = \Delta L = 1$ interaction:

$$\mathcal{L}_{GUT} = \frac{g^2}{2_X^2} \left[(u\gamma^\mu u)(e\gamma_\nu d) \quad , \quad (u\gamma^\mu d)(\nu\gamma_\mu d) \right] \quad (1.6)$$

This gives a baryon decay rate $\Gamma_B \propto m_X^{-4}$ and hence a baryon lifetime $\tau_B \sim m_X^4/m_B^5$. Detailed estimates in minimal SU(5) give [6]:

$$m_X \simeq (1 \text{ to } 2) \times 10^{15} \Lambda_{\overline{MS}} \simeq (1 \text{ to } 4) \times 10^{14} \text{ GeV} \quad (1.7)$$

and baryon lifetimes:

$$\tau_{p,n} \simeq 10^{-29 \pm 2} y \quad (1.8)$$

with preferred decay modes $p \to e^+\pi^0, \bar{\nu}\pi^+, n \to e^+\pi^-, \bar{\nu}\pi^0$. These estimates are in apparent conflict with the experimental lower limit $\tau(p \to e^+\pi^0) > 9 \times 10^{32} y$ [7].

1.2 - Testing GUT Models

In the absence of proton decay, one can test GUT models via their predictions for the low-energy gauge couplings, which are measured with precision at LEP and elsewhere. The energy-dependences of the Standard Model gauge couplings are given by the renormalization group equations:

$$Q \frac{\partial \alpha_i(Q)}{\partial Q} = -\frac{1}{2\pi}\left(b_i + \frac{b_{ij}}{4\pi}\alpha_j(Q)\right)\left[\alpha_i(Q)\right]^2 + \ldots \quad (1.9)$$

where the one-loop coefficients in a non-supersymmetric GUT are:

$$b_i = \begin{pmatrix} 0 \\ -22/3 \\ -11 \end{pmatrix} + N_g \begin{pmatrix} 4/3 \\ 4/3 \\ 4/3 \end{pmatrix} + N_H \begin{pmatrix} 1/10 \\ 1/6 \\ 0 \end{pmatrix} \quad (1.10)$$

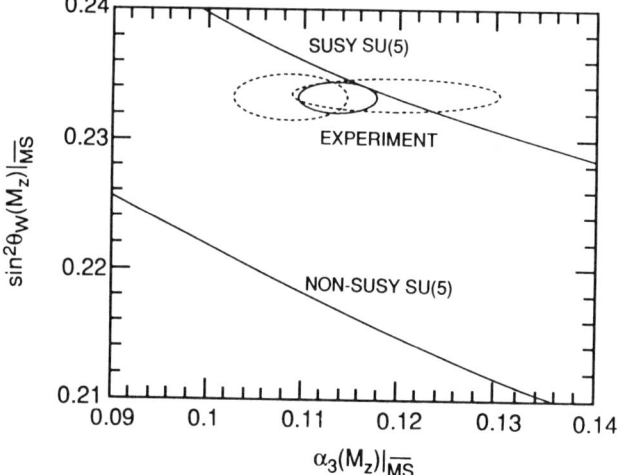

Fig. 1. *GUT predictions in the $(\alpha_3, \sin^2\theta_W)$ plane, both with and without supersymmetry, compared with experimental data.*

and the two-loop coefficients are:

$$b_{ij} = \begin{pmatrix} 0 & 0 & 0 \\ 0 & -136/3 & 0 \\ 0 & 0 & -102 \end{pmatrix} + N_g \begin{pmatrix} 19/15 & 3/5 & 44/15 \\ 1/5 & 49/3 & 4 \\ 11/30 & 3/2 & 76/3 \end{pmatrix} + N_H \begin{pmatrix} 9/50 & 9/10 & 0 \\ 3/10 & 13/6 & 0 \\ 0 & 0 & 0 \end{pmatrix} \quad (1.11)$$

independently of the specific GUT model.

The formulae (1.9,1.10,1.11) can be used together with the observed values of α_{em} and $\alpha_3(m_{Z^0})$ to predict $\sin^2\theta_W(M_Z)|_{\overline{MS}}$ with the result shown in fig. 1, which was unsuccessful even before the advent of LEP data. In the absence of supersymmetry one finds [6] :

$$\sin^2\theta_W(m_Z)|_{\overline{MS}} = 0.208 + 0.004(N_H - 1) + 0.006 \ln\left(\frac{400 \text{ MeV}}{\Lambda_{\overline{MS}}(N_q = 4)}\right) \quad (1.12)$$

which yields 0.214 ± 0.002 for $N_H = 1$ and $\Lambda_{\overline{MS}} = 100$ to 200 MeV, to be compared with the experimental value of 0.233 ± 0.001 [8]. Another way of expressing this disagreement is to show that the data extrapolate to unification of α_3, α_2 and α_1 with a single value at some GUT scale m. The prediction for $\sin^2\theta_W$ could in principle be rescued by increasing the number N_H of Higgs doublets to 6, but in this case m_X is still too low and the proton still decays too quickly. But we will now argue that the Mass Problem requires a more drastic modification of the particle spectrum.

1.3 - The Mass Problem

The generation of quark, lepton and gauge boson masses requires the spontaneous breaking of gauge symmetry via some $< 0|X_{I,I_3}|0 > \neq 0$. The ratio $\rho \equiv m_W^2/m_Z^2\cos^2\theta_W \simeq 1$ suggests that whatever X is, it mainly has $I = 1/2$, and this is also what is required to give fermion masses. The next question is whether X is elementary or composite. The former is the option exercised in the original version of the Standard Model, whereas a fermion condensate $< 0|X = \overline{F}F|0 > \neq 0$ has closer affinities to QCD, where $< 0|\bar{q}q|0 > \neq 0$, and to theories of superconductivity. The fermions F could either be new ones bound together by some new, very strong force: "technicolour"[1] , or perhaps just the t quark of the Standard Model [9], if it is sufficiently heavy. Both of these composite alternatives to an elementary Higgs boson are difficult to reconcile with the precision electroweak data: technicolour theories resemble in some respects theories with a very heavy Higgs boson, which is disfavoured by the data [8] , and naïve top quark condensation models require the top quark to weigh more than 200 GeV, which is also disfavoured by the data [8] . In view of these problems, we focus on the alternative of an elementary Higgs boson.

The problem here is that when one calculates radiative corrections to its mass, or equivalently that of the W, one finds quadratic divergences:

$$\delta m_{H,W}^2 \simeq g^2 \int^\Lambda \frac{d^4p}{(2\pi)^4}/p^2 \simeq 0\left(\frac{\alpha}{\pi}\right)\Lambda^2 \quad (1.13)$$

where Λ is a cutoff representing the appearance of new physics beyond the Standard Model. If Λ is identified with either $m_{GUT} \sim 10^{15}$ GeV or $m_P \sim 10^{19}$ GeV, one finds $\delta m_{H,W}^2 \gg m_{H,W}^2$. Such a situation is technically unnatural. It requires an absurd fine-tuning of the bare Higgs mass so that it cancels out almost exactly the radiative correction, so as to have the correct physical value:

$$m_{H,W}^2 = m_0^2 + \delta m^2 \simeq -(10^{15} \text{ GeV})^2 + \left[(10^{15} \text{ GeV})^2 + (10^2 \text{ GeV})^2\right] \simeq (10^2 \text{ GeV}^2) \tag{1.14}$$

This procedure looks very artificial, and physicists would prefer the radiative corrections to be "naturally" small: $|\delta m_{H,W}^2| \leq m_{H,W}^2$, just like corrections to fermion masses:

$$\delta m_f = 0\left(\frac{\alpha}{\pi}\right) m_f \, \ell n\left(\frac{\Lambda}{m_f}\right) \leq m_f \tag{1.15}$$

even for $\Lambda \sim m_P$.

Supersymmetry removes the quadratic divergences [2] and makes the radiative corrections "naturally" small as desired. This trick is based on the observation that boson and fermion loops have opposite signs:

$$\delta m_{H,W}^2 \simeq 0\left(\frac{\alpha}{\pi}\right) \left[(\Lambda^2 + m_B^2) - (\Lambda^2 + m_F^2)\right] \tag{1.16}$$

and will cancel if the boson and fermion couplings are chosen to be equal. This is guaranteed by supersymmetry. The minimal supersymmetric extension of the Standard Model contains extra particles with the same internal quantum numbers as those in the Standard Model, but have spins differing by half a unit: $\ell \rightarrow$ sleptons $\tilde{\ell}, q \rightarrow$ squarks $\tilde{q}, g \rightarrow$ gluinos $\tilde{g}$, Higgs $\rightarrow$ higgsino $\tilde{H}$, $W \rightarrow$ wino $\widetilde{W}, Z \rightarrow$ zino $\widetilde{Z}, \gamma \rightarrow$ photino $\tilde{\gamma}$. It is described by a Lagrangian. They would then be renormalized to different values in the effective low-energy theory [10]:

$$\mathcal{L} = \mathcal{L}_{\text{SUSY}} + \mathcal{L}_{\text{SUSY}\times} \tag{1.17}$$

where the supersymmetric part $\mathcal{L}_{\text{SUSY}}$ has the same gauge interactions as in the Standard Model, and the Yukawa interactions $(f_i f_J)\partial^2 P/\partial \phi_i \partial \phi_j$, are all derived from a superpotential $P(\phi_i)$:

$$P = \sum_L \lambda_L L_L L_L^c H_1 + \sum_{q,p^c} \lambda_p q_L p_L^c H_2 + \sum_{q,n^c} \lambda_n q_L n_L^c H_1 + \mu H_1 H_2 \tag{1.18}$$

The first 3 terms give masses to the leptons L, charge-2/3 quarks p and charge-1/3 quarks n respectively. The $\lambda_{p,n}$ are matrices in flavour space whose diagonalization yields the Kobayashi-Maskawa mixing angles. Note that two Higgs doublets $H_{1,2}$ are needed, which mix via the last term in (1.18). Associated with the superpotential P and gauge couplings g_α is an effective potential:

$$V = \sum_i (F_i)^2 + \frac{1}{2} \sum_{\alpha=1}^{3} g_\alpha^2 \sum_a \left| \phi_i^* (T^a)^i_j \phi^j \right|^2 \qquad (1.19)$$

where:
$$F_i^* \equiv \partial P / \partial \phi^i \qquad (1.20)$$

To get a realistic spectrum with $m_{\tilde{e}} \gg m_e$, etc., one must incorporate supersymmetry breaking:

$$\mathcal{L}_{SUSY\cancel{x}} \ni -\sum_i m_{0_i}^2 |\phi_i|^2 - \sum_\alpha M_\alpha \tilde{V}_\alpha \tilde{V}_\alpha + \dots \qquad (1.21)$$

where the m_{0_i} are scalar masses and the M_α ($\alpha = 1, 2, 3$) gaugino masses which presumably originate in some supergravity or superstring theory. In this case, they may be universal at some large renormalization scale:

$$m_{0_i}|_{Q \simeq m_P} \equiv m_0 \quad , \quad M_\alpha|_{Q \simeq m_P} \equiv m_{1/2} \qquad (1.22)$$

They would then be renormalized to different values in the effective low-energy theory [10]:

$$m_0^2|_{Q \simeq m_W} = m_0^2 + C_i m_{1/2}^2 \quad , \quad M_\alpha|_{Q \simeq m_W} = \frac{\alpha_\alpha}{\alpha_{GUT}} m_{1/2} \qquad (1.23)$$

where:
$$C_{\tilde{q}} \simeq 6, \quad C_{\tilde{\ell}_L} \simeq 0.5, \quad C_{\tilde{\ell}_R} \simeq 0.15, \quad \frac{\alpha_3}{\alpha_{GUT}} \simeq 3 \text{ etc.} \qquad (1.24)$$

Using the simplified parametrization in terms of m_0 and $m_{1/2}$, one can compare the constraints from different experiments [11]:

$$e^+e^- : m_{\tilde{q}, \tilde{\ell}, \tilde{W}^\pm / \tilde{H}^\pm} \geq 45 \text{ GeV} \quad , \quad \bar{p}p : m_{\tilde{q}, \tilde{g}} \geq 100 \text{ GeV} \qquad (1.25)$$

in a meaningful way. Present constraints in the $(m_0, m_{1/2})$ plane are shown in fig. 2 together with contours of heavier masses of sleptons, squarks and gluinos. The figures are plotted for two representative values of the ratio $\tan \beta \equiv v_2/v_1$ of Higgs vacuum expectation values. Constraints in the $(\mu, m_{1/2})$ plane [12] are compiled in fig. 3 for the $\tan \beta = 2$ case. Important constraints come from LEP searches for $Z^0 \to \widetilde{W}^\pm / \widetilde{H}^\pm$ pairs and $Z^0 \to$ pairs of $\widetilde{Z}/\tilde{\gamma}/\widetilde{H}_{1,2}$ contributions, as well as from the FNAL $\bar{p}p$ collider search for gluinos. Also shown are contours of $m_{\widetilde{W}^\pm / \widetilde{H}^\pm}$, m_{LSP} and $m_{\tilde{g}}$.

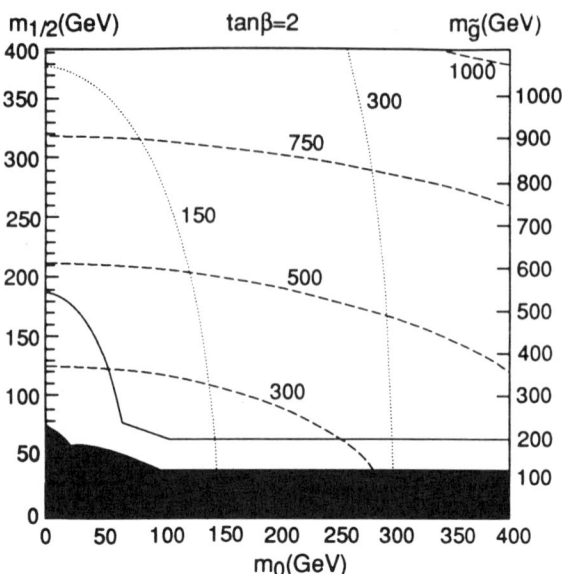

Fig. 2. Present limits and future sensitivity in the $(m_0, m_{1/2})$ plane, for the value $\tan\beta = 2$. The dotted (dashed) lines are contours of $m_{\tilde{e}}(m_{\tilde{q}})$.

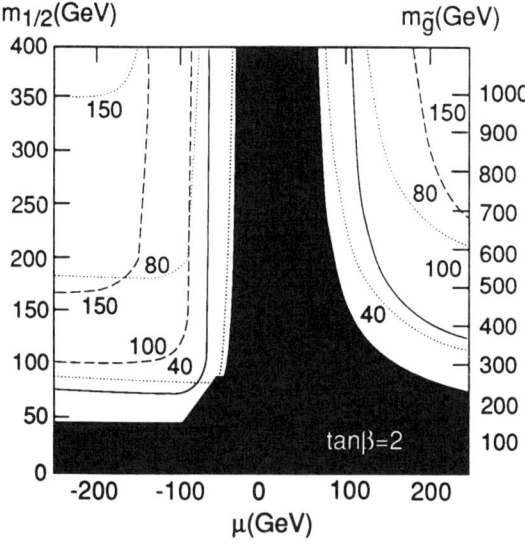

Fig. 3. Present limits and future sensitivity in the $(\mu, m_{1/2})$ plane, for the value $\tan\beta = 2$. The dotted (dashed) lines are contours of $m_{\text{LSP}}(m_{\tilde{W}^{\pm}})$, and the solid line represents the possible discovery potential of LEP.

1.4 - Higgs bosons in supersymmetric models

Supersymmetric models must contain an even number of Higgs doublets, in order to give masses to all the quarks and leptons, and to cancel triangle anomalies. The minimum number is 2 doublets, as discussed above. In this case there are five physical Higgs bosons, 3 neutral (2 CP-even: h,H, and one CP-odd: A) and two charged ($H^{\pm}$). At the tree-level, their masses and couplings are all described by two parameters, which are often taken to be (m_H, m_A) or ($m_A, \tan \beta$), and they obey interesting and powerful mass formulae:

$$m_h^2 + m_H^2 = m_A^2 + m_Z^2 ,$$

$$m_a^2 = m_{H^{\pm}}^2 - m_{W^{\pm}}^2 , \quad m_{h,H}^2 = \frac{1}{2} \left(m_A^2 + m_Z^2 \mp \sqrt{(m_A^2 + m_Z^2)^2 - 4 m_Z^2 m_A^2 \cos^2 2\beta} \right) \quad (1.26)$$

We deduce from these formulae that at least one neutral Higgs boson weighs less than m_Z, whilst at least one is heavier than m_Z. It used to be thought on the basis of these tree-level results that LEP 2 was very likely to be able to find at least the lightest neutral supersymmetric Higgs boson h.

However, it has recently been realized [13] that there are very important radiative corrections to the supersymmetric Higgs boson masses, of which the largest are those associated with the top quark and associated particle ($\tilde{t}, \tilde{b}$) loops. These can be approximated to within a few GeV by computing the one-loop effective potential:

$$m_{phys}^2 \simeq \left. \frac{\partial^2 V}{\partial \phi^2} \right|_{\phi_i = v_i \equiv <0|\phi_i|0>} \qquad V \simeq V_1(Q) = V_0(Q) + \delta V_1(Q) \quad (1.27)$$

where:

$$\delta V_1(Q) = \frac{1}{64 \pi^2} \, \text{Str} \, M^4 \left(\ell n \, \frac{M^2}{Q^2} - \frac{3}{2} \right) \quad (1.28)$$

where Q is the renormalization scale, M is the mass matrix, and $\text{Str} \equiv \sum_{B,F} (-1)^F$. The t quark diagrams contain two factors of the Yukawa coupling λ_t, and hence are proportional to m_t^2. The one-loop effective potential would be quadratically divergent in the absence of supersymmetry, which becomes an $m_{\tilde{q}}^2$ dependence when sparticles are introduced. In the limit $m_t, m_{\tilde{q}} \gg m_W$ with $m_t/m_{\tilde{q}}$ fixed, one has [13]:

$$\left. \frac{\partial V_1}{\partial \phi^2} \right|_{0_i = u_i} = \frac{3 g^2}{8 \pi^2} \frac{m_t^4}{m_W^2} \ell n \, m_{\tilde{t}}^4 / m_t^4 + 0(m_t^2) \quad (1.29)$$

i.e. a quartic dependence on m_t. The resulting shift in m_h can be tens of GeV, and it is quite possible for m_h to be increased above m_A, and even above m_Z.

This makes detection at LEP more problematic, and may move detection into the realm of the LHC.

1.5 - Supersymmetric SU(5) GUT

The supersymmetrization [14] of the SU(5) GUT is motivated by the hierarchy and naturalness arguments mentioned earlier. The standard GUT multiplets become supermultiplets: matter representations $\bar{5}, \bar{F}$ and $10, T$, and Higgs representations: $\underline{5} + \underline{\bar{5}}, H + \bar{H}$, and $\underline{24}, \Phi$. The Higgs part of the superpotential contains terms of the following form:

$$P \ni \left(\mu + \frac{3\lambda}{2}M\right)\bar{H}H + \nu\bar{H}\Phi H + f(\Phi) \qquad (1.30)$$

where $f(\Phi)$ is chosen so that the tree-level effective potential vanishes when:

$$<0|\Phi|0> = M \begin{pmatrix} 1 & & & \\ & 1 & & 0 \\ & & 1 & \\ 0 & & & -3/2 \\ & & & & -3/2 \end{pmatrix} \qquad (1.31)$$

The coefficient of the $\bar{H}H$ term is chosen to cancel out the $\bar{H}\Phi H$ term to 13 decimal places, giving a small residual $\bar{H}H$ mixing term $\propto \mu$. Although the origin of this fine-tuning is very difficult to understand, the radiative corrections are under control.

The approach to grand unification is given by renormalization group equations of the form (1.9) that have the same coefficients in all supersymmetric GUTs that have no light particles beyond those in the minimal supersymmetric extension of the Standard Model:

at the one-loop level:

$$b_i = \begin{pmatrix} 0 \\ -6 \\ -9 \end{pmatrix} + N_g \begin{pmatrix} 2 \\ 2 \\ 2 \end{pmatrix} + N_H \begin{pmatrix} 3/10 \\ 1/2 \\ 0 \end{pmatrix} \qquad (1.32)$$

and at the two-loop level:

$$b_{ij} = \begin{pmatrix} 0 & 0 & 0 \\ 0 & -24 & 0 \\ 0 & 0 & -54 \end{pmatrix} + N_g \begin{pmatrix} 38/15 & 6/5 & 88/15 \\ 2/5 & 14 & 8 \\ 11/15 & 3 & 68/3 \end{pmatrix} + N_H \begin{pmatrix} 9/30 & 9/10 & 0 \\ 3/10 & 7/2 & 0 \\ 0 & 0 & 0 \end{pmatrix}$$
$$(1.33)$$

The comparison with data is quite delicate, so we discuss each of the low-energy inputs in some detail.

The correct value of $\alpha_{em}(m_{Z^0})$ is not as well determined as the value of the fine structure constant in the Thomson Limit, because of renormalization due to vacuum polarization diagrams. The main uncertainty is due to light quark loops, which must be evaluated using data on $e^+e^- \to$ hadrons and a dispersion integral over E_{cm}. This gives:

$$1/\alpha_{em}(m_{Z^0}) = 127.9 \pm 0.2 \tag{1.34}$$

with the largest contribution to the error coming from 3 GeV $\leq E_{cm} \leq$ 10 GeV. Better data in this range would be useful for tests of GUTs.

It is possible to estimate $\alpha_3(m_{Z^0})$ by extrapolation from lower-energy data. Deep inelastic and prompt photon production data give:

$$\alpha_3(m_{Z^0}) = 0.109 \pm 0.007 \tag{1.35}$$

where the scale error is included, whilst bottomonium decays yield:

$$\alpha_3(m_{Z^0}) = 0.109 \pm 0.005 \tag{1.36}$$

with an unknown systematic error. Medium-energy $e^+e^- \to$ hadrons total cross-section data yield a less precise value:

$$\alpha_3(m_{Z^0}) = 0.140 \pm 0.020 \tag{1.37}$$

Our best estimate of $\alpha_3(m_{Z^0})$ from $Z^0 \to$ jets data is [15]:

$$\alpha_3(m_{Z^0}) = 0.118 \pm 0.004 \pm 0.005 \tag{1.38}$$

where the first error is due to hadronization, and the second to uncertainties in higher-order contributions. We [15] believe that (1.36) is best extracted by renormalizing QCD at m_{Z^0} and exponentiating all the higher-order double infrared logarithms which yield corrections $O(\pi\alpha_s)^n$, avoiding the 2-jet region (where there are other large higher-order corrections) and the 4-jet region (where no radiative corrections have been calculated). On the basis of (1.35,1.36,1.37 and 1.38) our best guess is that [15]:

$$\alpha_3(m_{Z^0}) = 0.113 \pm 0.004 \tag{1.39}$$

This value is also compatible with a more recent estimate from τ decays[16].

Next we discuss the experimental value of $\sin^2\theta_W$, which requires a careful analysis of precision electroweak data including radiative corrections. It is convenient for this analysis to use the "mass-shell" definition [17]:

$$\sin^2\theta_W \equiv 1 - \frac{m_W^2}{m_Z^2} \tag{1.40}$$

in terms of which the vector boson masses are given at one-loop order by [18]:

$$m_W^2 \sin^2\theta_W = m_Z^2 \cos^2\theta_W \sin^2\theta_W = \frac{A}{1-\Delta r} \qquad (1.41)$$

where:

$$A = \frac{\pi\alpha}{\sqrt{2}G_\mu} = (37.2802/3 \text{ GeV})^2 \qquad (1.42)$$

and

$$\Delta r = \Delta\alpha - \cos^2\theta_W \, \Delta\rho + \widehat{\Delta r} \qquad (1.43)$$

The first term in the decomposition (1.43) is the change in α_{em} already mentioned. The second term also appears in the ratio of low-energy neutral and charged current cross-sections, and has a quadratic dependence on m_t [19]:

$$\Delta\rho \simeq \frac{3G_\mu \, m_t^2}{8\pi^2\sqrt{2}} \qquad (1.44)$$

The residual term has an additional logarithmic sensitivity to m_t:

$$\widehat{\Delta r} \ni \frac{\sqrt{2}G_\mu \, m_W^2}{8\pi^2}\left(\cos^2\theta_W - \frac{1}{3}\right) \ell n \, \frac{m_t^2}{m_W^2} \qquad (1.45)$$

whilst Δr also has a logarithmic sensitivity to the Higgs mass [18]:

$$\Delta r \ni \frac{\sqrt{2}G_\mu m_W^2}{16\pi^2}\left\{\frac{11}{3}\left(\ell n \, \frac{m_H^2}{m_W^2} - \frac{5}{6}\right)\right\} \qquad (1.46)$$

in the limit $m_H \gg m_W$.

Because of these radiative corrections, the apparent values of $\sin^2\theta_W$ extracted from different experiments have different dependences on m_t and m_H. Agreement between the extracted values of $\sin^2\theta_W$ is only possible for limited ranges of m_t and m_H. Figure 4 shows χ^2 curves as functions of m_t, for an assumed value $m_H = m_Z$, and fig. 5 shows the corresponding χ^2 curves as functions of m_H, for an assumed value $m_t = 130$ GeV. From this analysis, we find [8]:

$$m_t = 92 \text{ to } 147 \text{ GeV} \qquad (1.47a)$$

and

$$m_H = 1.4 \text{ to } 160 \text{ GeV} \qquad (1.47b)$$

The $\Delta\chi^2 = 1$ and 2.7 contours in the (m_H, m_t) plane are shown in fig. 4. We see that a "light" Higgs mass ($m_H < 300$ GeV) is apparently more likely than a "heavy" Higgs mass ($m_H > 300$ GeV). Moreover, there is very good consistency with the range $m_H \sim m_Z \pm 40$ GeV expected [13] for the lighter CP-even neutral

Higgs in the minimal supersymmetric extension of the Standard Model discussed in the previous section.

For comparison with GUTs, the most appropriate prescription for defining $\sin^2\theta_W$ is the $\overline{MS}$ definition [20], which is related to the "mass-shell" definition by:

$$\sin^2\theta_W(m_{Z^0})|_{\overline{MS}} = \left(1 - X_{\overline{MS}}\right)\sin^2\theta_W \qquad (1.48)$$

where:

$$X_{\overline{MS}} = \frac{\cos^2\theta_W}{\sin^2\theta_W}\,\text{Re}\left[\frac{A_{ZZ}(m_Z^2)}{m_Z^2} - \frac{A_{WW}(m_W^2)}{m_W^2}\right] \qquad (1.49)$$

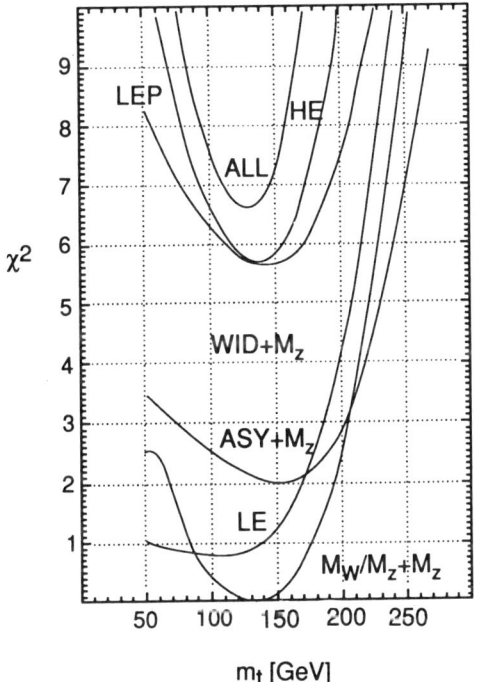

Fig. 4. The χ^2 functions, versus m_t, for different sets of the precision electroweak data.

and the $A_{ij}(Q^2)$ are vacuum polarization functions. The global fit [8] to precision electroweak data gives:

$$\sin^2\theta_W(m_{Z^0})|_{\overline{MS}} = 0.2327 \pm 0.0007 \qquad (1.50)$$

with the principal uncertainty that in m_t (1.47a).

You have all seen the plots showing that the low-energy couplings extrapolate perfectly using (1.32, 1.33) to meet at a supersymmetric GUT scale [21], [22], [23], [24]. It is certainly true, as seen in fig. 1, that the supersymmetric GUT prediction works much better than the non-supersymmetric GUTs. However, this agreement is not quite perfect [23], [24] as seen also in fig. 1, and the approximate agreement certainly cannot[23], [24] be used to set very stringent constraints on the scale of supersymmetry breaking. Parametrizing the light sparticle thresholds as in

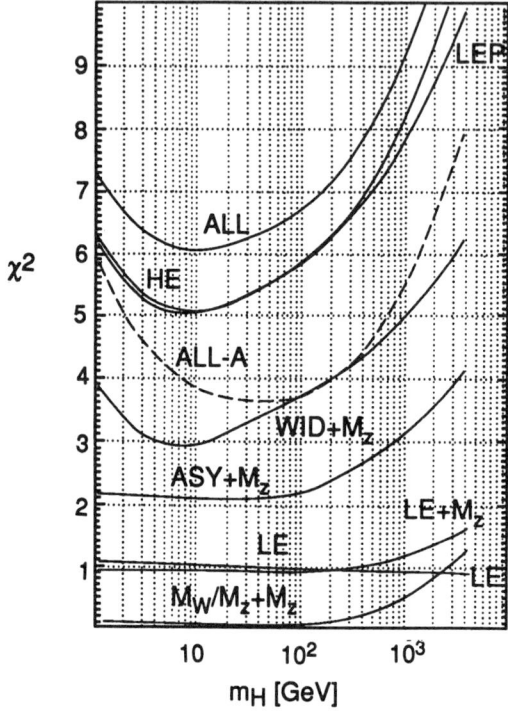

Fig. 5. The χ^2 functions, versus m_H, for different sets of the precision electroweak data.

equations (1.21,1.22), and keeping the Higgs mixing parameters μ and $m_H \simeq m_A$ as free parameters, we find [23] :

$$\sin^2\theta_W(m_{Z^0})|_{\overline{MS}} \simeq 0.2029 + 7\frac{\alpha_{em}(m_{Z^0})}{15\,\alpha_3}$$

$$+ \frac{\alpha_{em}(m_{Z^0})}{20\pi}\left[-3\ell n\left(\frac{m_t}{m_Z}\right) + \frac{28}{3}\ell n\left(\frac{m_{\tilde{g}}}{m_Z}\right) - \frac{32}{3}\ell n\left(\frac{m_{\widetilde{W}}}{m_Z}\right) - \ell n\left(\frac{m_h}{m_Z}\right)\right.$$

$$- 4\ell n\left(\frac{\mu}{m_Z}\right) + \frac{5}{2}\ell n\left(\frac{m_{1/2}\sqrt{6+y}}{m_Z}\right) - 3\ell n\left(\frac{m_{1/2}\sqrt{0.5+y}}{m_Z}\right) + 2\ell n\left(\frac{m_{1/2}\sqrt{0.15+y}}{m_Z}\right)$$

$$\left. - \frac{19}{36}\ell n\left(\frac{m_{1/2}\sqrt{6+y+w}}{m_Z}\right) - \frac{35}{36}\ell n\left(\frac{m_{1/2}\sqrt{6+y-w}}{m_Z}\right)\right]$$

(1.51)

where $y \equiv m_0/m_{1/2}$ and w parametrizes the stop mass matrix. The formula (1.51) can be inverted to give [23]:

$$\ln\left(\frac{m_{\frac{1}{2}}}{m_Z}\right) = \frac{15\pi}{\alpha_{em}(m_{Z_0})}\left[0.2029 + \frac{7\alpha_{em}(m_{Z^0})}{15\alpha_3(m_{Z^0})} - \sin^2\theta_W(m_{Z^0})|_{\overline{MS}}\right.$$
$$\left. + f(y,w) - \frac{1}{4}\ln\left(\frac{m_t}{m_Z}\right) - \frac{3}{4}\ln\left(\frac{m_h}{m_Z}\right) - 3\ln\left(\frac{\mu}{m_Z}\right)\right] + 8.839$$

where:

$$f(y,w) = \frac{15}{16}\ln(6+y) - \frac{19}{96}\ln(6+y+\omega) - \frac{35}{96}\ln(6+y-w) - \frac{9}{8}\ln(0.5+y)$$
$$+ \frac{3}{4}\ln(0.15+y)$$
(1.52)

Putting in central values and 1σ errors for $\alpha_{em}(1.34), \alpha_3(m_{Z^0})(1.39)$ and $\sin^2\theta_W(m_{Z^0})|_{\overline{MS}}$ (1.50), assuming the range (1.47a) for m_t, which is largely correlated with the value of $\sin^2\theta_W(m_{Z^0})|_{\overline{MS}}$, and allowing $\mu = 100$ GeV $\times 3^{0\pm1}$, $m_H = 300$ GeV $\times 3^{0\pm1}$ we find [23]:

$$\ln\left(\frac{m_{\frac{1}{2}}}{m_Z}\right) = 19.8 \mp 7.3 \mp 6.9 \mp 0.3 \,^{-3.3}_{+0.3} \mp 0.2 \mp 0.8$$
$$\sin^2\theta_W \quad \alpha_3 \quad \alpha_{em} \quad m_h \quad y \quad \mu$$
(1.53)

Adding all the errors in quadrature, we find:

$$\ln\left(\frac{m_{\frac{1}{2}}}{m_Z}\right) = 20 \pm 10 \quad (1.54)$$

which is a very broad range indeed. In fact, the formula does not apply when $m_{1/2} < m_Z$, and we find in this case another allowed range $m_{1/2} < 65$ GeV. Note that the range between this and (1.52) is also allowed at the $2-\sigma$ level. We conclude that although the latest precision data are consistent with supersymmetric GUTs, this consistency cannot yet be used to a set stringent upper bound on the supersymmetric-breaking scale $m_{1/2}$.

2 - STRING MODEL-BUILDING

2.0 - ... and the biggest problem of all?

This may be that of reconciling quantum mechanics with general relativity. The "gravity" of the problem may be "weighed" by considering the general form of a graviton exchange amplitude:

$$A_G \propto G_N E_1 E_2 \quad (2.1)$$

where E_1 and E_2 are the energies of the scattering particles and $G_N \equiv 1/m_P^2 : m_P \simeq 1.2 \times 10^{19}$ GeV is Newton's constant. The runaway high-energy behaviour (2.1) yields uncontrollable divergences when one tries to calculate quantum corrections:

$$\delta A_G \propto (G_N \Lambda^2)^p \ : \ p \in \mathcal{Z}^+ \tag{2.2}$$

It is reasonable to hope that progress in this reconciliation would lead us to as big a theoretical revolution as the previous reconciliation of Maxwell's and Newton's equations, which led to Special Relativity, and of Special Relativity and Gravity, that led to General Relativity. As in those two cases, new geometric ideas may be needed, and this is certainly the point of view of string theorists.

2.1 - Is the TOE made of string?

According to this hypothesis, what we have previously thought of as point particles are in fact loops of strings, and a particle's world-line is replaced by the world-sheet described by a loop of string. Because the string is not point-like, it has an infinite number of degrees of freedom and hence of excited states:

$$m_W^2 = \text{``0''} + n \times O\left(\frac{1}{G_N}\right) \ : \ n \ni \mathcal{Z} \tag{2.3}$$

where "0" represents the lowest harmonics that are massless in some approximation and correspond to the particles we see, and all of the higher harmonics have masses $0(G_N^{-1/2}) = 0(10^{19}$ GeV$)$. It used to be thought that consistency with quantum mechanics required string theory be formulated in more space-time dimensions: 26 for the bosonic string and 10 for a superstring. The surplus dimensions beyond 4 would be curled up or compactified with characteristic dimensions $L_P \sim 1/m_P \sim 10^{-33}$ cm. It is now known that these extra dimensions can in fact be replaced by combinations of internal degrees of freedom. There is now considerable freedom in the formulation of consistent string theories, and the challenge is to make a definite prediction that can be either proved or disproved by experiment.

The string action can be represented in the general form:

$$A = \frac{1}{2\pi\alpha'} \int d^2z \bigg[(G_{\mu\nu}(X)\sqrt{g}\, g_{ij} + B_{\mu\nu}(X)\epsilon_{ij}) \partial^i X^\mu \partial^j X^\nu \\ + \text{(internal degrees of freedom)} \bigg] \tag{2.4}$$

where the X^μ are space-time coordinates, $G_{\mu\nu}$ is the space-time metric, $B_{\mu\nu}$ is an antisymmetric tensor field, and the internal degrees of freedom are represented by a two-dimensional field theory on the world-sheet parametrized by the coordinates z. Bosonic fields on the world-sheet should obey periodic boundary conditions:

$$\phi(x + 2\pi) = \phi(x) \tag{2.5}$$

whilst real fermions could be periodic or antiperiodic:

$$\psi(x + 2\pi) = \pm\psi(x) \tag{2.6}$$

Because the string is closed, left- and right-moving degrees of freedom are independent: recall the independent solutions $\phi(x \pm t)$ of the two-dimensional wave equation:

$$\left(\frac{\partial^2}{\partial t^2} - \frac{\partial^2}{\partial x^2}\right)\phi(x,t) = 0 \tag{2.7}$$

Solutions to the equation (2.7) subject to the boundary conditions (2.5) have an infinite discrete set of frequencies: $\omega = n$: $\phi(x,t) \propto \exp[in(x \pm t)]$. At the quantum level, there are annihilation and creation operators a_n, a_n^+ corresponding to each one of these modes, with excitation levels $\delta m^2 \sim n/G_N$.

The internal world-sheet degrees of freedom are subject to various quantum consistency contraints, of which the first is conformal invariance. This corresponds to the classical reparametrization invariance of the world-sheet, i.e. holomorphic coordinate changes:

$$\omega = f(z) \tag{2.8}$$

This corresponds to an infinite set of transformations generated by an infinite-dimensional algebra:

$$[L_n, L_m] = i(n - m)L_{n+m} \quad (n \epsilon \mathcal{Z}) \tag{2.9}$$

This conformal algebra is not in general valid after including quantum corrections: there is normally an anomaly on the right-hand side proportional to:

$$c(n^3 - n)\,\delta_{n+m,0} \tag{2.10}$$

The coefficient c, called the central charge, gets contributions from reparametrization ghosts (cf gauge theories) as well as from physical degrees of freedom. The ghosts contribute -26 for bosonic strings and -15 for superstrings. Hence to cancel them one needs:

$$c = \begin{cases} 26 & \text{(bosonic strings)} \\ 15 & \text{(superstrings)} \end{cases} \tag{2.11}$$

to be consistent at the quantum level.

Each real free boson contributes $\delta c = 1$, including the space-time coordinates X^μ. Therefore, if there were no other degrees of freedom, the number of space-time coordinates would have to equal 26, as in the original bosonic string. However, there could also be internal degrees of freedom that do not necessarily describe flat space-time coordinates, and could parametrize some curved manifold of "compactification".

Each real free fermion contributes $\delta c = 1/2$, for example in the case of supersymmetry in D dimensions there are D spartners ψ^μ of the bosonic coordinates X^μ, giving a total:

$$\delta c = D + \frac{D}{2} = \frac{3D}{2} \qquad (2.12)$$

Hence, if there were no other internal degrees of freedom, the critical dimension of a superstring would be 10.

In point of fact, there could also be internal degrees of freedom such as interacting or constrained bosons or fermions corresponding, for example, to statistical mechanical models such as the unitary discrete series with:

$$\delta c = 1 - \frac{6}{m(m+1)} : m = 3, 4 \ldots \qquad (2.13)$$

Several of these would be combined with other internal degrees of freedom to build up to $c = 26$ or 15.

Conformal invariance takes care of locally-generated divergences. There are also possible divergences associated with incorrect counting of higher-order diagrams. Conventional perturbation theory becomes a topological expansion equivalent to a sum over all closed surfaces, equivalent to spheres with n handles. Each equivalent surface should be counted just once and only once. However, there are many ways of introducing the modular parameters that characterize each such higher-order surface. The requirement of modular invariance is that one must choose the internal degrees of freedom on the world-sheet (e.g. the boundary conditions for fermions) in such a way that each choice of modular parameters is treated equivalently, and one can finally divide the functional integral by the volume of the modular group that inter-relates them.

There are different types of string theory that are consistent with these requirements of conformal and modular invariance. One is the original bosonic string, whose critical dimension is D=26, but which could live in fewer large space-time dimensions if the surplus dimensions were compactified. The main phenomenological defect of this model is that it does not contain fermions. The superstring has critical dimension D=10, but even when compactified to a lower number of dimensions it cannot accommodate parity violation in a phenomenologically acceptable way. The first phenomenologically promising string theory was the original heterotic string [25], with a supersymmetric left-moving sector formulated directly in 10 space-time dimensions, and a right-moving sector which had internal degrees of freedom parametrizing an SO(32) or $E_8 \times E_8$ group manifold, which together with the 10 space-time dimensions gave a central charge of 26. Because of the asymmetric treatment of left- and right-movers, the theory could violate parity when compactified to 4 dimensions, e.g. on a Calabi-Yau manifold [26].

More generally, one can formulate heterotic string theories directly in 4 dimensions, with for example $(X^\mu, \psi^\mu, 18 \text{ fermions})_L$ and $(X^\mu, 44 \text{ fermions})_R$ [27]. There is no need to use the language of compactification, and the internal degrees of freedom offer great liberty in the choice of symmetry group, which is in principle broader than that available from simple manifold compactification of the 10-dimensional $E_8 \times E_8$ heterotic string. The following are some of the main trends in the choice of 4-dimensional gauge group.

Geometrical: models have been proposed based on Calabi-Yau manifold compactifications [26], which yield a 4-dimensional gauge group $\mathcal{G}_4 \subset E_6$, such as $[SU(3)]^3$, and on orbifold compactifications [28], which can lead to $\mathcal{G}_4 = SU(3) \times SU(2) \times U(1)^n$.

Algebraic: direct products of unitary discrete series models with N=2 supersymmetry have been studied, and shown to be related to Calabi-Yau compactification, and models have been constructed using free fermions on the world-sheet with $\mathcal{G}_4$ as a GUT gauge group [4] , or $\mathcal{G}_4 = SU(3) \times SU(2) \times U(1)^n$ [29].

Which type of 4-dimensional gauge group $\mathcal{G}_4$ should one seek from string? Here the basic choice is: to GUT or not to GUT? The disadvantages of models with $SU(3) \times SU(2) \times U(1)^n$ gauge groups are that proton decay tends to be either too rapid or too slow to be observed, whilst there is no mechanism to obtain naturally small neutrino masses. On the other hand, GUTs offer both these possibilities, and so we have chosen [4] to look for string models with GUT-like gauge groups.

The chief obstacle to this programme is that (almost) all GUTs, from the SU(5) model discussed earlier on upwards, require adjoint or larger Higgs representations. Thus SU(5) needs at least a $\underline{24}$ to break down to $SU(3) \times SU(2) \times U(1)$, and larger representations such as $\underline{75}, \underline{50}, \underline{\overline{50}}$ are discussed in order to obtain a natural splitting of the light Higgs doublet and its colour triplet partners. The corresponding jobs in SO(10) require $\underline{45}$ and perhaps $\underline{54}, \underline{120}, \underline{126}, \underline{210}$ representations. This is a problem because the effective field theories obtained from string theories do not in general contain adjoint or larger Higgses. One example of this statement is the Calabi-Yau class of compactifications of the 10-dimensional heterotic string, which give $\underline{27}$, $\underline{\overline{27}}$ and $\underline{1}$ representations of E_6, but not the adjoint $\underline{78}$. This observation has been elevated into a theorem [30] applying to all supersymmetric field theories arising from 4-dimensional heterotic strings in which the gauge symmetry is realized by a level-one representation of a Kac-Moody current algebra.

The only known solution to this conundrum is offered by the only known GUT that does not need an adjoint or larger Higgs...

2.2 - Flipped supersymmetric SU(5) GUT[4]

This GUT model has the gauge group SU(5), with SU(3) and SU(2) embedded in the SU(5) factor, and $U(1)_Y$ a linear combination of the external U(1) factor and the same internal U(1) subgroup as in conventional SU(5). The matter fields of the minimal flipped SU(5) model are assigned to $T_i \equiv (\underline{10}, 1/2), \overline{F}_i \equiv (\underline{\bar{5}}, -3/2)$ and $L_i^c = (\underline{1}, 5/2)(i = 1, 2, 3)$ representations of SU(5), with the external U(1) hypercharges indicated:

$$T_1 = \frac{1}{\sqrt{2}} \begin{pmatrix} 0 & d_r^c & -d_y^c & \vdots & d_r & u_r \\ -d_b^c & 0 & d_r^c & \vdots & d_y & u_y \\ d_y^c & -d_r^c & 0 & \vdots & d_b & u_b \\ \cdots & \cdots & \cdots & \vdots & \cdots & \cdots \\ -d_r & -d_y & -d_b & \vdots & 0 & \nu^c \\ -u_r & -u_y & -u_b & \vdots & -\nu^c & 0 \end{pmatrix}_L, \overline{F}, = \begin{pmatrix} u_r^c \\ u_y^c \\ u_b^c \\ \nu_e \\ e^- \end{pmatrix}_L, L_1^c = e^c$$

(2.14)

Note, by comparison with conventional SU(5), the flipped particle assignments $\nu \leftrightarrow e, u \leftrightarrow d$ are made possible by the extra U(1), which means that $\text{Tr}Q_{em} \neq 0$ in general. The magic of this model lies in the Higgs fields: $H \equiv (\underline{10}, 1/2) \ni (d_H^c, (d_H, u_H), \nu_H^c)$ and $\overline{H} \equiv (\underline{\overline{10}}, 1/2)$ suffice to break $SU(5) \times U(1) \to SU(3) \times SU(2) \times U(1)_Y$ via their ν_H^c and $\bar{\nu}_{\overline{H}}^c$ components: $< 0|\nu_H^c, \bar{\nu}_{\overline{H}}^c|0> \neq 0$. Then $h \equiv (\underline{5}, -1) \ni (D, h^-, h^0)$ and $\bar{h} \equiv (\underline{\bar{5}}, 1) \ni (\overline{D}, h^+, \bar{h}^0)$ suffice to break $SU(2)_L \times U(1)_Y \to U(1)_{em}$. In addition there must be some singlet fields $\phi_{0,i} = (\underline{1}, 0)$.

The superpotential of the minimal flipped SU(5) model is[4] :

$$\begin{aligned} P =& \lambda_1 TTh + \lambda_2 T\overline{F}\bar{h} + \lambda_3 \overline{F}L^c h \\ & + \lambda_4 HHh + \lambda_5 \overline{HH}\bar{h} \\ & + \lambda_6 T\overline{H}\phi + \lambda_7 h\bar{h}\phi_0 + \lambda_8 \phi\phi\phi_0 + \lambda_9 \phi_0^3 \end{aligned}$$ (2.15)

The first three terms give masses to the quarks and leptons, the next two terms split naturally the Higgs doublets and their colour triplet partners, the sixth term plays a rôle in the neutrino mass matrix, the seventh removes an unwanted axion field, and the last two terms ensure that $<0|\phi|0>$ is not large. The superpotential (2.15) is the most general respecting the reflection symmetry $H \leftrightarrow \overline{H}$. It gives a potential with flat directions corresponding to possible large v.e.v's:

$$V = <0|H|0> = \overline{V} \equiv <0|\overline{H}|0> \qquad (2.16)$$

There are no flat directions corresponding to v.e.v.'s for $h, \bar{h}$ and ϕ, implying that they can only acquire v.e.v.'s in the presence of supersymmetry breaking. Naturalness tells us that this should be small (≤ 1 TeV), so that we also expect v.e.v.'s of this order of magnitude for $h, \bar{h}$ and ϕ. Thus the scale of SU(5) × U(1) breaking is expected to be much greater than that of SU(2) × U(1) breaking.

The fourth and fifth terms in (2.15) give large Dirac mass terms coupling the D and $\overline{D}$ (triplet partners of the light doublet Higgses) with the d^c_H and $\bar{d}^c_{\overline{H}}$ states in the $\underline{10}$ and $\overline{\underline{10}}$ Higgs representations:

$$m_{d^c_H D} = \lambda_4 V \quad , \quad m_{\bar{d}^c_{\overline{H}} \overline{D}} = \lambda_5 \overline{V} \qquad (2.17)$$

There are no corresponding mass terms for their Higgs doublet partners. Hence there is naturally large doublet/triplet mass splitting, a holy grail for GUT model-builders.

The neutrino mass matrix has a characteristic seesaw form:

$$(\nu, \nu^c, \phi)_i \begin{pmatrix} 0 & m_u & 0 \\ m_u & 0 & \lambda_6 \overline{V} \\ 0 & \lambda_6 \overline{V} & \lambda_8 x \end{pmatrix}_{ij} \begin{pmatrix} \nu \\ \nu^c \\ \phi \end{pmatrix}_j \qquad (2.18)$$

leading to one massive approximate Dirac mass eigenstate:

$$m^c_\nu \simeq \lambda_6 \overline{V} \qquad (2.19)$$

in each generation, and one very light, approximately left-handed, Majorana state:

$$m^2_{\nu_L} \simeq \frac{m^2_W \lambda_8 x}{(\lambda_6 \overline{V})^2} \sim \frac{m^3_W}{m_{GUT}} \qquad (2.20)$$

In general, flipped SU(5) models also have neutrino mixing with neutralinos, but this is model-dependent and will not be discussed further here.

Proton decay in flipped SU(5) is predominantly via vector boson exchange, which gives an effective d=6 interaction of similar form to that in conventional SU(5), but with numerical differences in the coefficients [31]:

$$\mathcal{L}_{\text{eff}} = \frac{g_5^2}{2m_X^2} \left[(d_R \gamma^\mu d_L)(u_R \gamma_\mu \nu_L) + (d_R \gamma^\mu u_L)(u_R \gamma_\mu e_L) \right] \quad (2.21)$$

This gives $p \to e^+ \pi^0, \bar{\nu} \pi^+$ and $n \to e^+ \pi^-, \bar{\nu} \pi^0$, but with different branching ratios from conventional SU(5):

$$\Gamma(p \to \bar{\nu} \pi^+) = \Gamma(n \to e^+ \pi^-) = 2\Gamma(n \to \nu \pi^0) = 2\Gamma(p \to e^+ \pi^0) \quad (2.22)$$

The estimated nucleon lifetime is about $10^{35 \pm 2}$y. The d=5 higgsino exchange that is dangerous in conventional SU(5) is very strongly suppressed in flipped SU(5), as a result of its natural doublet/triplet mass splitting, whilst d=6 heavy Higgs exchange is expected to be smaller than d=6 vector exchange.

This flipped SU(5) GUT has many attractive features, but would not have got so much attention if it had not been for the possibility of deriving it from string, to which we now turn.

2.3 - Flipped SU(5) GUT derived from string[4]

We start from the fermionic formulation of heterotic strings directly in 4 dimensions [27]. The numbers of left- and right-moving fermions are fixed by conformal invariance, and the choice of model depends on the choice of fermionic boundary conditions, which must be made consistently with modular invariance. Physical states are made out of the fermionic excitations, projecting out using the boundary conditions. Internal quantum numbers are obtained from the fermionic indices. A suitable choice of boundary conditions gives a gauge group of the form $[SU(5) \times U(1)] \times U(1)^n$.

I will not discuss the details of the model construction here, but just present the physical spectrum of light observable particles in one string version of flipped SU(5)[4]:

$$(M_1)_{(-\frac{1}{2},0,0,0)} \; ; \; (M_2)_{(0,-\frac{1}{2},0,0)} \; ; \; (T_3)_{(0,0,\frac{1}{2},-\frac{1}{2})} + (\overline{F}_3 + L_3^c)_{(0,0,\frac{1}{2},\frac{1}{2})}$$

$$(T_4)_{(-\frac{1}{2},0,0,0)} \; ; \; (F_4)_{(\frac{1}{2},0,0,0)} \; ; \; (\overline{L}_4^c)_{(\frac{1}{2},0,0,0)}$$

$$(\overline{T}_5)_{(0,\frac{1}{2},0,0,)} \; ; \; (\overline{F}_5)_{(0,-\frac{1}{2},0,0,)} ; (L_5^c)_{(0,-\frac{1}{2},0,0,)}$$

$$h_{45} \equiv (\underline{5}, -1)_{(-\frac{1}{2},-\frac{1}{2},0,0)} \; ; \; \phi_{45} \equiv (\underline{1}, 0)_{(\frac{1}{2},\frac{1}{2},1,0)} \; ;$$

$$\phi_i \equiv (\underline{1}, 0)_{(\frac{1}{2},-\frac{1}{2},0,0)} \; ; \; (i = 1 \text{ to } 4); \quad (2.23)$$

$$\phi^+ \equiv (\underline{1}, 0)_{(\frac{1}{2}, -\frac{1}{2}, 0, 1)} \; ; \; \phi^- \equiv (\underline{1}, 0)_{(\frac{1}{2}, -\frac{1}{2}, 0, -1)} \; ;$$

$$h_1 \equiv (\underline{5}, -1)_{(1,0,0,0)} \; ; \; h_2 \equiv (\underline{5}, -1)_{(0,1,0,0,)} \; ; \; h_3 \equiv (\underline{5}, -1)_{(0,0,1,0)};$$

$$\Phi_{23} \equiv (1,0)_{(0,-1,1,0)} \; ; \; \Phi_{31} \equiv (\underline{1}, 0)_{(1,0,-1,0)} \; ; \; \Phi_{12} \equiv (\underline{1}, 0)_{(-1,1,0,0)};$$

$$\Phi_I \equiv (1, 0)_{(0,0,0,0)} (I = 1 \text{ to } 5)$$

where $M \equiv T + \overline{F} + L^c$ and the subscripts denote hypercharges for the 4 extra $U(1)$ factors. It is also possible in this model to calculate all the Yukawa couplings, which include for the observable particles:

$$W = g\sqrt{2} \Bigg[F_1 F_1 h_1 + F_2 F_2 h_2 + F_4 F_4 h_1 + F_4 \bar{f}_5 h_{45} + F_3 \bar{f}_3 h_3$$

$$+ \bar{f}_1 L_1^c h_1 + \bar{f}_2 L_2^c h_2 + \bar{f}_5 L_5^c h_2 + \frac{1}{\sqrt{2}} \left(F_4 \overline{F}_5 \phi_3 + f_4 \bar{f}_5 \phi_2 + \overline{L}_4^c L_5^5 \phi_2 \right)$$

$$+ \overline{F}_5 \overline{F}_5 \bar{h}_2 + f_4 \overline{L}_4^c \bar{h}_1 + \overline{F}_5 f_4 h_{45} + h_1 \bar{h}_2 \Phi_{12} + h_2 \bar{h}_3 \Phi_{23} + h_3 \bar{h}_1 \Phi_{31} + h_3 \bar{h}_{45} \bar{\phi}_{45} + \text{h.c.}$$

$$+ \frac{1}{2} \left(\phi_{45} \bar{\phi}_{45} \Phi_3 + \phi^+ \bar{\phi}^+ \Phi_3 + \phi^- \bar{\phi}^- \Phi_3 + \phi_i \bar{\phi}_i \Phi_3 \right)$$

$$+ \left(\phi_1 \bar{\phi}_2 + \bar{\phi}_1 \phi_2 \right) \Phi_4 + \left(\Phi_{12} \Phi_{23} \Phi_{31} + \Phi_{12} \phi^+ \phi^- + \Phi_{12} \phi_i \phi_i + \text{h.c.} \right) \Bigg]$$
(2.24)

Note that all the non-zero couplings are simple multiples of the gauge coupling g. It is also possible [32] to calculate non-zero non-renormalizable superpotential terms. For example, the full set of non-vanishing observable-sector quadrilinear couplings is:

$$c_1 T_1 \overline{F}_1 \bar{h}_{45} \phi_1 + c_2 T_2 \overline{F}_2 \bar{h}_{45} \bar{\phi}_4 \quad (2.25)$$

where:
$$c_1 = -3.07 g^2, c_2 = i c_1 \quad (2.26)$$

These and higher-order superpotential couplings contribute to the masses of the lighter fermions that do not get their masses from the trilinear Yukawa couplings.

The model contains *a priori* 4 additional $U(1)$ factors beyond those in the original flipped $SU(5)$ model, of which one linear combination is anomalous.

$$\text{Tr}(Y_A) \neq 0 : Y_A \equiv -2U(1)_1 - U(1)_2 + 2U(1)_3 - U(1)_4 \quad (2.27)$$

The orthogonal $U(1)$'s are free of gauged and mixed gravitational anomalies. Via a Fayet-Iliopulos term [33], the Y_A gauge boson acquires a mass by eating the imaginary part of the dilaton supermultiplet that appears in all string models. Nevertheless, the vacuum energy can vanish, so that supersymmetry is unbroken at this stage, for a suitable choice of singlet v.e.v.'s $< 0|\phi|0 >$ that break all (or all except one) of the 4 extra U(1) factors. Thus the effective gauge group at energies less than m_P becomes the desired SU(5) × U(1)[4] . The model has an ambiguity in the choice of these v.e.v.'s, so at the moment one just chooses the pattern that is phenomenologically the most realistic, although ultimately the pattern should be fixed by the supersymmetry-breaking terms in the effective potential.

The breaking of SU(5) × U(1) is via v.e.v.'s:

$$\overline{V} \equiv < 0|\overline{T}_5|0 > = V \equiv < 0|T|0 > \neq 0 \; : \; T = \sum_{i=1}^{3} \alpha_i T_i, \sum_{i=1}^{3} |\alpha_i|^2 = 1 \qquad (2.28)$$

whilst $< 0|T_4|0 > = 0$ because of a $\overline{T}_5 - T_4$-singlet coupling. There are linear combinations of the 5 h and $\overline{5}$ $\overline{h}$ fields that do not acquire large masses, and so could acquire v.e.v.'s in the presence of supersymmetry breaking. Precisely which combination of the $\overline{h}$ remains light depends on an (as yet) undetermined ratio of large v.e.v.'s for singlet fields.

2.4 - Superstring "prediction" for m_t

The above is just one of the many possible 4-dimensional string models, that can be considered as corresponding to different vacua with different v.e.v.'s $< 0|\phi_i|0 > \neq 0$. In general, one can predict non-zero Yukawa couplings in string models, that are of order of the gauge coupling:

$$\lambda = 0(1) \times g \qquad (2.29)$$

when renormalized at the Planck scale. For example, in the flipped SU(5) GUT [4] , in general:

$$\lambda = \sqrt{2} \times g \qquad (2.30)$$

for all non-zero couplings, although many vanish. The physical top quark is in general a mixture between states with and without Yukawa couplings, so it has:

$$\lambda_t = \sqrt{2}g \, \cos\theta_t \qquad (2.31)$$

where the value of the mixing angle θ_t remains to be determined. Likewise, one has:

$$\lambda_b = \lambda_\tau = \sqrt{2}g \, \cos\theta_b \qquad (2.32)$$

at the Planck scale, which yields the successful prediction:

$$m_b \simeq 2.8 m_\tau \simeq 5 \text{ GeV} \tag{2.33}$$

for the physical b quark mass, if and only if there are just 3 generations.

The formula (2.31) gives an absolute upper bound for the physical top quark mass of about 190 GeV, and the range (1.47a) indicated by electroweak radiative corrections appears completely natural. Indeed, the puzzle is not "why is the top quark so heavy?" but rather "why are the other quarks so light?" The value $m_t = 123$ GeV corresponds to $\lambda_t = 1/\sqrt{2}$ in a model with just one Higgs v.e.v., whereas $m_e = 0.51$ MeV corresponds to $\lambda_e \simeq 1.5 \times 10^{-6}$. The latter is technically "natural", in the sense that radiative corrections are under control ($\delta m_f \leq m_f$), but what is its origin? In the context of the string-derived flipped SU(5) GUT discussed earlier, we have proposed [4] a possible origin in the non-renormalizable superpotential interactions, which provide effective Yukawa couplings of order:

$$\lambda_f \simeq 0(1) \times g^{n+1} \times \left(\frac{<0|\phi|0>}{m_P}\right)^n \tag{2.34}$$

for some positive integer n.

These may be able to provide the necessary hierarchy of fermion masses.

2.5 - Effective string unification scale

In a field-theoretical GUT, both the grand unification scale m_X and the gauge coupling g_X at that scale are arbitrary free parameters. It should be possible to calculate both of them in string theory. So far, we know how to calculate the effective string unification scale, at which all the extrapolated low-energy gauge couplings appear to become equal. This information is contained in the one-loop string formula [34]:

$$\frac{16\pi^2}{g_i^2(\mu)} = k_i \frac{16\pi^2}{g^2} + b_i \ln \frac{M^2}{\mu^2} + \Delta_i \tag{2.35}$$

where g_i are the different low-energy couplings renormalized at some scale μ, the k_i are their corresponding Kac-Moody levels, the b_i are their one-loop β-functions:

$$M = \left(\frac{2e^{1-\gamma_E} 3^{-3/2}}{\pi \alpha'}\right)^{\frac{1}{2}} \simeq 1.03 \times g \times 10^{18} \text{ GeV} \tag{2.36}$$

where α' is the string tension and g the string coupling, in the Pauli-Villars regularization scheme that is most natural for string, γ_E is the familiar Euler constant, and the Δ_i are threshold correction factors obtained from subtracted one-loop string integrals:

$$\Delta_i = \int_\Gamma \frac{d^2\tau}{\tau_2} (B_i(\tau,\bar\tau) - b_i) \qquad (2.37)$$

where Γ is the fundamental domain, and:

$$B_i(\tau,\bar\tau) \equiv -\text{Tr}\left(Q_5^2 Q_i^2 q^H \bar q^{\bar H}\right) \qquad (2.38)$$

where Q_5 is the helicity operator, the Q_i are the group generators, $q = e^{i\pi\tau}$, and $H(\bar H)$ is the Hamiltonian for the left (right)-moving in the partition function:

$$\mathcal{Z} = \text{Tr}(q^H \bar q^{\bar H}) \qquad (2.39)$$

It is evident from (2.37) that the threshold correction factors Δ_i have potential infrared divergences as $\tau_2 \to \infty$. These are cancelled by the β-function terms in (2.37). From the point of view of string theory, μ is an infrared cutoff, where high-energy physics now rejoins condensed-matter physics!

The dependences of the Δ_i on the moduli parameters for theories with 3 N=2 supersymmetric sectors have been calculated previously [35]:

$$\Delta_i = \sum_{\alpha=1}^{3} \left[-\hat b_i^\alpha \text{ Re } \ell n\{T_\alpha \eta^4(T_\alpha)\} + c_i^\alpha \right] \qquad (2.40)$$

where the $\hat b_i^\alpha$ are the contributions of the 3 sectors to the b_i, the T_α are the moduli, η is the conventional Dedekind function, and the c_i^α are the residual non-moduli corrections. The flipped SU(5) model derived from string has:

$$T_\alpha = i(\alpha = 1,2,3) \qquad (2.41)$$

which minimize the moduli contribution to the effective string unification scale. The next question is: how large are the non-moduli corrections c^{α_i} [36]?

We recall that in a generic free fermion model, the physical spectrum is obtained from all combinations of fermionic excitations by making various projections. It is easy to convince oneself that sectors with N=4 supersymmetry do not contribute because of a θ-function identity, whilst N=1 sectors do not contribute because they depend on odd θ-functions. The only possible contributions come from N=2 sectors, which are model-dependent. In the flipped SU(5) model, the scale at which the U(1) and SU(5) couplings appear to become equal is governed by:

$$\delta\Delta \equiv \Delta_1 - \Delta_5 \qquad (2.42)$$

which receives contributions only from fermions which are neither periodic nor antiperiodic, i.e. intrinsically complex. The result is [36]:

$$\delta\Delta = 24.13 \qquad (2.43)$$

whose implications we now discuss.

In general, equation (2.35) tells us that

$$\ell n\left(\frac{m_{SU}}{M}\right) = \frac{1}{2}\left(\frac{\delta\Delta}{\delta b}\right) \qquad (2.44)$$

where

$$\delta b = b_1 - b_5 : b_1 = \frac{43}{2}, \ b_5 = 1 \qquad (2.45)$$

in our case. Using the numerical value (2.36) for M, we therefore find [36]:

$$m_{SU} = 1.86 \times g \times 10^{18} \text{ GeV} \qquad (2.46)$$

Several comments are in order. The first is that the non-moduli corrections are quite small, implying that m_{SU} is close to what would have been obtained from an easy moduli estimate, neglecting the c_i^α in (2.40). If so, m_{SU} would be similar in other models with 3 N=2 sectors, and larger in most other models. This means in particular that $m_{SU} \simeq 10^{18}$ GeV is larger than the unification scale $m_X \simeq 10^{16}$ GeV calculated in minimal supersymmetric GUTs.

This suggests a possible observable effect on $\sin^2\theta_W$ [37]. If there is a range of energies between m_X and m_{SU} where the gauge group is SU(5) × U(1), the different evolutions of g_5 and g_1 with scale mean that from $g_1 = g_5$ at m_{SU} one gets $g_1 < g_5 = g_3 = g_2$ at $m_X \leq \mu < m_{SU}$, implying that $\sin^2\theta_W(\mu)$ is less than its usual GUT symmetry value of 3/8:

$$\sin^2\theta_W(\mu) = \frac{15}{16 + 24[\alpha_5(\mu)/\alpha_1(\mu)]} \qquad (2.47)$$

In the flipped SU(5) string model, one has at the one-loop level:

$$\sin^2\theta_W(\mu) = \frac{3}{8 + \frac{27}{\pi}\alpha_5 \ell n\left(\frac{m_{SU}^2}{\mu^2}\right)} \qquad (2.48)$$

The effect on $\sin^2\theta_W$ is less than 5% if:

$$\mu \geq 0.6 m_{SU} \sim 5 \times 10^{17} \text{ GeV} \qquad (2.49)$$

which is still much larger than the estimate of m_X obtained by extrapolating from low energies using just the minimal set of particles and sparticles.

A larger value of m_X can be accommodated in string theory if SU(5) non-singlet fields participate in the cancellation of the anomalous U(1), and the value of m_X can be increased if there are more low-energy particles [37]. These should include at least one "light" vector-like pair of states transforming as ($\underline{3}$,2) and ($\underline{\bar{3}}$,2) under SU(3) × SU(2), together with at least one other "light" pair of supermultiplets. Although such states are absent in the flipped SU(5) string model derived earlier, they can be found in closely-related string flipped SU(5) models. Thus flipped SU(5) is surely able to reconcile the top-down, a priori string calculation of $\sin^2\theta_W$ with the bottom-up estimate of m_X obtained by extrapolating the low-energy couplings. It even offers the possibility that $\sin^2\theta_W$ might be somewhat lower than the minimal supersymmetric GUT prediction, which is even consistent [8] with the data!

References

[1] E. Farhi and L. Susskind, *Physics Reports* **74C** (1981) 2777.
[2] E. Witten, *Nucl. Phys.* **B188** (1981) 513;
     L. Maiani, Proc. Summer School on Particle Physics, Gif-sur-Yvette (IN2P3, Paris, 1980) 1.
[3] H. Georgi and S.L. Glashow, *Phys. Rev. Lett.* **32** (1974) 438;
     H. Georgi, H.R. Quinn and S. Weinberg, *Phys. Rev. Lett.* **33** (1974) 451.
[4] I. Antoniadis, J. Ellis, J.S. Hagelin and D.V. Nanopoulos, *Phys. Lett.* **B194** (1987) 231, **B205** (1988) 459, **B208** (1988) 209 and **B231** (1989) 65.
[5] J. Ellis and D.V. Nanopoulos, *Nature* **292** (1981) 436.
[6] J. Ellis, M.K. Gaillard, D.V. Nanopoulos and S. Rudaz, *Nucl. Phys.* **B176** (1980) 61.
[7] Particle Data Group, *Phys. Lett.* **B236** (1990) 1.
[8] J. Ellis, G.L. Fogli and E. Lisi, CERN preprint TH. 6273/91 (1991);
     See also: D. Schaile, CERN preprint PPE/91/187 (1991).
[9] Y. Nambu, in Proc. XI Int. Symp. on Elementary Particle Physics, eds. Z. Ajduk, S. Pokorski and A. Trautman (World Scientific, Singapore, 1989);
     A.Miranski, M. Tanabashi and K. Yamauraki, *Mod. Phys. Lett.* **A4** (1989) 1043 and *Phys. Lett.* **B221** (1989) 177;
     W.A. Bardeen, C.T. Hill and M. Linder, *Phys. Rev.* **D41** (1990) 1647.

[10] A.B. Lahanas and D.V. Nanopoulos, *Phys. Reports* **145** (1987) 1.

[11] M. Davier, Proceedings of the LP-HEP conference, Geneva, 1991.

[12] J. Ellis, G. Ridolfi and F. Zwirner, *Phys. Lett.* **B237** (1990) 423

[13] Y. Okada, M. Yamaguchi and T. Yanagida, *Prog. Theor. Phys. Lett.* **85** (1991) 1;
J. Ellis, G. Ridolfi and F. Zwirner, *Phys. Lett.* **B257** (1991) 83;
H.E. Haber and R. Hempfling, *Phys. Rev. Lett* **66** (1991) 1815.

[14] S. Dimopoulos and H. Georgi, *Nucl. Phys.* **B193** (1981) 150 ;
N. Sakai, *Z Phys.* **C11** (1982) 153.

[15] J. Ellis, D. Nanopoulos and D.A. Ross, *Phys. Lett.* **B267** (1991) 132.

[16] A. Braaten, S. Narison and A. Pich, CERN Preprint TH.6070 (1991) and references therein.

[17] A. Sirlin, *Phys. Rev.* **D22** (1980) 971.

[18] M. Consoli and W. Hollik, in "Z Physics at LEP", eds. G. Altarelli, R. Kleiss and C. Verzegnassi, CERN Report 89-08 (1989).

[19] M.J.G. Veltman, *Nucl. Phys.* **B123** (1977) 89 ;
M.S. Chanowitz, M.A. Furman and I. Hinchliffe, *Phys. Lett.* **B78** (1978) 285.

[20] A. Sirlin, *Phys. Lett.* **B232** (1989) 123;
S. Fanchiotti and A. Sirlin, *Phys. Rev.* **D41** (1990) 319;
G. Degrassi, S. Fanchiotti and A. Sirlin, *Nucl. Phys.* **B351** (1991) 49;
G. Degrassi and A. Sirlin, *Nucl. Phys.* **B352** (1991) 342.

[21] J. Ellis, S. Kelley and D.V. Nanopoulos, *Phys. Lett.* **B249** (1990) 442, **B260** (1991) 131;
P. Langacker, Univ. of Pennsylvania preprint UPR-0435T (1990).

[22] U. Amaldi, W. de Boer and F. Fürsteneau, *Phys. Lett.* **B260** (1991) 447.

[23] J. Ellis, S. Kelley and D.V. Nanopoulos, CERN Preprint TH.6140/91 (1991).

[24] F. Anselmo, L. Cifarelli, A. Petermann and A. Zichichi, preprint CERN-PPE/91-123, 15 July 1991 and *Il Nuovo Cimento* **104A** (1991).

[25] D.J. Gross, J. Harvey, E. Martinec and R. Rohm, *Phys. Rev. Lett.* **54** (1985) 502; *Nucl. Phys.* **B256** (1985) 253; **B267** (1986) 75.

[26] P. Candelas, G.T. Horowitz, A. Strominger and E. Witten, *Nucl. Phys.* **B258** (1985) 46.

[27] I. Antoniadis, C. Bachas and C. Kounnas, *Nucl. Phys.* **B289** (1987) 87; I. Antoniadis and C. Bachas, *Nucl. Phys.* **B298** (1988) 586; H. Kawai, D.C. Lewellen and S.H.-H. Tye, *Phys. Rev. Lett.* **57** (1986) 1832; *Phys. Rev.* **D34** (1986) 3794; *Nucl. Phys.* **B288** (1987) 1; R. Bluhm, L. Dolan and P. Goddard, *Nucl. Phys.* **B309** (1988) 330.

[28] For a review, see: J.A. Harvey, Princeton preprint PUPT-1082 (1987).

[29] A Faraggi, D.V. Nanopoulos and K. Yuan, *Nucl. Phys.* **B335** (1990) 347.

[30] H. Dreiner, J. Lopez, D.V. Nanopoulos and D. Reiss, *Phys. Lett.* **B216** (1989) 283.

[31] J. Ellis, J.S. Hagelin, S. Kelley and D.V. Nanopoulos, *Nucl. Phys.* **B311** (1988).
[32] S. Kalara, J. Lopez and D.V. Nanopoulos, *Phys. Lett.* **B245** (1990) 421 and *Nucl. Phys.* **B353** (1991) 650.
[33] M. Dine, N. Seiberg and E. Witten, *Nucl. Phys.* **B289** (1987) 589.
[34] V. Kaplunovsky, *Nucl. Phys.* **B307** (1988) 145.
[35] L.J. Dixon, V. Kaplunovsky and J. Louis, *Nucl. Phys.* **B355** (1991) 649.
[36] I. Antoniadis, J. Ellis, R. Lacaze and D.V. Nanopoulos, CERN preprint TH. 6136/91 (1991).
[37] I. Antoniadis, J. Ellis, S. Kelley and D.V. Nanopoulos, CERN preprint TH.6169/91 (1991).

CHAIRMAN: J. Ellis

Scientific Secretaries: A. Hasan, Z. Pluciennik, H.J. Schulze

DISCUSSION I

– *Hasan:*

1) What is the SUSY breaking scale? Does it depend on the specific model chosen?

2) Does SUSY solve the hierarchy problem?

– *Ellis:*

1) I think the explanation of the SUSY breaking scale comes from superstring theory and I do not think we are far enough along in our understanding of superstring theory to know what fixes this magnitude and why it is as big as it apparently is.

2) There are two aspects of the hierarchy problem: one is to fine tune the parameters to make them small compared to the Planck scale. The other is to maintain the fine-tuning despite radiative corrections. SUSY has nothing to say about why the parameters were small initially, but it guarantees that the radiative corrections will not introduce large changes. It guarantees the stability of the hierarchy.

– *Hernandez:*

Considering that the one-loop corrections to Higgs masses in minimal SUSY change the bounds on the parameters considerably, how can one be safe if one stops at the one-loop level and not consider higher-order corrections?

– *Ellis:*

It has something to do with what Weinberg called accidental symmetry, according to which the tree-level term is small for special reasons. A familiar example is the Coleman-Weinberg limit of the Standard Model, where $m_h = 0$ at the tree-level, becomes about 10 GeV at the one loop level, and changes by only a few hundred MeV at the two-loop level. The one-loop correction is big, but it does not mean that the higher loop corrections are going to be bigger than the one-loop correction. Several groups have calculated the two-loop corrections to supersymmetric Higgs masses and found that they were small.

– *Hernandez:*

If the top quark is finally found in the Tevatron range, when do you think minimal SUSY will be most definitely excluded?

– Ellis:

If the top mass lies in the range of 100-200 GeV, there is a fair chance to find the minimal SUSY Higgs boson at LEP. If the Higgs is not found at LEP, one will have to wait for LHC. LHC and SSC will be able to explore a reasonable region of SUSY parameter space but will not be able to ultimately prove or disprove SUSY.

– Neubert:

Two questions concerning the extrapolation of coupling constants to the GUT-scale:

1) How big are the changes as one goes from one-loop to two-loop β-functions? If they are large, could even higher-order corrections "rescue" the non-SUSY Standard Model by improving the matching?

2) How sensitive are the results to the assumptions about SUSY-parameters?

– Ellis:

1) The two-loop corrections in a non-SUSY Standard Model change the value of the heavy boson masses m_X by a factor of a few and change $\sin^2\theta_W$ by 1%.

2) The variation of the sparticle masses, Higgs masses and the mixing parameter μ in a reasonable range does not change the results very much. The biggest error comes in our estimation from the value of $\sin^2\theta_W$ and α_3, as discussed in my first lecture.

– Gourdin:

In the Standard Model the top quark is probably the only fermion to have a natural mass, which means the Yukawa coupling constant is of the order of one. For the other fermions we have a hierarchy for λ of 5 orders of magnitude. For SUSY models we are faced with similar problems for the fermions and sfermions. How can SUSY give an explanation for those two hierarchies?

– Ellis:

SUSY does not explain the hierarchy problem: it tolerates hierarchy. The question of the Yukawa coupling constants is a flavour problem which SUSY does not address. A possible answer could be found in superstring theory, as discussed in my second lecture.

– Wadhwa:

1) Could you comment on the $\sin^2\theta_W$ value calculated by LEP compared to that predicted by GUT?

2) The value of the Higgs mass depends strongly on the radiative corrections. What is the upper limit for the radiative corrections?

– *Ellis:*

1) I said that the value of $\sin^2\theta_W$ measured at LEP is 0.233 which is completely different from the value 0.214 obtained in a non-supersymmetric GUT. However, in a SUSY model with SUSY breaking scale m_W one gets just the right value.

2) The variation of the Higgs mass with the top mass in the range 100-160 GeV determined from electroweak data is shown in my first lecture. The shift can be up to 100 GeV values if the squark mass ~ 1 TeV.

– *Weselka:*

Is the problem of baryon stability solved in the minimal supersymmetric extension of the Standard Model? If so, what is the prediction for the proton lifetime?

– *Ellis:*

In a non-Susy model the proton lifetime $\tau \sim m_X^4$ and the value of m_X is $\sim 2 \times 10^{14}$ GeV. In the case of SUSY model $m_X \sim 2 \times 10^{16}$ GeV but there are additional contributions which go like $\tau \sim m_X^2 m_W^2 \sim C$. The coefficient C is very small and with a bit of luck if m_X is of the order of $10^{16} - 10^{17}$ GeV you are just about OK.

– *Khoze:*

Could you please make some comments about the light stop scenario?

– *Ellis:*

In realistic models we have $m_{\tilde{t}} > m_t$. In a light stop scenario one could have very different decay modes, e.g. $t \to \tilde{t} + \tilde{g}$ or $\tilde{t} + \tilde{\gamma}$.

– *Sivaram:*

1) Would there be any manifestation of SUSY at one level of parity violation in atomic spectroscopy, e.g. explaining the discrepancy with the SM for Thallium?

2) Is proton decay inevitable in all GUTS or SUSY GUTS, or is there any way it can be suppressed almost completely?

– *Ellis:*

1) There is no discrepancy for Cesium, the element where parity violation is measured most precisely and the atomic physics calculations are the most reliable. If there is any discrepancy for Thallium it could not be explained by SUSY and would most likely be due to problems with the atomic physics calculations.

2) There are certain models which make the proton lifetime longer. However, there is no gauge symmetry principle which prevents the proton from decaying, and it is generally believed that all global quantum numbers such as baryon number are inevitably violated.

– *Kaur:*

Different experiments measure α_3 from data using different ways of calculating the radiative corrections, second and higher-order QCD corrections, and fixing the renormalization scale. Is it really justified to compare the results (on equal footing) to get an average value?

– *Ellis:*

I cannot really answer your question. We have combined with equal weights all the determinations which seemed to us reliable.

– *Rothstein:*

It is possible to explain charge quantisation through the necessity of anomaly cancellation, do you have a prejudice against this?

– *Ellis:*

Anomaly cancellation is necessary for mathematical consistency, but it does not answer how the SM knows that the anomalies should be cancelled. I would prefer a more profound and physical explanation of charge quantization.

– *Gallo:*

If the top is not found up to $100 \div 300$ GeV, what is your opinion about it?

– *Ellis:*

If the top is not found below 200 GeV then the fantastic agreement between LEP results and the SM seems to be fortuitous and absolutely inexplicable.

– *Wu:*

You mentioned the possible large radiative correction effect on the mass of the lightest Higgs particle in the minimum SUSY model. What is the radiative correction to the width of this particle?

– *Ellis:*

In the region of the parameter space where the CP-even Higgs is heavier than the CP-odd one, the radiative corrections could increase the amplitude for the decay of CP-even → CP-odd Higgs pairs and it could completely change the LEP phenomenology, and invalidate the previous searches for light Higgs bosons.

– *Pluciennik:*

Did you perform a complete renormalization program in the SSM in calculating the large shifts of masses by one-loop corrections? I am asking this question because people in Warsaw found smaller shifts performing this calculation.

– *Ellis:*

We calculated the Higgs mass using the effective potential formalism. Another group used a diagrammatic method and obtained similar results, so I am surprised by what you say, and do not believe it.

– *Khalatyan:*

At present there exist many programs for collider physics (LHC, SSC). They have almost the same research problems. What do you think about the possibility to increase energy up to 200 TeV?

– *Ellis:*

There will be important physics questions which the LHC and SSC will not be able to illuminate (e.g. discovering of sparticles in the range 1-10 TeV). If we want to have the option of going to a 200 TeV machine we have to start talking about it now. Also LHC and SSC already benefit from the research made for Eloisatron.

CHAIRMAN: J. Ellis

Scientific Secretaries: A. Hasan, Z. Pluciennik, H.J. Schulze

DISCUSSION II

– *Hasan:*

1) I refer to the equation:

$$m_\nu = \left(\frac{m_W^3}{m_{GUT}^2}\right)$$

in your lecture. In light of the fact that a few experiments have supported 17 KeV neutrino, could you comment on M_{GUT} scale?

2) What are the ghosts and how do they enter the superstring theory?

– *Ellis:*

1) I am very sceptical about the 17 KeV neutrino and I would not discard a theory on the basis of the 17 KeV neutrino. I remind you that the first claim to observe a 17 KeV neutrino was made 6 years ago, and was then refuted by several more precise experiments. The recent claims conflict with these earlier limits.

2) If you quantize a gauge theory or a string theory which has a large number of gauge symmetries, you have to introduce ghosts to ensure that the functional integration is performed in a gauge-invariant way.

– *Neubert:*

1) Your prediction for a neutrino mass in flipped SU(5)-GUT is $m_\nu \sim m_W^3/m^2_{GUT} \simeq 0$. Would a small neutrino mass, therefore, rule out this model, or could there be higher-order corrections (i.e. in the superpotential) which generate $m_\nu \neq 0$?

2) What are the arguments that 2-dimensional objects, i.e. strings, might be enough to build the theory of everything?

– *Ellis:*

1) This is true in the minimal realization of the flipped SU(5) model. In non-minimal realizations, for instance, in the version derived from string theory, neutrino masses can be larger.

2) Some theories with higher-dimensional objects, like fivebranes for instance, are just an alternative description of a string theory. I am not aware of any candidate for the TOE which is not a string theory.

– *Hernandez:*

Could you explain in what way a string theory fixes the number of generations, if it does?

– *Ellis:*

In the original geometrical compactification of string theory, the number of generations is connected to the Euler characteristic of the Calabi-Yau manifold, but the formula for the number of generations is more complicated in other compactification schemes. In our specific flipped SU(5) model building we only looked for models with three generations. I do not know if it is possible to construct in our particular framework a model with a different number of generations. If you only could have 3 generations in this framework, it would be really exciting.

– *Danielsson:*

You have been talking about string theories based on critical strings. During the past few years there has been a lot of progress towards constructing non-critical string theories, since more and more has been learnt about 2-dimensional quantum gravity. There are even some non-perturbative results. Could you comment on the importance of these results? Do you think that the critical strings are all we need?

– *Ellis:*

It is possible to analyze the critical strings from the point of view of non-critical string theory. There is a very interesting way of thinking about the time dimension which is the Liouville mode of a non-critical string.

– *Danielsson:*

Yesterday the anthropic principle was mentioned and you got very upset. Could you explain why? Is it really so disgusting?

– *Ellis:*

My understanding of at least one formulation of the anthropic principle is as follows: there are scientific questions which are not worthwhile to ask because if things were not the way they are, human life would not exist and we could not ask the question. This is the very form of the anthropic principle which I do not like. I want to ask all kinds of scientific questions. An example is the baryon number of the Universe. Before 1967, no-one knew how to generate it, and the anthropic principle could have prevented us from trying to explain it. Then in 1967 Sakharov showed that the baryon number of the Universe could in principle be calculated in terms of microscopic physics.

– *Potters:*

How do you break SUSY?

– *Ellis:*

The gauge group in a string model splits into an observable and a hidden part. The hidden part contains strong, confining non-Abelian gauge factors and matter particles which could condense in the vacuum. They could satisfy the requirement for SUSY breaking and this breaking could propagate into the observable sector.

– *Rothstein:*

Is it possible to "predict" the CP violating phases in the Yukawa sector?

– *Ellis:*

There are many sources of CP violation including phases from large dimensions ($D > 4$) effective operators, as well as from spontaneous CP violation. However details are model-dependent.

– *Junk:*

There is a sector in the particle spectrum of flipped SU(5) which is called the crypton sector. What are the characteristics of these particles?

– *Ellis:*

Cryptons are metastable bound states of the confined matter fields in the hidden sector. They are one of the candidates for dark matter. Their masses are in the range of $10^{10} - 10^{16}$ GeV.

– *Gourdin:*

1) In the flipped SU(5) model the relevant gauge group is SU(5) × U(1) and it is not simple. We have 2 coupling constants. You were speaking about one; what is the second one?

2) You were making qualitative statements concerning the Yukawa coupling constants for fermion mass. Can you do more quantitative ones?

– *Ellis:*

1) The two coupling constants become equal at the string unification scale which I calculated in my second lecture.

2) The physical top in our model is a linear combination of the charge 2/3 quarks sitting in different 10-dimensional representations of SU(5). We cannot make quantitative predictions, because we do not know analytically which combination it is.

– *Pluciennik:*

Is there a prediction in your model for the ratio of the V.E.V's of the two Higgs doublets?

– *Ellis:*

No, I cannot predict this ratio.

– *Duff:*

The absence of adjoint Higgs required the caveat that the Kac-Moody algebra be level 1. Can you tell us why nature abhors levels > 1?

– *Ellis:*

I do not know what nature does, but I know what the physicists do. And I do not know any consistent string theory in which the Kac-Moody algebra level is greater than 1. We, Lewellen, Ibanez and others have tried and failed to construct one.

– *Weselka:*

Superstring theory is at least 10-dimensional. Obviously our observed space-time is approximatly flat and 4-dimensional. Have you any ideas why this is so, why there are 3 flat space dimensions and the others are curled up?

– *Ellis:*

I do not understand why we have three space dimensions and one time. As I said before, one can try to understand time as the Liouville mode of a non-critical string theory. My own hunch is that our space-time is 4-dimensional because only in 4 dimensions do there exist interesting renormalizable, as distinct from non-renormalizable, and as distinct from finite, field theories.

– Peccei:

You argued nicely for SUSY GUTs from the apparent unification of α_3, α_2 and α_1 at M $\sim 10^{16}$ GeV, favouring this over ordinary GUTs which do not naïvely unify after the coupling constant extrapolation. However, in your favorite flipped model the picture you finished with was just like GUTs with disunification at 10^{16} GeV and unification at 10^{18} GeV. Does this not bother you? Have you flipped?

– Ellis:

Clearly I would have preferred m_{GUT} to be closer to m_{SU}, but I have to follow what Nature tells us. It is in fact possible to increase m_{GUT} closer to m_{SU} by including extra particles in the effective low-energy theory. These can be chosen in such a way that the successful prediction of $\sin^2\theta_W$ can be retained, but this is not automatic. One way to retain it automatically would be to have an SU(3) × SU(3) × SU(3) gauge group between m_{SU} and m_{GUT}.

– Sivaram:

1) You had the string unification scale at $\sim 10^{18}$ GeV. At that energy the dimensionless gravitational constant $(GM^2/\hbar C)$ is $\sim 10^{-2}$. Is the GUTs coupling also of this magnitude? Do you expect any modification in gravity between 10^{15} and the Planck scale at 10^{19} GeV?

2) Implications of flipped SU(5) × U(1) for dark matter (neutralinos etc.)? Your neutrino masses are too light $\sim 10^{-20}$ eV.

– Ellis:

1) M_{SU} is of the order of lightest higher string states. This is where those additional particles start coming into the game. The gravitational strength is comparable to the gauge coupling at this scale. Because such a large number of states come in at this scale the gauge coupling rapidly overtakes the gravitational strength and becomes asymptotically non-free.

2) In the minimal field-theoretical realization of the flipped SU(5) model, the neutrinos are too light to be the dark matter. There are two types of candidate for the dark matter: one is the LSP and the others are cryptons.

– Etzion:

ρ and $\sin^2\theta_W$ have already been measured in various ways: νe scattering, Z line shape, e^+e^- decay asymmetries etc. These measurements gave different values from which it was possible to deduce for example limits on top mass, technicolour models etc. Is there a way to tell something from these results about string theories, or do we still need to wait to reach the energies of 10^{16} or 10^{18} GeV^2, or wait for more statistics?

– *Ellis:*

Our model offers a possibility of getting $\sin^2\theta_W$ which is lower than in the minimal SUSY GUT. Comparing those results with experiment one might at least say that one is not discouraged.

DO WEAK INTERACTIONS BECOME STRONG AT HIGH ENERGY?

R.D. Peccei
Department of Physics
University of California
Los Angeles, CA 90024

Abstract

Weak coupling theories in a semiclassical approximation lead to rapidly growing cross-sections coming from multiparticle production. In these lectures, I discuss two circumstances where such an approximation provides a sensible starting point: B + L violation in the standard electroweak theory and high order estimates in $\lambda\phi^4$. Even though a rapidly growing cross-section may obtain for a limited energy range, such behaviour eventually must level off. Such a leveling off is hinted at by studying corrections to the semiclassical limit, and a general formalism for performing these corrections, due to Khlebnikov, Rubakov and Tyniakov, is explained. Because no closed form of the correction exists, however, the situation remains unclear. For the $\lambda\phi^4$ theory, because of its Borel summability, one can show that the total cross-section at weak coupling is exponentially small. For the electroweak theory, however, a similar strong assertion cannot be made. Nevertheless, unitarity arguments, emphasized particularly by Zakharov and by Maggiore and Shifman, strongly hint that also B + L violation in the standard model will not grow strong at high energy, reaching at most a cross-section of order $\frac{1}{s}e^{-2\pi/\alpha_w}$. Some of the caveats surrounding these conclusions are emphasized.

1. Semiclassical Approximations and Phase Space Growth

A clear goal of the hadronic supercolliders [LHC, SSC and Eloisatron] which hopefully will become operational at, or near, the start of the next millenium, is to probe the origin of the spontaneous symmetry breakdown of the electroweak theory. Because the partonic sub energy in these machines will far exceed the order parameter $v = (\sqrt{2}G_F)^{-1/2} \simeq 250$ GeV of the electroweak theory:

$$\sqrt{\hat{s}} \gg v \ , \tag{1}$$

the prospects of achieving this goal are excellent. It has been recently realized, however, that there is another interesting energy scale which might possibly be explored. This scale is characterized by energies of order M_w/α_w, with $\alpha_w = g_2^2/4\pi \simeq 4\alpha$ being the fine structure constant of the weak $SU(2)$ group. Since $M_w = g_2 v/2$, this second scale is related to the order parameter v **divided** by the $SU(2)$ coupling constant: v/g_2. It has been suggested by Ringwald[1] and Espinosa[2] that at energies above this scale:

$$\sqrt{\hat{s}} > \frac{M_w}{\alpha_w} \qquad (2)$$

novel nonperturbative phenomena may arise characterized by:

i) Copious production of W and Z bosons (as well as Higgs boson) with a multiplicity growing rapidly with energy

$$<n> \sim \alpha_w^{1/3} \left(\frac{\hat{s}}{M_w^2}\right)^{2/3} \qquad (3)$$

ii) Violation of total fermions number by 6 units:

$$\Delta(B+L) = 6 \qquad (4)$$

The purpose of these lectures is to examine the reasoning which leads to the identification of this new energy scale and to these rather amazing predictions. After this, then we will try to understand if the phenomena identified by Ringwald and Espinosa really obtains. That is, we want to ask and try to answer the question: can weak interactions really become strong? Before entering into details, however, it might be helpful to understand why the scale M_w/α_w is the natural scale where one may perhaps expect some new phenomena to occur. If the energy is sufficiently high, multiparticle production is allowed kinematically to occur.* So to produce n W's one needs to have at least

$$\sqrt{\hat{s}} > nM_w \qquad (5)$$

In perturbation theory, these processes are very suppressed since they originate at $0(\alpha_w^n)$ and α_w is a small parameter. However, combinatoric effects may vitiate this reasoning at sufficiently high value of n. In fact the interesting regime is when $n \sim \frac{1}{\alpha_w}$, where an $n!$ factor can compensate a factor of α_w^n. So indeed for energies of the order of M_w/α_w [cf Eq. (2)] one may expect surprises associated with copious production of W's and Z's.

These kinematical arguments, however, are not enough to make the case. For these processes really to become important it is necessary that the amplitudes for production not be rapidly damped, so that the natural phase space growth with energy,

---

*Whether it actually occurs is a dynamical question.

which favors multiparticle production, obtains. Remarkably, there is a simple circumstance which guarantees this to happen. This occurs when one can make a semiclassical approximation for the multiparticle Green's functions. We shall demonstrate this interesting point and derive some useful formulas in this Section. In the next Sections we will take up the much more difficult issue of trying to justify when such a semiclassical approximation is applicable and what are its first corrections.

For our purposes, it is sufficient to consider here only scalar fields ϕ. If we want to compute the physical amplitude for an n-leg process, $A_n(p_1, p_2, ..., p_n)$, we must consider the Green's function

$$G_n(x_1, x_2, ..., x_n) = <0|T(\phi(x_1), \phi(x_2), ..., \phi(x_n))|0> \quad . \tag{6}$$

Then the amplitude in question is given by

$$(2\pi)^4 \delta^4(\sum_i p_i) A_n(p_1, p_2, ..., p_n) = \prod_{i=1}^n (p_i^2 + m^2) \int d^4x_i e^{ip_i x_i} G_n(x_i, x_2, ..., x_n)|_{p_i^2 = -m^2} \tag{7}$$

A perturbative evaluation of this amplitude, say in a $\lambda \phi^4$ field theory, would give an amplitude A_n which, in general, is damped in the momentum transfers $p_i \cdot p_j$, since the building blocks are both the point-like vertices as well as the damped bare propagators,

$$\Delta(p) = i \int d^4(x_1 - x_2) <0|T(\phi(x_1)\phi(x_2))|0> e^{-ip(x_1 - x_2)} \quad . \tag{8}$$

The situation is totally different, however, if one tries to evaluate the amplitude A_n by making a semiclassical approximation for the Green's function G_n. Semiclassically, the idea is to replace the operator quantum field $\phi(x)$ by some classical field configuration. In general, such a classical field will depend not only on x, but also on where the classical field is centered and on other parameters describing its extent, possible direction in some internal space, etc. Let us denote by $\{\rho\}$ the collection of all parameters for the classical field, except its location, z. Then the semiclassical replacement intended for G_n replaces the quantum field $\phi(x)$ by the classical field $\phi_{cl}(x - z; \{\rho\})$:

$$\phi(x) \to \phi_{cl}(x - z; \{\rho\}) \tag{9}$$

and the Green's function becomes just a product of these classical fields, with some weighted integral - with weight $e^{-S[\{\rho\}]}$ - over the parameters characterizing these fields:

$$G_n(x_1, x_2, ..., x_n) = \int d^4z d\{\rho\} \, e^{-S[\{\rho\}]} \Pi_i \phi_{cl}(x_i - z; \{\rho\}) \tag{10}$$

The integral over z above insures that energy and momentum is conserved. Indeed, the Fourier transform of G_n needed to obtain the amplitude A_n [cf Eq. (7)], because of the integral over d^4z above, automatically provides the energy momentum conserving δ-function:

$$\Pi_i \int d^4x_i e^{ip_i x_i} G_n(x_1, x_2, ..., x_n) = (2\pi)^4 \delta^4(\sum_i p_i) \int d\{\rho\} e^{-S\{\rho\}} \Pi_i \int d^4x_i e^{ip_i x_i} \phi_{cl}(x_i; \{\rho\}) \tag{11}$$

Eq. (11), in contrast to the propagator (8), does **not** contain coordinate differences. Thus no dependence on $p_i \cdot p_j$ will arise in the amplitude A_n in the semiclassical approximation. Let us write the individual Fourier transforms of the classical field ϕ_{cl} as

$$\int d^4x_i e^{ip_i x_i} \phi_{cl}(x_i; \{\rho\}) \equiv \frac{Z[p_i^2; \{\rho\}]}{p_i^2 + m^2} \tag{12}$$

Calling the on shell residue $Z[\{\rho\}] \equiv Z[p^2;\{\rho\}]|_{p^2=-m^2}$, one sees that in the semiclassical approximation A_n is just a point-like amplitude:

$$A_n(p_1, p_2, ..., p_n) = \int d\{\rho\}\, e^{-S[\{\rho\}]}(Z[\{\rho\}])^n \tag{13}$$

Having established that, semiclassically, one has point-like amplitudes it is easy to see how a rapid growth of multiparticle production ensues, essentially as a result of the growth in the n-particle phase space. The phase space for producing n identical particles - assumed to be relativistic, for simplicity - grows like $\hat{s}^n$:

$$\begin{aligned}\Phi_n &= \frac{1}{n!}\Pi_i \int \frac{d^3 p_i}{(2\pi)^3 2 E_i} (2\pi)^4 \delta(\sqrt{\hat{s}} - \sum_i E_i) \delta^3(\sum_i p_i) \\ &\xrightarrow[\substack{|\vec{p}_i|\gg m \\ n\,\text{large}}]{} \frac{1}{(n!)^3}\left[\frac{\hat{s}}{16\pi^2}\right]^n \frac{1}{\hat{s}}\end{aligned} \tag{14}$$

Typically, the semiclassical amplitudes A_n grow like $n!$ for large n

$$A_n = k\frac{n!}{\sigma^n}. \tag{15}$$

where σ is a typical scale associated with the classical field ϕ_{cl}. Thus one obtains partial cross sections which are Poisson distributed

$$\sigma_n \simeq \frac{k^2}{\hat{s}} \frac{1}{n!} \left(\frac{\hat{s}}{16\pi^2 \sigma^2}\right)^n, \tag{16}$$

with a multiplicity which grows as a power of the energy[†] and a total cross section which grows **exponentially** with the energy:

$$\sigma_{tot} = \sum_n \sigma_n \simeq \frac{k^2}{\hat{s}} \exp[\frac{\hat{s}}{16\pi^2 \sigma^2}] \tag{17}$$

Various remarks are in order:

i) The above demonstrates clearly that irrespective of the strength of the coupling, in the semiclassical approximation one is lead to rapidly increasing particle multiplicities and a total cross section which is growing exponentially fast with energy. Thus, in this approximation, weak interactions becomes eventually strong at sufficiently high energy.[3]

ii) Since unitarity does not permit one to have a cross section indefinitely growing exponentially with energy, eventually the semiclassical approximation - even if it was a good approximation in some energy regime - must break down. From this point of view, whether weak interactions become strong depends on how far the semiclassical growth is allowed to go on.[4] If this growth goes on till one is near the unitarity limit, and then a proper unitarization of the amplitudes kills the unphysical behaviour, then one may well have a weakly interacting theory which becomes, effectively, strongly interacting at high energy. This is the exciting possibility that the work of Ringwald[1] and Espinosa[2] suggested for the electroweak theory.

[†]Here $<n> \sim \frac{\hat{s}}{\sigma^2}$.

iii) Before worrying about what really happens to amplitudes evaluated semiclassically at high energy, one must establish whether this approximation makes sense for any process at some energy. As we shall see in the next two Sections, there are circumstances where indeed this is the case.

Before closing this Section, it proves useful for our later development to establish a semiclassical formula for the total cross section, using the result (13) for the n-leg amplitude. The formula for the total cross section

$$\sigma_{tot} = \frac{1}{2\hat{s}} \sum_n \frac{1}{n!} \Pi_i \int \frac{d^3 p_i}{(2\pi)^3 2E_i} |A_{n+2}|^2 (2\pi)^4 \delta^4(P - \sum_i p_i) \tag{18}$$

can be easily summed since the amplitude squared $|A_{n+2}|^2$ has a factorized form in the semiclassical approximation considered:

$$|A_{n+2}|^2 = \int d\{\rho\} d\{\rho'\} e^{-S[\{\rho\}]} e^{-S[\{\rho'\}]} (Z[\{\rho\}] Z^*[\{\rho'\}])^n (Z^*[\{\rho\}] Z[\{\rho'\}])^2 \quad , \tag{19}$$

and one can similarly "factorize" the energy - momentum conserving δ-function:

$$(2\pi)^4 \delta^4(P - \sum_i p_i) = \int d^4 x \, e^{i[P - \sum_i p_i]x} \quad . \tag{20}$$

Using (19) and (20) one can rewrite Eq. (18) as

$$\sigma_{tot} = \frac{1}{2\hat{s}} \int dt d^3\vec{x} d\{\rho\} d\{\rho'\} (Z^*[\{\rho\}] Z[\{\rho'\}])^2 e^{iPx} e^{-S[\{\rho\}]} e^{-S[\{\rho'\}]}$$

$$\sum_n \frac{1}{n!} \left[\int \frac{d^3 p_i}{(2\pi)^3 2E_i} e^{-ip_i x} Z[\{\rho\}] Z^*[\{\rho'\}] \right]^n \quad . \tag{21}$$

Performing the trivial sum one establishes the following formula for σ_{tot}:

$$\sigma_{tot} = \frac{1}{2\hat{s}} \int dt d^3\vec{x} d\{\rho\} d\{\rho'\} e^{\Gamma[\{\rho\}, \{\rho'\}, \vec{x},t; \hat{s}]} (Z^*[\{\rho\}] Z[\{\rho'\}])^2 \tag{22}$$

where, in the cm system $(\vec{P} = 0, P^0 = \sqrt{\hat{s}})$,

$$\Gamma = -i\sqrt{\hat{s}} t - S[\{\rho\}] - S[\{\rho'\}] + \int \frac{d^3 p}{(2\pi)^3 2E} e^{i(Et - \vec{p}\cdot\vec{x})} Z[\{\rho\}] Z^*[\{\rho'\}] \tag{23}$$

This formula is a particular case of a more general functional formula derived by Khlebnikov, Rubakov and Tyniakov[5] for the total cross section. At high energies, this expression can be further reduced by doing the integrals over $\vec{x}$ and t by a saddle point technique. The stationary point of the exponent Γ occur at $\vec{x}_{saddle} = 0$ and at t given by the equation

$$-i\sqrt{\hat{s}} + Z[\{\rho\}] Z^*[\{\rho'\}] i \int \frac{d^3 p}{(2\pi)^3} \frac{e^{iEt}}{2} = 0 \tag{24}$$

The integral above is defined by giving t a small positive imaginary part. In the relativistic limit one has

$$i \int \frac{d^3 p}{(2\pi)^3 2} e^{iEt} = \frac{1}{2\pi^2 t^3} \quad . \tag{25}$$

Thus the saddle point occurs at imaginary time

$$t_{saddle} = i \left[\frac{Z[\{\rho\}] Z^*[\{\rho'\}]}{2\pi^2 \sqrt{\hat{s}}} \right]^{1/3} \tag{26}$$

In the high energy limit the integral over $\vec{x}$ and t just gives back Γ evaluated at the saddle points. Hence

$$\sigma_{tot} \sim \frac{1}{\hat{s}} \int d\{\rho\} d\{\rho'\} e^{-S[\{\rho\}]} \exp\frac{3}{2}\left[\frac{Z[\{\rho\}]Z^*[\{\rho'\}]\hat{s}}{2\pi^2}\right]^{1/3} e^{-S[\{\rho'\}]} \qquad (27)$$

This expression for the total cross section is shown pictorially in Fig. 1. Creating and destroying the multiparticle states introduces factors of $e^{-S[\{\rho\}]}$ and $e^{-S[\{\rho'\}]}$, respectively, reflecting the original weight of the semiclassical configurations. The pointlike amplitudes, along with the opening up of the multiparticle phase space, introduce a "bond function" - in the language of Maggiore and Shifman[6] - which is growing exponentially with energy. It is this function which - upon doing the integrals over the other parameters $\{\rho\}, \{\rho'\}$ of the semiclassical configurations - will give the exponential growth of σ_{tot} with energy

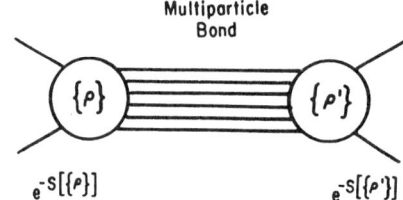

Figure 1. Schematic form of the total cross section formula in the semiclassical approximation.

In the next Section we will establish what $S[\{\rho\}]$ and $Z[\{\rho\}]$ are in the electroweak theory and detail under what circumstances such a semiclassical approximation is a sensible starting point. Although there will be some slight technical changes, since we will be dealing with both scalar fields and gauge fields, much of what we will do is just a simple extension of the formulas we have just derived.

2. θ-Vacua, Anomalies and $B+L$ Violation in the Electroweak Theory

Non Abelian gauge theories have a vacuum state which is more complicated than that in ordinary field theories. As a result, the vacuum functional of these theories - which is the generator of all the Green's functions - is a superposition of different pieces, characterized by the value of a certain integral over particular field configurations. For non zero values of this integral, it will become clear that semiclassical configurations play an important role, for one does expansions about these nontrivial fields. Although this is slightly removed from the main topic of my lectures, it is worthwhile making a brief excursion into these matters, so as to render these notes more self contained.

The correct vacuum state of a non Abelian gauge theory is the, so called, θ - vacuum. The origin of the θ - vacuum is perhaps simplest to understand by considering an $SU(2)$ gauge theory in the $A_a^0 = 0$ gauge.[7] In this gauge, the vacuum configuration is a pure gauge field with only spatial components, depending solely on $\vec{x}$ and not on t. These pure gauge fields $A_a^i(\vec{x})$ are assumed to vanish as $\vec{x} \to \infty$ and, because of this, they can be classified depending on how their $\vec{x} \to \infty$ behaviour is obtained. If we write the gauge fields as 2×2 matrices

$$A^i(\vec{x}) \equiv \frac{T_a}{2} A^i_a(\vec{x}) \;, \tag{28}$$

then the behaviour of the gauge transformation matrix at space infinity

$$\Omega_n(\vec{x}) \xrightarrow{\vec{x} \to \infty} e^{2\pi i n} \quad [n = \text{integer}] \;, \tag{29}$$

allows one to classify these fields by the index n:

$$[A^i(\vec{x})]_n = \frac{i}{g_2} \Omega_n(\vec{x}) \nabla^i \Omega_n^{-1}(\vec{x}) \;. \tag{30}$$

This integer is the index of the map of $SU(2) \sim S_3 \to S_3$ and it is just the number of windings of this map. As such, n is related to the Jacobian of the transformation and one can show that[8]

$$n = \frac{ig_2^3}{24\pi^2} \int d^3\vec{x} \; \text{Tr} \; \epsilon_{ijk} [A^i(\vec{x})]_n [A^j(\vec{x})]_n [A^k(\vec{x})]_n \tag{31}$$

The vacua associated with the pure gauge fields $[A^i(\vec{x})]_n$ (the n-vacua) are clearly not gauge invariant, since a gauge transformation can transform $[A^i(\vec{x})]_n$ into $[A^i(\vec{x})]_{n+1}$:[‡]

$$[A^i(\vec{x})]_{n+1} = \Omega_1(\vec{x}) [A^i(\vec{x})]_n \Omega_1^{-1}(\vec{x}) + \frac{i}{g_2} \Omega_1(\vec{x}) \nabla^i \Omega_1^{-1}(\vec{x}) \;. \tag{32}$$

That is, if we let $|n>$ stand for an n - vacua state, then

$$\Omega_1 |n> = |n+1> \tag{33}$$

The θ - vacuum is superposition of n-vacua which, by construction, is gauge invariant. It is defined by

$$|\theta> = \sum_n e^{-in\theta} |n> \tag{34}$$

and it is easy to check that, for example,

$$\Omega_1 |\theta> = e^{i\theta} |\theta> \tag{35}$$

Because the θ-vacuum is a superpositon of n-vacua, the vacuum functional, which describes the transition probability of the vacuum from $t = -\infty$ to $t = +\infty$, is split into distinct sectors,[7]

$$_+<\theta|\theta>_- = \sum_{n,m} e^{i(m-n)\theta} {}_+<m|n>_- = \sum_\nu e^{i\nu\theta} \left\{ \sum_n {}_+<n+\nu|n>_- \right\} \tag{36}$$

That is, one must sum over transitions of fixed $\nu = n_+ - n_-$, and then sum over distinct ν sectors with a weight $e^{i\nu\theta}$. A pictorial representation of various different transitions from the vacuum at $t = -\infty$ to that at $t = +\infty$ is shown in Fig. 2.

Using Eq. (31), the integer ν in the $A^0_a = 0$ gauge is given by

$$\nu = n_+ - n_- = i\frac{g_2^3}{24\pi^2} \int d^3\vec{x} \epsilon_{ijk} \; Tr[A^i(\vec{x})]_n [A^j(\vec{x})]_n [A^k(\vec{x})]_n \Big|_{t=-\infty}^{t=+\infty} \tag{37}$$

[‡]Eq. (32) uses that $\Omega_n(\vec{x}) = [\Omega_1(\vec{x})]^n$.

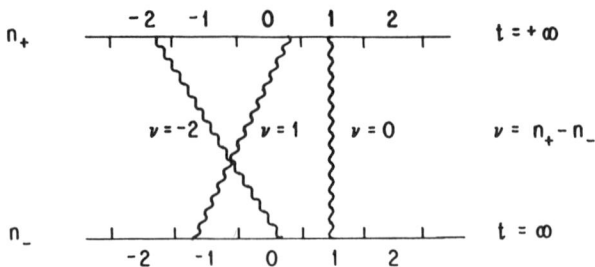

Figure 2. Transitions making up the vacuum amplitude. Shown are various paths of different ν.

Using Bardeen's identity[9]
$$\partial_\mu K^\mu = F_a^{\mu\nu} \tilde{F}_{a\mu\nu} , \qquad (38)$$
where
$$K_\mu = \epsilon_{\mu\alpha\beta\gamma} A_a^\alpha [F_a^{\beta\gamma} - \frac{g_2}{3}\epsilon_{abc} A_b^\beta A_c^\gamma] \qquad (39)$$
and $\tilde{F}_a^{\mu\nu}$ is the dual of $F_a^{\mu\nu}$:
$$\tilde{F}_a^{\mu\nu} = \frac{1}{2}\epsilon^{\mu\nu\alpha\beta} F_{a\alpha\beta} , \qquad (40)$$
one can readily show that (37) is precisely the $A_a^0 = 0$ expression corresponding to the invariant
$$\nu = \frac{g_2^2}{32\pi^2} \int d^4x F_a^{\mu\nu}(x)\tilde{F}_{a\mu\nu}(x) \qquad (41)$$
Thus, schematically, the vacuum functional written as a path integral decomposes into a sum over distinct sectors, characterized by a fixed value of the integral of $F\tilde{F}$, Eq. (41):
$$+ <0|0>_- = \sum_\nu \left[\int \delta A_\mu e^{iS_{\text{eff}}}\right]_\nu \qquad (42)$$
where
$$S_{eff} = S + \theta \frac{g_2^2}{32\pi^2} \int d^4x F_a^{\mu\nu} \tilde{F}_{a\mu\nu} \qquad (43)$$

Ordinary perturbation expansions in gauge theories are clearly connected with the $\nu = 0$ sector, since one expands about $F = \tilde{F} = 0$. To see the $\nu \neq 0$ sectors one must focus on processes which, as t goes from $-\infty$ to $+\infty$, necessarily involve field configurations, where $\int d^4x F\tilde{F} \neq 0$. In as much as these transitions involve nontrivial field configurations, they provide justification for using semiclassical methods for evaluating appropriate Green's functions. In the electroweak theory as first pointed out by 't Hooft,[10] $\nu \neq 0$ processes are connected with $B + L$ violation, as a result of the Adler Bell Jackiw anomaly.[11] So one can hope that for these processes a semiclassical approach to the evaluation of the scattering amplitudes may be a sensible first approximation.

In the standard electroweak theory of Glashow, Salam and Weinberg, at the classical level, the total fermion number current
$$J_{B+L}^\mu = \sum_{i=\text{quarks, leptons}} \bar{f}_i \gamma^\mu f_i \qquad (44)$$

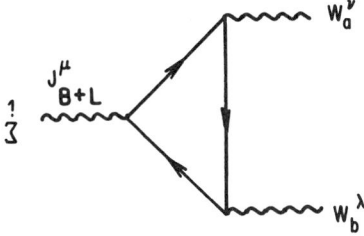

Figure 3. Triangle graph containing the f_{iL} loops which give rise to the nonconservation of $B+L$ in the electroweak theory.

is conserved. However, because the left-handed fermions, f_{iL}, are part of $SU(2)$ doublets, while the right-handed fermions, f_{iR} are $SU(2)$ singlets, quantum mechanically the $B+L$ current has an anomaly. The anomaly arises because only the left-handed fermions in J^μ_{B+L}

$$J^\mu_{B+L} = \sum_i \bar{f}_i \gamma^\mu f_i = \sum_i \left(\bar{f}_{iL}\gamma^\mu f_{iL} + \bar{f}_{iR}\gamma^\mu f_{iR}\right) \tag{45}$$

contribute to the triangle graph of Fig. 3. Because there is no f_{iR} loop, effectively J^μ_{B+L} is a chiral current and has an Adler Bell Jackiw[11] anomaly.

In the electroweak theory, as the result of the graph of Fig. 3, one finds that the divergence of the $B+L$ current is a measure of $F\tilde{F}$:

$$\partial_\mu J^\mu_{B+L} = 2N_g \left[\frac{g_2^2}{32\pi^2} F^{\mu\nu}_a \tilde{F}_{a\mu\nu}\right] \tag{46}$$

where N_g is the number of generations.§ As a result of Eq. (46), any standard model process which involves a change in the fermionic charge must also **necessarily** involve a change of the gauge field n vacua as one goes from $t = -\infty$ to $t = +\infty$. The total change of $B+L$ as one makes this transition is directly related to ν. Indeed

$$\Delta(B+L) = \int d^4x \partial_\mu J^\mu_{B+L} = 2N_g \frac{g_2^2}{32\pi^2} \int d^4x F^{\mu\nu}_a \tilde{F}_{a\mu\nu} = 2N_g \nu \tag{47}$$

I want to mention here since I shall use it shortly below, a related aspect of chirality and the index ν, connected to the spectrum of the Dirac equation in Euclidean space in the presence of background gauge fields. In Euclidean space, the Dirac operator

$$\not{D}_E = \gamma_\mu [\partial_\mu - ieA_\mu] \tag{48}$$

is a Hermitian operator and so one can study the spectrum of its eigenvalues. There is a famous result, due to Atiyah and Singer,[12] which characterizes this eigenvalue spectrum and its eigenstates as follows. The eigenvalue problem

$$\not{D}_E \phi_n = \lambda_n \phi_n \tag{49}$$

---

§We shall assume, in what follows, that $N_g = 3$ as deduced from neutrino counting at the Z.

has both vanishing ($\lambda_n = 0$) and non vanishing ($\lambda_n \neq 0$) eigenvalues. The non vanishing eigenvalues always have two eigensolutions of opposite chirality, ϕ_{nL} and ϕ_{nR}. For the zero eigenvalues, however, there is a mismatch in the number of chiral eigenstates. In particular, if the background field corresponds to a configuration which has index ν,¶ then the difference between the number of left-handed and right-handed eigenstates with zero eigenvalue is given by ν:

$$\nu = n_L - n_R \tag{50}$$

To have a $B + L$ violating process in the electroweak theory, in the language of the $A_a^0 = 0$ gauge, the gauge field configurations must go from an n-vacuum state at $t = -\infty$ to, say an $(n+1)$ - vacuum state at $t = +\infty$. 't Hooft[10] suggested that this kind of transition could be estimated precisely as one would a simple quantum mechanical tunnelling process. The sketch in Fig. 4, perhaps, renders this idea more vividly. Semiclassically, therefore - in total analogy to the WKB approximation in

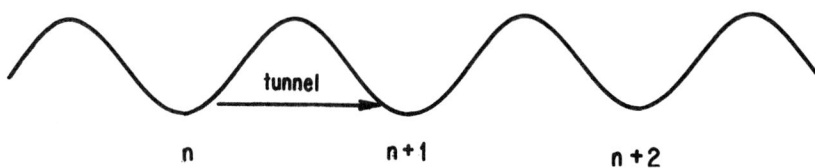

Figure 4. $B + L$ transitions require gauge fields to "tunnel" from n-vacuum to $(n+1)$ - vacuum.

quantum mechanics - the transition amplitude in the gauge sector that accompanies a process where a change of $B + L$ takes place should be given simply by[10]:

$$_+<n+1|n>_- \sim e^{-S_E^{min}[\nu=1]} \tag{51}$$

Here $S_E^{min}[\nu = 1]$ is the minimal Euclidean action for configurations with $\nu = 1$. These configurations, and the associated minimal Euclidean action for the case of a pure $SU(2)$ gauge theory, were actually identified previous to 't Hooft's paper by Belavin, Polyakov, Schwartz and Tyupkin.[13] They are the famous **instantons** - self dual solutions of the Yang Mills equations of motion:

$$F_a^{\mu\nu} = \tilde{F}_a^{\mu\nu} \tag{52}$$

Because of the self dual nature of these solutions, the minimal action is trivial to evaluate since

$$S_E^{min}[\nu = 1] = \frac{1}{4}\int d^4x_E F_a^{\mu\nu} F_a^{\mu\nu} = \frac{1}{4}\int d^4x_E F_a^{\mu\nu} \tilde{F}_a^{\mu\nu} = \frac{8\pi^2}{g_2^2} = \frac{2\pi}{\alpha_w} \tag{53}$$

¶In Euclidean space ν has an analogous definition to Eq. (41), except that one uses Euclidean field strengths and an Euclidean integration measure.

The result (53), unfortunately, is rather discouraging since $\frac{2\pi}{\alpha_w}$ is a very large number in the electroweak theory [$\frac{2\pi}{\alpha_w} \sim 180$]! So, just looking at the purely gauge field factor that always must accompany a $B + L$ violating process, the amplitude for these processes is so suppressed to make their experimental detection totally out of the question. Despite knowing this fact, however, Helen Quinn and I, soon after 't Hooft's paper appeared, looked at the process of fermion number violation in the electroweak theory.[14] Because we knew the rate was so suppressed, our interest was in trying to understand how one could really get in the electroweak theory Green's functions with all fermions either coming out or coming in. We finally succeeded to understand how this happens, by appealing to the Atiyah Singer theorem[12] alluded to above. The formula which we derived, whose structure I will elaborate on below, not surprisingly gives rise to point-like $B + L$ violating amplitudes. Unfortunately (or, perhaps, fortunately!), the formula we wrote down did not make this immediately obvious and, being sufficiently discouraged by the $e^{-2\pi/\alpha_w}$ factor, we did not pursue the matter further. So one had to await more than 10 years for Ringwald and Espinosa to discover that there was something interesting to further investigate in these processes!

To set the stage for the calculation of Ringwald[1] and Espinosa,[2] let us imagine for simplicity that there is only one generation of quarks and leptons and let us try to compute the $\Delta(B+L) = 2$ (and hence $\nu = 1$) process $\bar{u} + \bar{d} \to u + e$. For these purposes we need a Green's function with only fermion fields and **no antifermion** fields. Because the instantons are classical solutions in Euclidean space we will do all our computations in Euclidean space and eventually rotate the result for the Green's function back to Minkowski space. We want to compute the Green's function

$$G_{uude} = <0|T(\psi_u(x_1)\psi_d(x_2)\psi_u(x_3)\psi_e(x_4))|0> \tag{54}$$

This Green's function does not vanish identically only because there is an asymmetry in the number of zero modes ($\lambda_n = 0$ eigenvalues) of the Dirac equation in the $\nu = 1$ sector. This follows directly from the Atiyah Singer theorem.[12]

It turns out, as 't Hooft showed, that the Euclidean Dirac equation in the $\nu = 1$ sector has only one (normalizable) solution, corresponding to a left-handed eigenmode and no right-handed eigensolution. 't Hooft gave in his paper already the form of the left-handed eigensolution of the Euclidean Dirac equation

$$\slashed{D}_E \psi_{0L}(x) = \gamma_\mu [\partial_\mu - ig_2 W_\mu^{\text{inst}}(x)]\psi_{0L}(x) = 0 \tag{55}$$

in the presence of an instanton located at z of size ρ

$$W_\mu^{\text{inst}}(x-z;\rho) = \frac{i}{g_2} \frac{\rho^2}{(x-z)^2} \frac{[\delta_{\mu\nu} - \sigma_\mu^\dagger \sigma_\nu](x-z)_\nu}{[\rho^2 + (x-z)^2]} \tag{56}$$

It reads

$$[\psi_0(x-z;\rho)]_{a\alpha} = \sqrt{\frac{2\rho^3}{\pi^2}} i \frac{[(x-z)_\mu \sigma_\mu]_{ab}\epsilon_{b\alpha}}{|x-z|[\rho^2 + (x-z)^2]^{3/2}} \tag{57}$$

In Eqs. (56) and (57) σ_μ and $\sigma_\mu^\dagger$ are the Pauli matrix combinations

$$\sigma_\mu = (\vec{\sigma}, i) \; ; \sigma_\mu^\dagger = (\vec{\sigma}, -i) \tag{58}$$

and a, α are respectively $SU(2)$ and Weyl indices.

The Green's function G_{uude} is computed semiclassically by replacing each of the quantum fields $\psi(x)$ in Eq. (53) by the classical solutions $\psi_0(x-z;\rho)$:

$$\psi(x) \longrightarrow \psi_0(x-z;\rho) \tag{59}$$

with a weight given by the 't Hooft tunnelling factor of Eq. (51). Actually, strictly speaking, this is not quite correct, since there are no combined solutions of the Higgs and gauge fields equations of motion[15] and so there is no corresponding $S_E^{min}[\nu=1]$ in the full electroweak theory. However, as long as the size of the instanton is not too large compared to the inverse size of the order parameter of the electroweak theory, one can find a good approximate solution for the combined theory. This approximate solution still has the gauge configurations given by the $SU(2)$ instantons of Eq. (56), while the scalar doublet field takes the form

$$\Phi(x-z;\rho) = \frac{v}{\sqrt{2}}\left[\frac{(x-z)^2}{\rho^2+(x-z)^2}\right]^{1/2}\begin{pmatrix}1\\0\end{pmatrix} \tag{60}$$

The action for this approximate solution has an additional contribution to the 't Hooft piece, coming from the Higgs field configuration (60). One finds

$$S_E^{min}[\nu=1] \simeq \frac{2\pi}{\alpha_w} + \pi^2 v^2 \rho^2 \tag{61}$$

So, as long as $\rho v \ll 1$, the effect of the Higgs contributions is negligible. However, for large, ρ, the action becomes large indicating that there is really no solution[‖].

Putting everything together one arrives at a formula for G_{uude} which is essentially the one that was derived in the 1977 Peccei Quinn paper[14][16]:

$$G_{uude} \simeq \int d^4 z d\rho n_0(\rho) \exp -[\frac{2\pi}{\alpha_w} + \pi^2 v^2 \rho^2] \prod_{i=1}^{4} \psi_0(x_i - z;\rho) \tag{62}$$

Here $n_0(\rho)$ is a Jacobian factor associated with the integrals over the collective coordinates describing the instanton. It contains besides a factor of ρ^{-5}, which serves to restore the proper dimensions to Eq. (62), a further dimensionless factor arising from integrals over the $SU(2)$ orientations of the instanton[2]:

$$n_0(\rho) = 861.94 e^{-0.997 n_f} \frac{2^{10+6n_f} \pi^{6+4n_f}}{g_2^8} \frac{(\rho v)^{\frac{43-8n_f}{6}}}{\rho^5}, \tag{63}$$

where n_f is the number of fermions (here $n_f = 4$). Although Eq. (62) leads to a pointlike 4-Fermi amplitude for the process $\bar{u}+\bar{d} \to u+e$, because of the 't Hooft factor of $e^{-\frac{2\pi}{\alpha_w}}$ this amplitude remains hopelessly small.

To get any sizeable effects, as we have seen in Sec. 1, we must have sufficient energy for the growing phase space factor associated with **pointlike multiparticle production** to compensate the 't Hooft tunnelling factor. In the 3 generation standard model, the core $B+L$ violating process contains 12 fermions (9 quarks and 3 leptons) giving $\Delta(B+L) = 6$. Nevertheless, since the number of particles is fixed, this process itself is unimportant since it would take extraordinary energies to overcome the factor of $e^{-2\pi/\alpha_w}$. Only by considering processes in which besides the core violation of $B+L$ by 6 units, multi W, Z and Higgs are produced can one contemplate cancelling expeditiously the 't Hooft factor. Indeed, as our qualitative discussion of Sec. 1 indicated, this naively should occur at energies of the order of $\sqrt{\hat{s}} \sim M_w/\alpha_w$. This was the crucial observation made by Ringwald[1] and Espinosa[2] about 2 years ago.

[‖]In principle, really α_w in Eq. (61) is not fixed, but is a function of the scale ρ. To the extent that the size of ρ is set by $\frac{1}{v}$, in effect α_w is α_w evaluated at the Fermi scale.

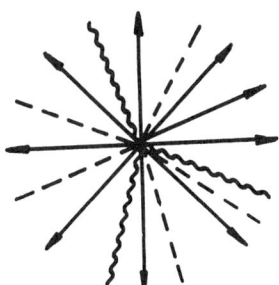

Figure 5. Point-like amplitude for $B + L$ violating processes in the electroweak theory, computed in semiclassical approximation.

In a semiclassical approximation the Green's function associated with the process $q + q \to 7\bar{q} + 3\bar{\ell} + n_H H + n_w W$ is given by the expression

$$G \simeq \int d^4z \, d\rho n_0(\rho) e^{-S_E[\rho]} \prod_{i=1}^{12} \psi_0(x_i - z; \rho)$$
$$\prod_{j=1}^{n_H} \Phi(x_j - z; \rho) \prod_{k=1}^{n_w} W_\mu^{\text{inst}}(x_k - z; \rho) \tag{64}$$

where

$$S_E[\rho] = \frac{2\pi}{\alpha_w} + \pi^2 v^2 \rho^2 \tag{65}$$

This Green's function will give point-like amplitudes for this process, as sketched in Fig. 5. The fermion factors above, since they are limited dynamically to 12, are irrelevant for the rapid growth of the total $B + L$ violating cross section with energy. So, in what follows, I will focus only on the W and Higgs factors which contribute to the relevant multiparticle amplitudes.**

To compute the contribution of the multiparticle production of W's and Higgses to the total $B + L$ violating cross section we need, according to our discussion in Sec. 1, to compute the residues of the Fourier transform of the semiclassical solution. For the present purposes, it is consistent to ignore both the Higgs and W masses. Thus we need to find the coefficient of $1/p^2$ in the relevant Fourier transform. Since

$$\int d^4x \, e^{ipx} \frac{v}{\sqrt{2}} \left[\frac{x^2}{x^2 + \rho^2} \right]^{1/2} = \frac{2\pi^2 \rho^2 v}{p^2} , \tag{66}$$

one identifies the Higgs field residue as

$$Z_H(\rho) = 2\pi^2 \rho^2 v \tag{67}$$

For the W field the residue "carries" both momentum and isospin information, since one is dealing with a spin 1, $SU(2)$ triplet object. In terms of the 't Hooft η symbols,[10] defined by the equation

$$i\sigma_a \eta_{a\mu\nu} = \sigma_\mu^\dagger \sigma_\mu - \delta_{\mu\nu} , \tag{68}$$

**I speak here loosely of W production. In fact, both W and Z are produced, correlated with the isospin properties of the instanton. For a more detailed discussion, see.[2]

the relevant Fourier transform to consider is

$$\int d^4x e^{ipx} \frac{2}{g_2} \frac{\epsilon_\mu \eta_{a\mu\nu} \rho^2 x_\nu}{x^2(x^2+\rho^2)} = \frac{i\epsilon_\mu p_\nu \eta_{a\mu\nu}}{p^2} \frac{4\pi^2 \rho^2}{g_2} \quad , \tag{69}$$

which identifies the W field residue as

$$Z_a(p;\rho) = i\epsilon_\mu p_\nu \eta_{a\mu\nu} Z_w(\rho) \tag{70}$$

with

$$Z_w(\rho) = \frac{4\pi^2 \rho^2}{g_2} \tag{71}$$

The amplitude for producing n_H Higgses and n_w W's will contain n_H factors of Z_H and n_w factors of Z_a:

$$A_{n_H,n_w} \sim \int d\rho n_0(\rho) e^{-S_E[\rho]} (Z_H(\rho))^{n_H} (Z_a(p_i;\rho))^{n_w} \tag{72}$$

Using this result, it is straightforward to obtain an expression for the total $B+L$-violating cross section, analogous to that given in Eq. (22). One finds in this way that

$$\sigma_{B+L \text{ viol}} \sim \frac{1}{\hat{s}} \int dt\, d^3\vec{x}\, d\rho d\rho'\, e^\Gamma \tag{73}$$

where the exponent Γ is now given by

$$\Gamma = -i\sqrt{\hat{s}}t - S_E[\rho] - S_E[\rho'] + Z_H(\rho) Z_H^*(\rho') \int \frac{d^3p}{(2\pi)^3 2E} e^{iEt - i\vec{p}\cdot\vec{x}}$$
$$+ Z_w(\rho) Z_w^*(\rho') \int \frac{d^3p}{(2\pi)^3 2E} e^{iEt - i\vec{p}\cdot\vec{x}} [3E^2 + \vec{p}^{\,2}] \quad . \tag{74}$$

The extra kinematical factors in the Fourier integral of the W contributions arise from doing the W polarization sum, taking into account the polarization dependent residue (70).††

In the high energy limit, the $B+L$-violating cross section can be immediately estimated by doing the integrals in Eq. (73) by a saddle point approximation. It is easy to see again that $\vec{x}_{\text{saddle}} = 0$. Neglecting both the masses of the Higgs and W bosons, which as we shall see later on is a self-consistent approximation, the Fourier transforms associated with multi Higgs and multi-W production are simply

$$\int \frac{d^3p}{(2\pi)^3 2E} e^{iEt} = -\frac{1}{4\pi^2 t^2}; \quad \int \frac{d^3p}{(2\pi)^3} E e^{iEt} = \frac{3}{\pi^2 t^4} \tag{75}$$

Thus Γ at the $\vec{x}=0$ saddle point reads

$$\Gamma_{\text{saddle}}(\rho,\rho',t;\sqrt{\hat{s}}) = -i\sqrt{\hat{s}}t - \frac{4\pi}{\alpha_w} - \pi^2 v^2(\rho^2 + \rho'^2) - \frac{\pi^2 v^2 \rho^2 \rho'^2}{t^2} + \frac{96\pi^2 \rho^2 \rho'^2}{g_2^2 t^4} \quad , \tag{76}$$

where the last two terms are the multi-Higgs and multi-W contributions, respectively. Because of the momentum dependent factor in the W Fourier transform, the W production contribution is proportional to t^{-4} and will be the dominant contribution to Γ_{saddle} for large sub-energies $\sqrt{\hat{s}}$. However, beyond a certain critical energy - which we will identify below - this no longer will be true.

---

††This factor is not given correctly in the original paper of Ringwald.[1] For a discussion of some of the subtelties involved see.[5]

Neglecting, for the nonce, the multi-particle Higgs contribution in Γ_{saddle}, the t saddle point of this function is again found to occur for imaginary time:

$$t_{\text{saddle}} = i \left[\frac{384\pi^2 \rho^2 \rho'^2}{g_2^2 \sqrt{\hat{s}}} \right]^{1/5} \tag{77}$$

Evaluating Γ at both the t and $\vec{x}$ saddle points, the exponent takes the form

$$\Gamma_{\text{saddle}}(\rho, \rho'; \sqrt{\hat{s}}) = -S_E[\rho] - S_E[\rho'] + \frac{5}{4} \left[\frac{384\pi^2 \rho^2 \rho'^2 \hat{s}^2}{g_2^2} \right]^{1/5} \tag{78}$$

In the language of Fig. 1, the "bond function" for multi W production grows like $\hat{s}^{2/5}$, rather than $\hat{s}^{1/3}$ as it did in the scalar case [cf Eq. (27)].

The formula for the $B + L$ - violating cross section obtained,

$$\sigma_{B+L \text{ viol.}} \simeq \frac{1}{\hat{s}} \int d\rho \, d\rho' \, exp \left\{ -\frac{4\pi}{\alpha_w} - \pi^2 v^2 (\rho^2 + \rho'^2) + \frac{5}{4} \left[\frac{384\pi^2 \rho^2 \rho'^2 \hat{s}^2}{g_2^2} \right]^{1/5} \right\} , \tag{79}$$

at high energy can again be evaluated by saddle points techniques. The saddle point now occur at equal values for the sizes of the classical solutions and a simple calculation gives

$$\rho_{\text{saddle}} = \rho'_{\text{saddle}} = \frac{\sqrt{6}}{2M_w} \left[\frac{E}{E_0} \right]^{2/3} \tag{80}$$

Here $E = \sqrt{\hat{s}}$ is just the incident sub energy, while E_0 is an energy scale which, not surprisingly, is related to M_w/α_w:

$$E_0 = \frac{\sqrt{6}\pi M_w}{\alpha_w} \tag{81}$$

At the saddle point, the factor in the exponential is just

$$\Gamma_{\text{saddle}}(E) = -\frac{4\pi}{\alpha_w} \left[1 - \frac{9}{8} \left(\frac{E}{E_0} \right)^{4/3} \right] , \tag{82}$$

whence one finds for the $B + L$ violating cross section the result

$$\sigma_{B+L \text{ viol}} \sim \frac{1}{E^2} \exp -\frac{4\pi}{\alpha_w} \left[1 - \frac{9}{8} \left(\frac{E}{E_0} \right)^{4/3} \right] . \tag{83}$$

Various comments are in order:

i) One sees from Eq. (83) that the 't Hooft damping factor of $e^{-4\pi/\alpha_w}$ is **overcome** at an energy

$$E = \sqrt{\hat{s}} \simeq E_0 = \sqrt{6}\pi \frac{M_w}{\alpha_w} \simeq 18 \, TeV \quad . \tag{84}$$

The crucial point that compensation of the tunnelling factor can occur at $E \simeq E_0$ is the principal result obtained by Ringwald and Espinosa[1].[2]

ii) The specific form of Eq. (83) - in which the W multi-particle cross section give an $\exp E^{4/3}$ growth - was first obtained by Zakharov[17] and Porrati,[18] although the numerical coefficients in these papers are not quite those given in Eq. (83). To my knowledge Eq. (83), with all the correct factors, was first derived by Khlebnikov, Rubakov and Tyniakov.[5]

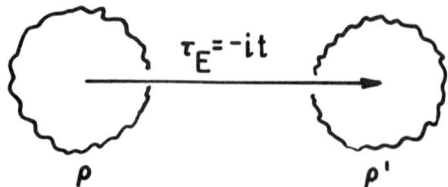

Figure 6. Conditions under which a semiclassical approximation for $\sigma_{B+L \text{ viol}}$ may be sensible. If $\rho/\tau_E \sim 0(1)$ and $\rho'/\tau_E \sim 0(1)$ quantum effects from the classical source overlap can become important.

iii) Using the value of ρ and ρ' at the saddle points, one sees that the t saddle point, given in Eq. (77), scales as $(E/E_0)^{1/3}$:

$$t_{\text{saddle}} = i \frac{\sqrt{6}}{M_w} \left[\frac{E}{E_0}\right]^{1/3} \quad (85)$$

From the above and the saddle point values for ρ and ρ' given in Eq. (80), one sees that the multi-Higgs contribution to Γ, which we have neglected, contributes a term to $\sigma_{B+L \text{ viol}}$ which grows as $\exp{(E/E_0)^2}$:

$$(\sigma_{B+L \text{ viol}})_{\text{Higgs}} \sim \frac{1}{E^2} \exp \frac{4\pi}{\alpha_w} \left\{ \frac{3}{32} \left[\frac{E}{E_0}\right]^2 \right\} \quad (86)$$

Thus, as long as $E \ll E_0$, indeed the dominant growth of the $B+L$ - violating cross section is due to multi-W production.

iv) For $E \simeq E_0$ one cannot trust Eq. (83), not only because of the extra Higgs contribution (86) which adds a further growing term to the $B+L$ - violating cross section, but because, as we shall see shortly, there are **quantum corrections** to the semiclassical formula that become of equal importance.

v) One can understand qualitatively why the semiclassical formula (83) may suffice for $E \ll E_0$ by replacing the time t in the formal expression for the cross section, Eq. (73), in terms of its Euclidean continuation $\tau_E = -it$ - with τ_E representing the separation between classical sources of sizes ρ and ρ' which are emitting and absorbing W's. As long as the sizes of the classical sources are small compared to their distance apart, as shown schematically in Fig. 6, a semiclassical approximation ought to be reasonable. This condition, using Eqs. (80) and (85) is satisfied if $E \ll E_0$

$$\left(\frac{\rho}{\tau_E}\right)_{\text{saddle}} = \left(\frac{\rho'}{\tau_E}\right)_{\text{saddle}} = \frac{1}{2}\left[\frac{E}{E_0}\right]^{1/3}. \quad (87)$$

3. Large Order Perturbation Theory a la Lipatov

Besides fermion number violating processes in the electroweak theory, there is another weakly coupled theory where amplitudes appear to become strong at high energy: $\lambda\phi^4$ theory. It is interesting to discuss this other example here, since it provides another instance of a theory where, in certain circumstances, semiclassical configurations dominate. However, we shall see that there is an important physical difference between $\lambda\phi^4$ theory and the electroweak theory, since only the former theory is Borel summable.[19]

In $\lambda\phi^4$ for weak coupling ($\lambda \ll 1$) one naively expects the N-leg amplitude to be highly suppressed due to the high powers of λ. For instance, for tree graphs the N-leg amplitude will be proportional to $\lambda^{\frac{N}{2}-1}$. This naive expectation, however, is not quite correct since one can show that for large N the expansion parameter is instead λN^2 and not λ, due to the combinatoric growth of the graphs contributing to the amplitude. Thus

$$A_N \sim N!\lambda^{\frac{N}{2}} \sim (\lambda N^2)^{\frac{N}{2}} , \qquad (88)$$

and there is a perturbative breakdown at large N.

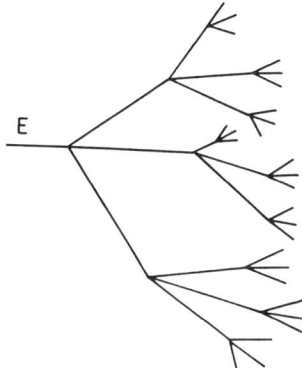

Figure 7. Branching graphs which dominate the tree graph amplitudes in $\lambda\phi^4$ theory.

The growth of N-leg amplitudes for $\lambda\phi^4$ theory has been examined recently by Cornwall[20] and Voloshin,[21] using semiclassical methods, and by Goldberg,[22] using graphical techniques. Let us consider, following Goldberg,[22] the special multiple branching graphs of Fig. 7 in the special kinematical configuration where all the final particles are essentially at rest, so that $E \simeq mN$. In these circumstances, each propagator in Fig. 7 gives just a factor of $1/m^2$ and the evaluation of the amplitude is just a counting problem. In fact, the amplitude is just gotten from the result of a typical graph, $\lambda^{N/2}/(m^2)^{N/2}$, times a combinatorial factor of $N!$. This gives precisely the result (88) alluded to above!

It is particularly instructive for our purposes to rederive the growth of the tree amplitudes given in Eq. (88) by using a technique pioneered by Lipatov[23] to study perturbation theory at high order. As we shall see, Lipatov's method appeals to important semiclassical configurations in the problem, so it ties in directly with the considerations we have discussed earlier. The idea behind Lipatov's method is the following. In Euclidean space $\lambda\phi^4$ theory is well defined for $\lambda > 0$, corresponding to the positive definitive potential shown in Fig. 8a. If $\lambda > 0$, however, the theory given by the action

$$S_E = \frac{1}{2}(\partial_\mu \phi)^2 + \frac{1}{2}m^2\phi^2 + \frac{\lambda}{4!}\phi^4 \qquad (89)$$

is unstable, since the ground state shown in Fig. 8b decays by tunnelling.

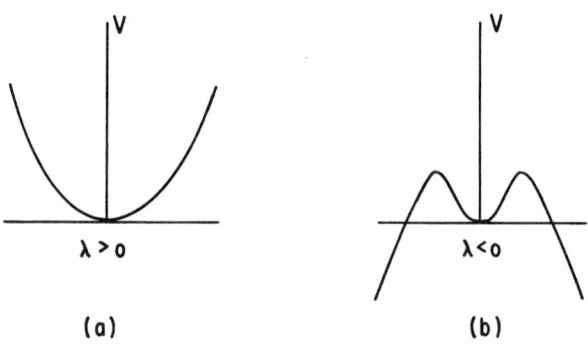

Figure 8. Potential in $\lambda\phi^4$ theory for the case when a)$\lambda > 0$ and b) $\lambda < 0$.

If $\lambda < 0$ then $\lambda' = -\lambda$ is a positive quantity. Because of the instability of the theory for $\lambda < 0$, all amplitudes in $\lambda\phi^4$ theory have an imaginary part in the λ' - plane along the positive real axis reflecting this instability. Thus, formally, one can write for any amplitude in the theory a dispersion relation of the form

$$A(\lambda) = -\frac{1}{\pi}\int_0^\infty d\lambda' \frac{ImA(\lambda')}{\lambda' + \lambda} \tag{90}$$

Obviously, Eq. (90) is only useful if one has some means to estimate $ImA(\lambda')$. If $\lambda' \to 0$ the potential of Fig. 8b becomes rather steep and narrow before it turns down towards negative infinity. Since $ImA(\lambda')$ is essentially the tunneling probability for the ground state to disappear, we see that for $\lambda' \to 0$ because the potential is so steep and narrow, it might be possible to compute $ImA(\lambda')$ by semiclassical methods. As usual, this necessitates finding a solution to the classical equations of motion. So if there are circumstances where the tip of the cut in Eq. (90) dominates, we see that semiclassical solutions of $\lambda\phi^4$ theory - for $\lambda < 0$ - play a role in determining the amplitudes of the theory.

For $m = 0$ it is easy to see that the field equations for the theory for $\lambda < 0$

$$-\partial^2\phi(x) = -\frac{\lambda'}{6}\phi^3(0) \tag{91}$$

have a classical solution: the Lipaton

$$\phi_{cl}(x - z; \rho) = \pm\sqrt{\frac{48}{\lambda'}} \frac{\rho}{(x-z)^2 + \rho^2} \tag{92}$$

As usual, in Eq. (92) z is the location of the Lipaton and ρ is its size. The action corresponding to Eq. (92) is easily seen to be

$$S_E = \frac{16\pi^2}{\lambda'} \tag{93}$$

When $m \neq 0$, as in the case of the electroweak interactions, there is no corresponding classical solution. However, as long as $\rho m \ll 1$ it is still possible to use the $m = 0$ classical solution with a slightly modified action[15]

$$S_E = \frac{16\pi^2}{\lambda'}\left\{1 + \frac{3}{2}(\rho m)^2 \ell n \frac{1}{m\rho}\right\} \tag{94}$$

One can use the semiclassical approximation (92) to evaluate amplitudes in the $\lambda\phi^4$ theory provided that the physical situation is such that the tip of the cut in the λ' plane dominates. As Lipatov first showed,[23] this obtains for amplitudes calculated in large order in perturbation theory. The tip of the cut is also important for large N tree graphs and the Lipatov technique has been used by Maggiore and Shifman[24] to calculate the N-leg amplitude, obtaining the result of Eq. (88).

For the N-leg amplitude one can write a power series expansion in λ in which the lowest order term, proportional to $\lambda^{N/2-1}$, is given by tree graphs, while terms containing a factor of $\lambda^{N/2-1+k}$ correspond to graphs containing k-loops:

$$A_N = a_N^{(0)} \lambda^{N/2-1} + a_N^{(1)} \lambda^{N/2} + \ldots + a_N^{(k)} \lambda^{N/2-1+k} + \ldots \quad (95)$$

Lipatov[23] studied the case of small N and large k - that is, scattering amplitudes of few final states calculated in large order in perturbation theory. Here, following Maggiore and Shifman,[24] we are more interested in the case of tree graphs at large N. Formally, in either cases, using Eq. (90), it is clear that for the coefficients $a_N^{(k)}$ the tip of the cut dominates since

$$a_N^{(k)} = (-1)^{N/2+k} \frac{1}{\pi} \int_0^\infty d\lambda' \frac{Im A_N(\lambda')}{(\lambda')^{N/2+k}} \quad (96)$$

In the semi classical approximation, the discontinuity across the cut on the positive λ' axis for the N-leg amplitude is given simply by

$$Im A_N(\lambda') \simeq \int d\rho \, e^{-S_E[\rho,\lambda']} [Z(\rho; \lambda')]^N \quad (97)$$

Here $Z(\rho; \lambda')$ is just the residue of the Lipaton field by which, semiclassically, one approximates the N-leg amplitude. Using the explicit solution (92) one finds for $Z(\rho, \lambda')$, defined by

$$\int d^4x \, e^{ipx} \, \phi_{cl}(x;\rho) = \frac{Z[\rho; \lambda']}{p^2} \quad , \quad (98)$$

the expression

$$Z[\rho; \lambda'] = \frac{16\sqrt{3}\pi^2 \rho}{\sqrt{\lambda'}} \quad (99)$$

Using Eqs. (96) and (97), it is straightforward to evaluate the coefficients $a_N^{(k)}$ by a saddle point technique. One has that

$$a_N^{(k)} = (-1)^{N/2+k} \frac{1}{\pi} \int d\rho \, d\lambda' \, e^{\Gamma(\rho;\lambda')} \quad (100)$$

where the exponent $\Gamma(\rho; \lambda')$ is given by

$$\Gamma = -\frac{16\pi^2}{\lambda'} \left[1 + \frac{3}{2}(m\rho)^2 \ln\frac{1}{m\rho}\right] - (N+k)\ln\lambda' + N\ln(16\sqrt{3}\pi^2\rho) \quad . \quad (101)$$

The saddle points of Γ occur at

$$\lambda' \simeq \frac{16\pi^2}{N+k}; \quad (\rho m)^2 = \frac{N}{N+k}\left[\frac{1}{3(2\ln\frac{1}{m\rho}-1)}\right] \quad (102)$$

The first of Eqs. (102) shows that the semiclassical approximation is justified since indeed for large N and/or k at the saddle point $\lambda' \to 0$. From the second of Eqs. (102) one can extract the value of ρ where the saddle occurs. In the large order example considered by Lipatov[23] (N fixed, k large) it is clear that $\rho_{\text{saddle}} \ll 1/m$ so that the approximate expression used, which is valid only for $\rho m \ll 1$, indeed turns out to be self consistently correct. For the case of tree amplitudes, where N is large and $k = 0$, on the other hand ρm is not arbitrarily small, but $\rho m \lesssim \frac{1}{\sqrt{3}}$. Although this does not totally guarantee self consistency, ρm is sufficiently small at the saddle that it is worthwhile proceeding and see what results one obtains.

Using the saddle point evaluation of the integral in Eq. (100), one finds[24] for the case of N fixed, but large k, the expression

$$a_N^{(k)} \simeq (-1)^k \left(\frac{k}{16\pi^2}\right)^k e^{-k} \qquad (103)$$

This is Lipatov's result,[23] which gives factorial growth for the amplitudes evaluated in large order in perturbation theory. In fact, since the N point amplitude for k loops is proportional to $\lambda^k a_N^{(k)}$,[‡‡] one finds for the amplitude A_N evaluated at k loops the formula

$$A_N \sim (-1)^k \left[\frac{\lambda k}{16\pi^2}\right]^k e^{-k} \qquad (104)$$

which corresponds to factorial growth at large order. Note that the amplitude (104) will start growing out of control for $k > k_{\text{crit}}$, where

$$k_{\text{crit}} = \frac{16\pi^2}{\lambda} . \qquad (105)$$

For tree amplitudes, on the other hand, one finds

$$a_N^{(0)} = (-1)^{N/2} \left(\frac{N}{m}\right)^N . \qquad (106)$$

This equation clearly shows factorial growth for the large N amplitudes. Since $(A_N)_{\text{tree}} = a_N^{(0)} \lambda^{N/2-1}$, one finds for the tree amplitudes at large N precisely the expression (88):

$$(A_N)_{\text{tree}} \sim \left(\frac{\sqrt{\lambda}N}{m}\right)^N . \qquad (107)$$

4. The KRT Formula and the Valley Approach

The formula for the amplitudes obtained semiclassically in the electroweak theory and in the $\lambda\phi^4$ theory, which leads to point-like amplitudes and cross sections which grow exponentially with energy, is an **approximation**. Thus before one takes this behaviour seriously and argues that weak interactions become strong at high energy, it is necessary to ascertain that corrections to the semiclassical approximation do not, in fact, damp this high energy growth. Two general procedures have been developed to account for the corrections to the semiclassical limit: The KRT formula[5] and the valley approach.[25] I would like here to briefly describe these techniques and see how the Ringwald Espinosa result gets modified.

Schematically, the semiclassical approximation for evaluating an n-leg amplitude can be represented by the diagram in Fig. 9a, in which each of the individual legs is attached to a classical solution. The corrections that the KRT formula or the valley approach take into account correspond to tying (some of) the external legs together by

---

[‡‡] We neglect, in what follows, subdominant terms in k for $k \gg N$.

means of propagators in the semiclassical background. A typical graph depicting those corrections is shown in Fig. 9b.

The KRT formula is a closed functional integral formula for the total cross section, deduced by means of coherent state methods by Khlebnikov, Rubakov and Tyniakov.[5] I shall not derive this result here directly, as it is a bit detailed. However, I hope to be able to motivate the final formula obtained by Khlebnikov, Rubakov and Tyniakov by analogy with the simple semiclassical calculation I described in Sec. 1 for scalar fields. Recall that the formula we obtained for the total cross section in this case, in the semiclassical approximation, read [cf. eq. (22)]:

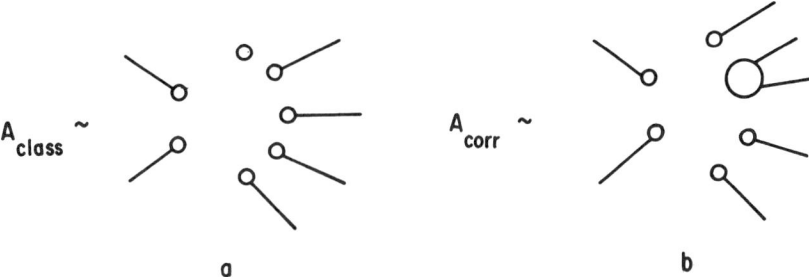

Figure 9. Schematic representation of the semiclassical approximation for : a) an n-leg amplitude ; b) a representative correction.

$$\sigma_{tot} = \frac{1}{2\hat{s}} \int d^3\vec{x} dt d\{\rho\} d\{\rho'\} \left[Z^{*2}[\{\rho\}] Z^2[\{\rho'\}] e^{-i\sqrt{\hat{s}}t} \right] \cdot \left[e^{-S[\{\rho\}]} e^{-S[\{\rho'\}]} \right]$$
$$\cdot \left[\exp\left\{ \int \frac{d^3p}{(2\pi)^3 2E} e^{i(Et-\vec{p}\cdot\vec{x})} Z[\{\rho\}] Z^*[\{\rho'\}] \right\} \right] \quad (108)$$

The structure of this equation is precisely that of the KRT formula, in which the total cross section is given as a functional integral over all field configurations ϕ and ϕ', rather than just over the coordinates $\vec{x}$ and t and the "sizes" $\{\rho\}$ and $\{\rho'\}$ of some semiclassical solutions. For each of the 3 terms in square brackets in Eq. (108) there is a corresponding term in the KRT formula. Specifically, the result of Khlebnikov, Rubakov and Tyniakov[5] is that the total cross section has the following functional integral representation.

$$\sigma_{tot} = \frac{1}{2\hat{s}} \int \delta\phi \delta\phi' \left[R^*(p_1) R^*(p_2) \right] \cdot \left[e^{-S[\phi]} e^{-S[\phi']} \right]$$
$$\cdot \left[exp\left\{ \int \frac{d^3p}{(2\pi)^3 2E} R(p) \right\} \right] \quad (109)$$

Here p_1 and p_2 are the momenta of the initial states and $R(p)$ is the truncated Fourier transform of the field configurations one integrates over:

$$R(p) = (p^2 + m^2) \left[\int d^4x e^{-ipx} \phi(x) \right] (p^2 + m^2) \left[\int d^4x' e^{+ipx'} \phi(x') \right] |_{p^2 = -m^2} \quad (110)$$

It is easy to see that the semiclassical formula (108) follows directly from the KRT formula by restricting the functional integral over ϕ and ϕ' just over the classical fields $\phi_{cl}(x-z;\{\rho\})$ and $\phi'_{cl}(x'-z';\{\rho'\})$. The resulting integral then reduces only to an ordinary integral over the collective coordinates of the classical solutions: $\{\rho\},\{\rho'\}$ and their relative separation in space time $\vec{x} = \vec{z} - \vec{z'}, t = z^0 - z'^0$. Since

$$(p^2+m^2)\int d^4x e^{-ipx}\phi_{cl}(x-z;\{\rho\})|_{p^2=-m^2} = Z[\{\rho\}]e^{-ipz} \tag{111}$$

indeed

$$R(p) = Z[\{\rho\}]Z^*[\{\rho'\}]e^{i(Et-\vec{p}\cdot\vec{x})} \tag{112}$$

and

$$R^*(p_1)R^*(p_2) = Z^{2*}[\{\rho\}]Z^2[\{\rho'\}]e^{-i\sqrt{s}t} , \tag{113}$$

so that the first and third square brackets in Eq. (108) are reproduced. Of course, in this approximation, $S[\phi] = S[\{\rho\}]$ and $S[\phi'] = S[\{\rho'\}]$ and the identification is complete.

Although the KRT formula reproduces the semiclassical result, it is clear that one can go beyond this. In particular, if one wants to consider corrections coming from fluctuations about ϕ_{cl} one can write ϕ as

$$\phi = \phi_{cl} + \chi \tag{114}$$

and considers the resulting integral over χ. Because ϕ_{cl} is an extremun of the action S, only terms of $0(\chi^2)$ or higher will appear in S. However, the R-term in Eq. (109) is not minimized by ϕ_{cl}, so that this term will give a term linear in χ. In effect, the total action relevant to σ_{tot} is not purely $S[\phi] + S[\phi']$, but it also includes the R-term [cf. Eq. (109)]. The combined system of the actions $S[\phi]$ and $S[\phi']$ plus the R-term has a different minimum than ϕ_{cl}, so the presence of this last term leads to a **distorsion** of the classical field configuration.[26]

Ideally, one would like to find the correct distorted field configuration. However, in practice this is not usually possible and one proceeds instead in a perturbative way. For these purposes it is convenient to introduce some notation.[26] If one defines the R operator by

$$R(p)|_{p^2=-m^2} \equiv <\phi|R|\phi'> , \tag{115}$$

then to $0(\chi^2,\chi'^2)$ the fluctuations χ and χ' obey the matrix equation

$$\begin{bmatrix} D & -R \\ -R^* & D' \end{bmatrix}\begin{pmatrix} \chi \\ \chi' \end{pmatrix} = \begin{bmatrix} R & 0 \\ 0 & R^* \end{bmatrix}\begin{pmatrix} \phi'_{cl} \\ \phi_{cl} \end{pmatrix} \tag{116}$$

Here D and D' are the Klein-Gordon operators in the presence of the background fields ϕ_{cl} and ϕ'_{cl}, respectively.

Eq. (116) can be solved provided one can compute the inverse propagators D^{-1} and D'^{-1} in the classical background fields. If one defines the double residue of these propagators by

$$Z_2[\{\rho\}] = (p^2+m^2)\left[\int d^4x d^4y e^{ipx}D^{-1}(x,y)e^{ip'y}\right](p'^2+m^2)|_{p^2=-m^2,p'^2=-m^2} \tag{117}$$

then it is easy to see that the first correction to the semiclassical formula, arising from the solution of Eq. (116) for the fluctuations χ and χ', takes the form

$$\text{corr} = \exp\{-\frac{1}{2}\int\frac{d^3p}{(2\pi)^3 2E}\int\frac{d^3p'}{(2\pi)^3 2E'}e^{i(E+E')t}e^{-i(\vec{p}+\vec{p'})\cdot\vec{x}}$$
$$\cdot[Z[\{\rho\}]Z_2^*[\{\rho'\}]Z[\{\rho\}] + Z^*[\{\rho'\}]Z_2[\{\rho\}]Z^*[\{\rho'\}]]\} \tag{118}$$

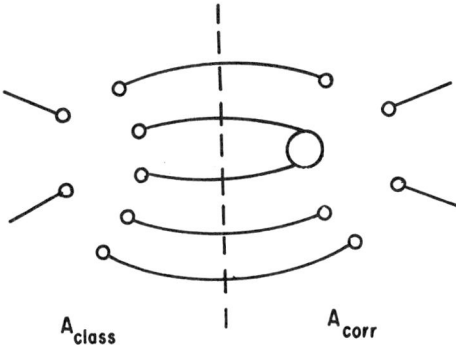

Figure 10. Correction to σ_{tot} which includes fluctuations about the semiclassical solution.

This correction has precisely the form that one would expect diagramatically. For instance, the first term in (118) corresponds to the contribution to σ_{tot} given in Fig. 10. I note also that as long as $E \ll E_0$ the correction term of Eq. (118) really gives a contribution to σ_{tot} which is subdominant to the classical term. The difference between (118) and the phase space integral of the semiclassical R term of Eq. (112) is the extra phase space integral over $\vec{p}'$ times the factor of $Z_2[\{\rho\}]$. Dimensionally this provides another factor of $(\rho/\tau)_{\text{saddle}} \sim (E/E_0)^{2/3}$ in the exponent, beyond the classical term.

The full corrections to the exponent function to $0[(E/E_0)^2]$ have now been computed in the electroweak theory.[27] The answer before doing the integrals over ρ and ρ' and t, but using $\vec{x}_{\text{saddle}} = 0$, is of the form

$$\sigma_{B+L \text{ viol}} \sim \frac{1}{\hat{s}} \int d\rho d\rho' dt \; e^{\Gamma_{class} + \Gamma_{corr}} \quad (119)$$

Here Γ_{class} is the exponent we computed earlier [c.f. eq. (76)]:

$$\Gamma_{\text{class}} = -iEt - \frac{4\pi}{\alpha} - \pi^2 v^2(\rho^2 + \rho'^2) + \frac{96\pi^2 \rho^2 \rho'^2}{g_2^2 t^4} - \frac{\pi^2 v^2 \rho^2 \rho'^2}{t^2} \quad (120)$$

Since at the saddle $\rho_{\text{saddle}} \sim \rho'_{\text{saddle}} \sim (E/E_0)^{2/3}$, while $t_{\text{saddle}} \sim i(E/E_0)^{1/3}$, the last term in Eq. (120), which is the multi-Higgs contribution, is already of $0[(E/E_0)^2]$. Thus no propagator corrections for this term are needed to the order we are considering and Γ_{corr} only contains contributions from the gauge field propagator in the background of the semiclassical solution (56).

Although these corrections are easily understood, in practice they turned out to be very difficult to calculate. Unless one properly subtracts away the zero mode contributions, the inverse propagator has double poles in p^2 and p'^2 and the residue Z_2 of eq. (117) - or its equivalent for gauge fields - blows up. This problem was eventually resolved by a variety of techniques[27] and, after some initial disagreement, there is now wide concordance on the form of Γ_{corr} which reads

$$\Gamma_{\text{corr}} = \frac{12\pi^2 M_w^2 \rho^2 \rho'^2}{g_2^2 t^2} + \frac{192\pi^2 \rho^2 \rho'^2 [\rho^2 + \rho'^2]}{g_2^2 t^6} \quad (121)$$

The first term above actually does not come from the propagator corrections but comes directly from the semiclassical solution when the phase space integral over R is done taking into account of the W mass. As expected, these kinematical corrections would tend to dampen the growth and the first term in Eq. (121) gives a negative contribution to Γ of $0[(E/E_0)^2]$.

To my knowledge, the full correction to $0[(E/E_0)^2]$ given in Eq. (121) was first calculated with all the right factor by Khoze and Ringwald[25] using the valley method. However, it was Zakharov[17] and subsequently Porrati[18] who first realized that also the gauge propagator corrections - besides the trivial kinematical mass corrections - damped the Ringwald Espinosa growth. Neglecting the W mass, the contribution from multi-W exchange from Eqs. (120) and (121) read

$$\Gamma_w = \frac{96\pi^2 \rho^2 \rho'^2}{g_2^2 t^4}\left[1 + \frac{2(\rho^2 + \rho'^2)}{t^2}\right] \sim \left(\frac{E}{E_0}\right)^{4/3}\left[1 - \left(\frac{E}{E_0}\right)^{2/3}\right] \quad (122)$$

where the second line is the value that Γ_w takes at the saddle. So the rapid growth proportional to $(E/E_0)^{4/3}$ found by Ringwald and Espinosa in the semiclassical approximation gets damped by the propagator corrections as $E \to E_0$. One must be careful, however, in being too cathegoric, since there are other positive contributions at $0[(E/E_0)^2]$ (e.g. those from multi-Higgs exchange) and higher order terms in E/E_0 which, of course, cannot be neglected as $E \to E_0$.

Using Eqs. (120) and the corrections (121) one can do the integrals over ρ, ρ' and t again by a saddle point technique. The saddles are now shifted by corrections of $0[(E/E_0)^{2/3}]$ from their "lowest order" values of Eqs. (80) and (85):

$$\rho_{\text{saddle}} = \rho'_{\text{saddle}} = \frac{\sqrt{6}}{2M_w}\left(\frac{E}{E_0}\right)^{2/3}\left[1 - \frac{1}{2}\left(\frac{E}{E_0}\right)^{2/3}\right] \quad (123)$$

$$t_{\text{saddle}} = i\frac{\sqrt{6}}{M_w}\left(\frac{E}{E_0}\right)^{1/3}\left[1 - \frac{3}{4}\left(\frac{E}{E_0}\right)^{2/3}\right] \quad (124)$$

and a simple calculation yields for the fermion number violating cross section the expression

$$\sigma_{B+L \; \text{viol}} \sim \frac{1}{\hat{s}} \exp - \frac{4\pi}{\alpha_w}\left[1 - \frac{9}{8}\left(\frac{E}{E_0}\right)^{4/3} + \frac{9}{16}\left(\frac{E}{E_0}\right)^2\right] . \quad (125)$$

Clearly the last term in the exponent damps the growth of the Ringwald Espinosa term as $E \to E_0$. However, in this limit one cannot really trust the result (125). Indeed, as $E \to E_0$, for example, taking into account the corrections to the saddle points given in Eqs. (123) and (124), now

$$\left(\frac{\rho}{\tau_E}\right)_{\text{saddle}} = \left(\frac{\rho'}{\tau_E}\right)_{\text{saddle}} \to 1 \quad , \quad (126)$$

so there is not a clear separation between the "emitting" and "absorbing" sources for the process.

I will spend the last Section of these lectures discussing in more detail the issue of what happens to $\sigma_{B+L \; \text{viol}}$ as $E \to E_0$, as this is the fundamental point one must clarify to gauge the physical import of all of these considerations. Before doing this, for completeness, however, I want to briefly describe the valley method by which one can also compute corrections to the semiclassical approximation. As I mentioned above, Khoze and Ringwald[25] used this method to obtain the correct $0[(E/E_0)^2]$ terms in $\sigma_{B+L \; \text{viol}}$ first, avoiding some of the difficulties encountered in implementing the KRT formula. The valley method, essentially makes use of the optical theorem to compute the total cross section and extracts $\sigma_{B+L\ viol}$ from the piece of the discontinuity of the 2 - to - 2 amplitude A_4 which is dominated by instanton - antiinstanton pairs. This is shown schematically in Fig. 11. Thus

$$\sigma_{B+L\ viol} \sim \frac{1}{\hat{s}} \text{Disc } A_4 \sim \frac{1}{\hat{s}} \text{Disc} \int \delta\Phi e^{-S[\Phi]} \Phi\Phi\Phi\Phi \quad , \quad (127)$$

where the "classical field" Φ which dominates the RHS of Eq. (127) contains a superposition of instanton and antiinstanton configurations, when these are widely separated from each other, plus a fluctuating field χ

$$\Phi_{cl}^{I\bar{I}} = \Phi_{cl}^{I} + \Phi_{cl}^{\bar{I}} + \chi \quad (128)$$

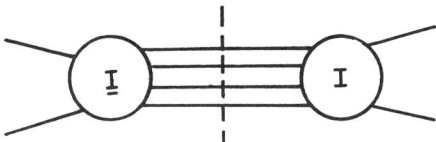

Figure 11. Schematic representation of $\sigma_{B+L\ viol}$ calculated via the valley method.

It is clear that the trick in the valley method is how to determine the fluctuating field χ, so that its fluctuations are under control. The idea here is to find a trajectory in the configuration space - the valley trajectory[28] - in which the field χ only makes excursions in directions **perpendicular** to the zero modes of Φ_{cl}^{I} and $\Phi_{cl}^{\bar{I}}$. This requirement on χ puts constraints on the action $S[\Phi_{cl}^{I\bar{I}}]$, which fixes the configuration. Having said this, I will not expand on the method further here, as it would take me too far afield. The interested reader, however, may consult the original papers of Khoze and Ringwald[25][29] or the recent review of Mattis[30] for more details. What I will do, nevertheless, is to show that, at least in lowest order in ρ/η_E (i.e. semiclassicaly), the valley method reproduces the formulas we derived earlier via the KRT formula.

Consider for these purposes a widely separated instanton - antiinstanton pair. Then the action $S[\Phi_{cl}^{I\bar{I}}]$ will be given by the sum of the separated actions for Φ_{cl}^{I} and $\Phi_{cl}^{\bar{I}}$, plus a $\Phi_{cl}^{I} - \Phi_{cl}^{\bar{I}}$ interaction term:

$$S[\Phi_{cl}^{I\bar{I}}] \simeq S[\Phi_{cl}^{I}] + S[\Phi_{cl}^{\bar{I}}] + \int d^4 x \Phi_{cl}^{I} \partial^2 \Phi_{cl}^{\bar{I}} + ... \quad (129)$$

The third term above in Fourier space gives a contribution

$$3^{\text{rd}} \text{term} = \int \frac{d^4 p}{(2\pi)^4} e^{-ipx} \frac{Z[\rho]Z^*[\rho']}{p^2 - i\epsilon} \quad . \quad (130)$$

To compute the discontinuity in Eq. (127) one expands the exponential e^{-S} in a power series, using Eq. (130) for the interaction term, making use of the usual Cutkosky rule[31]:

$$Im \frac{1}{p^2 - i\epsilon} = 2\pi \delta(p^2) \theta(p^0) \quad . \quad (131)$$

113

Using the above, each of the terms in the power series are easily seen to reassemble again in an exponential and one establshes that

$$\text{Disc } e^{-S[\Phi_{cl}^{II}]} \simeq e^{-S[\Phi^I]} e^{-S[\Phi^I]} \, exp \int \frac{d^3p}{(2\pi)^3 2E} e^{i(Et-\vec{p}\cdot\vec{x})} Z[\rho] Z^*[\rho'] \, , \qquad (132)$$

which is precisely the contribution to the total cross section one would have obtained in semiclassical approximation from the KRT formula [cf Eq. (109) and (112)].

A perturbative proof that the results from the valley method and those obtained by the KRT formula agree to all order in ρ/τ_E has been sketched by Arnold and Mattis[32] and by Mueller.[33] We will return briefly to this point in the next section, since there is also a direct connection between the difficulties in computing corrections to higher order in (E/E_0) in both methods. Nevertheless, it is possible that the valley method may have some advantages, since one may hope that one can perhaps fix $S[\Phi_{cl}^{II}]$ nonperturbatively, through the valley constraints, without resorting to an expansion in ρ/τ_E.

5. A Plethora of Damping Mechanism

We have seen in both the electroweak theory, for fermion violating processes, and in $\lambda\phi^4$ that a weakly coupled theory can become strong when one calculates amplitudes semiclassically. The problem is that the semiclassical answers one obtains can at best have only a limited range of validity, since they lead to cross sections which violate unitarity. For instance, the Ringwald Espinosa result for $\sigma_{B+L \, viol}$, given in Eq. (83), violates unitary for energies

$$E \gtrsim \tilde{E}_0 = \left(\frac{8}{9}\right)^{3/4} E_0 \simeq 16.5 \, TeV \, , \qquad (133)$$

where the factor in the exponent in this equation becomes positive. As we have seen, however, Eq. (83) gets important corrections precisely around this energy [c.f. Eq. (125)] and one is no longer sure if there are problems with unitarity.

In my opinion, nevertheless, the problem of unitarity is crucial to the whole issue of whether weak interactions become strong.[4] Unitarity constraints are always trivial as long as the amplitudes really remain weak, but cannot be ignored otherwise. We will see shortly that in $\lambda\phi^4$ theory, although multiparticle amplitudes seem to grow large for N beyond a certain critical value, in reality these amplitudes are bounded by an exponentially small quantity[20,34,24] So these amplitudes remain weak and unitarity considerations are irrelevant. For the electroweak theory, on the other hand, although it is possible that the corrections to the semiclassical approximation keep the growth of the cross section for fermion violating processes below the unitarity limit, it is more likely that unitarity corrections are the ones that control the growth of $\sigma_{B+L \, viol}$. In this case, it is important to ascertain whether unitarity keeps the theory weak or alternatively, it allows the theory to be strong - i.e. to remain near the unitarity limit. Obviously, the second alternative would be particularly exciting.*

The reason one understands $\lambda\phi^4$ theory better than the electroweak theory is connected to the fact that the former theory is Borel summable.[19] Thus one can interpret even rapidly diverging series in a clear fashion. Rather than trying to compute

---

*So that one should not think that such a possibility is remote, I remind the reader that $\pi\pi$ scattering has precisely this behaviour. The theory has small amplitudes at threshold which, however, grow linearly with s. Such a behaviour obviously eventually would violate unitary. One finds, however, that near the energy for the breakdown to occur, strong resonances, like the ρ meson, form restoring unitarity but keeping the amplitudes of the theory near their unitarity limit.

the total cross section in $\lambda\phi^4$ theory by summing over the partial cross sections for producing N particles,

$$\sigma_{tot} = \sum_N \sigma_N \ , \tag{134}$$

using for A_N the result (107), consider instead computing σ_{tot} via the optical theorem. It is clear that if one wants to pick out from Disc A_4 the contribution of multi-particle intermediate states with N particles, for N large one must study A_4 at large order $k \simeq N$. Fortunately, because of Lipatov's work,[23] which we reviewed briefly in Sec. 3, we know precisely what A_4 looks at large order [cf. Eq. (104), where we set now $k = N$].

$$A_4 \sim (-1)^N \left[\frac{\lambda N}{16\pi^2}\right]^N e^{-N} \tag{135}$$

This expression also gives factorial growth at large order. However, because $\lambda\phi^4$ is Borel summable, we know that beyond a certain N_{crit} [cf. Eq. (105)] the correct formula for A_4 is not that given in Eq. (135) but the integral expression associated with Eq. (96). That is

$$A_4 \sim (-1)^N \frac{\lambda N}{\pi} \int_0^\infty d\lambda' \frac{Im A(\lambda')}{[\lambda']^N} \tag{136}$$

For $N > N_{crit}$, this integral is in fact bounded by $A_4|_{N=N_{crit}}$. Since $N_{crit} = \frac{16\pi^2}{\lambda}$ one sees that the N loop amplitude contribution to A_4 is bounded by

$$[A_4]^{N-\text{loops}} \leq |\frac{\lambda^{N_{crit}}}{\pi} \int_0^\infty d\lambda' \frac{Im A(\lambda')}{(\lambda')^{N_{crit}}}| \leq e^{-\frac{16\pi^2}{\lambda}} \tag{137}$$

Thus, because the theory is Borel summable, even though the amplitudes for producing N particles seem to grow prodigiously for the large N, in fact the total cross section is well under control[20][34][24]:

$$\sigma_{tot} = \sum_N \sigma_N < \frac{1}{\hat{s}} e^{-\frac{16\pi^2}{\lambda}} \tag{138}$$

A nice bound like Eq. (138) is not applicable for the electroweak theory since this theory is not Borel summable. Furthermore, because one does not have some general principles to invoke, at the moment one cannot really tell what is the fate of the $B+L$ violating cross section at high energy. As I indicated above, three possibilities are open and one can find arguments (but no proofs!) in the literature to back each of these suppositions. I shall close my lectures by discussing each of these possibilities in turn, pointing out some of the weaknesses and strengths of the relevant calculations and speculations.

5.1. $B+L$ violating cross sections are intrinsically small

The contention made by a number of authors is that the original semiclassical instanton calculation of Ringwald and Espinosa badly over estimates the effect. The full answer is small and there is never any problem with unitarity. The physical idea behind these assertions is based on the intuitive picture that two pointlike high energy particles have an exponentially small probability to couple to an extended classical solution. Thus, so argue the proponents,[35] once one properly includes initial states corrections one should find that there is a small probability for $B+L$ violation and **not** the growing probability found by Ringwald[1] and Espinosa.[2]

To see why this might be the case consider, as do,[35] that the $B+L$ violating process occurs in three steps. In the first step, the initial states create the classical solution which dominates the $B+L$ amplitude (the "sphaleron"). In the second step, the sphaleron couples to the final states. Finally, in the third step the final state materializes into a large number of W's and Higgs. This three step process is shown schematically in Fig. 12, both for the amplitude and in terms of a tunnelling process. Step I corresponds to the probability of getting excited to a state of energy E which then tunnels in step II and materializes into real particles in step III.

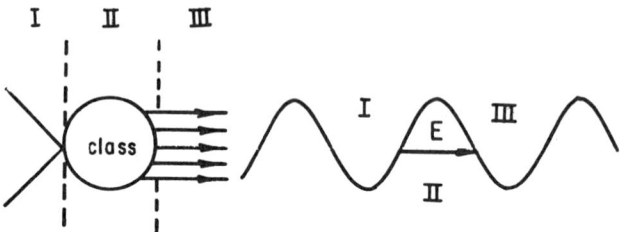

Figure 12. Three step process leading to $B+L$ violation at high energy.

It is clear that while the amplitude for processes II and III should be of order unity for E near E_0 - which here is taken to represent the energy at the top of the barrier - there is a tremendous mismatch between the initial pointlike state and the "sphaleron" solution. Typically, for an extended object of size M_w^{-1} one expects an exponentially suppressed amplitude of order e^{-E/M_w}. Whence, the expected amplitude for $B+L$ violation at energies of order of E_0 is given by

$$A = A_I A_{II} A_{III}|_{E \sim E_0} \sim e^{-E/M_w}|_{E \sim E_0} \sim e^{-\frac{1}{\alpha_w}}. \tag{139}$$

That is, also at high energies, the $B+L$ violating amplitude remains suppressed exponentially.

Although the above physical argument is both intuitive and appealing, I believe it actually misses the point. This is seen by examining in some detail a calculation of Mueller[36] in which he tried to estimate the effect of initial state branching in high energy $W-W$ scattering. Mueller basically computed the amplitude for two high energy W's to become many soft W's, which can then easily recombine into a semiclassical amplitude. His result is infrared sensitive and he used the size of the classical solution ρ as an infrared cut off. Rather remarkably these initial state corrections also appear to exponentiate and Mueller's answer takes the form

$$A_M \sim exp[-c\alpha_w \rho^2 E^2] \tag{140}$$

with c a numerical constant. With the size of the classical solution being given by M_w^{-1} one sees that for $E \sim E_0 \sim \frac{M_w}{\alpha_w}$, Eq. (140) gives precisely the exponential damping argued for by Eq. (139). However, there is an alternative way to look at (140). If we take ρ to be the size of the classical solution at the saddle point which dominates the $B+L$ cross section, then

$$\rho \simeq \rho_{\text{saddle}} = \frac{\sqrt{6}}{2M_w}\left(\frac{E}{E_0}\right)^{2/3} \tag{141}$$

and the initial state correction computed by Mueller corresponds simply to a further correction - of $0[(E/E_0)^{10/3}]$ - to the $B+L$ violating cross section of Eq. (125):

$$\sigma_{\text{initial state corr}} \sim \frac{1}{\hat{s}} \exp\left\{-\frac{4\pi}{\alpha_w}\left[\frac{9}{2}c\pi\left(\frac{E}{E_0}\right)^{10/3}\right]\right\} \tag{142}$$

From this point of view, the fact that near $E \simeq E_0$ corrections coming from the initial state give a damping like that given in Eq. (139) is irrelevant. This is just one of the many corrections of the same order which appear in the exponent. While some of the corrections, like (142), exponentially damp the answer, others may well enhance it. In fact, it appears pretty hopeless to try to compute corrections in a perturbative way for the exponential factor in the $B+L$ violating cross section. Let us write[5]

$$\sigma_{B+L \text{ viol}} \sim \frac{1}{\hat{s}} \exp\left\{-\frac{4\pi}{\alpha_w}F(E/E_0)\right\}. \tag{143}$$

Then what we definitely know is the first 3 terms in the function F in an expansion in $(E/E_0)^{2/3}$

$$F(E/E_0) = 1 - \frac{9}{8}\left(\frac{E}{E_0}\right)^{4/3} + \frac{9}{16}\left(\frac{E}{E_0}\right)^2 + 0\left[\left(\frac{E}{E_0}\right)^{8/3}\right] \tag{144}$$

Mueller's initial state correction being of $0[(E/E_0)^{10/3}]$ is not even the next term in the series. Furthermore, already the calculation of the terms of $0[(E/E_0)^{8/3}]$ is affected by even more harrowing technical difficulties than the $(E/E_0)^2$ term, connected to the proper subtraction of zero modes in the KRT method or, equivalently, for the valley connected to giving more precise specifications of the orthogonal directions to the valley trajectory.[37] Even if one cannot calculate F, it is possible that F remains positive for all values of E. In this case $\sigma_{B+L \text{ viol}}$ never violates unitarity, although if $F < 1$ the answer may be considerably larger than the 't Hooft estimate[10]: $\sigma_{B+L \text{ viol}} \sim \frac{1}{\hat{s}} e^{-\frac{4\pi}{\alpha_w}}$.

5.2. *Unitarity controls and dampens the behaviour of the $B+L$ violating cross section*

It is, however, equally possible that if one could somehow compute all the corrections to the semiclassical approximation, one would find that for $E > \tilde{E}_0$ [with $\tilde{E}_0$ probably of the $0(E_0)$] that F becomes negative. This apparent violation of unitarity signifies only that the distorted instanton which control the KRT formula (109) are not the whole answer. To regain unitarity, one must add to the instanton - antiinstanton contribution for $\sigma_{B+L \text{ viol}}$ also further terms involving further instanton - antiinstanton pairs[4][38][34],[6] as shown schematically in Fig. 13.

To understand how the presence of further instanton - antiinstanton pairs can radically alter the behaviour of the $B+L$ violating cross section, it is useful to consider a toy calculation of Maggiore and Shifman[6] [†]. Imagine computing the first term in Fig. 13 by only keeping the growing W-exchange "bond" of Eq. (78), which grows as $\hat{s}^{2/5}$. That is, approximate the instanton - antiinstanton contribution in Fig. 13 by

$$\sigma^{II}_{B+L \text{ viol}} \sim \frac{1}{\hat{s}}\int d\rho d\rho' e^{-S[\rho]} exp\left\{\frac{5}{4}[\frac{96\pi\hat{s}}{\alpha_w}^2]^{1/5}\rho^{2/5}\rho'^{2/5}\right\} e^{-S[\rho']} \tag{145}$$

This leads to the Ringwald Espinosa formula which, in terms of $\tilde{E}_0$ defined in Eq. (133), reads simply

$$\sigma^{II}_{B+L \text{ viol}} \sim \frac{1}{\hat{s}} exp\left\{-\frac{4\pi}{\alpha_w}\left[1 - \left(\frac{E}{\tilde{E}_0}\right)^{4/3}\right]\right\} \tag{146}$$

[†]This same result was obtained earlier, but in a less explicit fashion, by Zakharov[34]

In this approximation the function F becomes negative precisely at $E = \tilde{E}_0$.

Having identified the W-bond function which enters in Eq. (145), it is now trivial to write down the corresponding expression for the second term in Fig. 13. One has

$$\sigma_{B+L\ viol}^{(I\bar{I})^2} \sim \frac{1}{\hat{s}} \int d\rho_1 d\rho_2 d\rho_3 d\rho_4 \ e^{-S[\rho_1]} B(\rho_1, \rho_2) e^{-S[\rho_2]} B(\rho_2, \rho_3) e^{-S[\rho_3]} B(\rho_3, \rho_4) e^{-S[\rho_4]} \tag{147}$$

where the bond function $B(\rho, \rho')$ is given explicitly in Eq. (145). A saddle point evaluation of Eq. (147) gives for the $(I\bar{I})^2$ - correction

$$\sigma_{B+L\ viol}^{(I\bar{I})^2} \sim \frac{1}{\hat{s}} exp - \frac{8\pi}{\alpha_w}\left[1 - 2.05\left(\frac{E}{E_0}\right)^{4/3}\right]. \tag{148}$$

One sees that this contribution, although very small for $E << \tilde{E}_0$ [‡] actually reaches the unitarity limit ($F = 0$) **sooner** that the instanton - antiinstanton contribution:

$$E_{crit}|_{(I\bar{I})^2} \simeq 0.58 \tilde{E}_0 \tag{149}$$

At E_{crit} the cross secion (146) is still exponentially small $[\sigma_{B+L\ viol} \simeq \frac{1}{\hat{s}} e^{-\frac{2\pi}{\alpha_w}}]$ and is even smaller at the place where $\sigma^{I\bar{I}}$ and $\sigma^{(I\bar{I})^2}$ are comparable $[\sigma_{B+L\ viol} \simeq \frac{1}{\hat{s}} e^{-\frac{8\pi}{3\alpha_w}}]$.

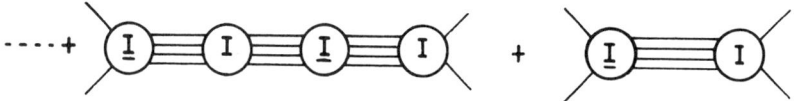

Figure 13. Contributions to $\sigma_{B+L\ viol}$ which help restore unitarity.

If one takes unitarity to require that the contribution of the multi- instanton amplitudes never exceed that of the single instanton - antiinstanton pair, then it is clear that the $B+L$ violating cross section can only be reliably computed up to some $E_{crit} < E_0$ where

$$\sigma_{B+L\ viol}(E_{crit}) \sim \frac{1}{\hat{s}} e^{-\frac{4\pi}{\alpha_w}K} \tag{150}$$

with $\frac{1}{2} \leq K \leq 1$. Indeed the number $K = 1/2$, suggested by Zakharov[34] and Maggiore and Shifman[6] has a simple interpretation, corresponding to each bond in the infinite expansion of Fig. 13 compensating precisely the adjoining factor of the classical action, leaving an overall factor of $e^{-S[\rho]} \simeq e^{-\frac{2\pi}{\alpha_w}}$ left over. Although the result (150) might be sensible, it is not clear to me if a "by hand" cutting off of the growing semiclassical amplitudes when these are still small makes sense. Furthermore, how to compute $\sigma_{B+L\ viol}$ for $E > E_{critical}$ is an open question. If the analogy with the parallel formula in $\lambda \phi^4$

[‡]The double tunnelling probability at low energy is just the square of the single tunnelling probability.

theory [Eq. (138)] were to hold, then (150) would also be a bound for $E > E_{critical}$. However, there are no compelling reasons to believe that the electroweak case is in any way analogous to the $\lambda \phi^4$ case.

5.3. *Unitarity does not affect the ultimate strength of the $B+L$ violating cross section*

It is of course possible that unitarity plays a relatively minor role in the way the $B+L$ violating cross section eventually behaves at high energy. The most optimistic assumption[3] is that once $\sigma_{B+L\ viol}$ hits the unitarity limit, the $B+L$ violating processes remain strong and the cross section asymptotes to an essentially constant value, up to $\log \hat{s}$ terms:

$$\sigma_{B+L\ viol} \sim (\pi R^2_{B+L\ viol}) \tag{151}$$

Roughly this is what happens in strong interaction physics with $R \sim \Lambda^{-1}_{QCD}$. Here $R_{B+L\ viol}$ should be related to the sphaleron mass, $R_{B+L\ viol} \sim \alpha_w/M_w$. Although a behaviour like (151) may emerge in some model field theories,[3] it is questionable whether it could really obtain in real life. Indeed some of the general arguments of the last subsection[34],[6] as well as the expectations from the most straightforward unitarization procedure[20] argue against it.

There is another circumstance, however, where perhaps the role of unitarity may not be as significant. This corresponds to the case in which although the exponent function F vanishes, it does so in an asymptotically smooth way. That is, for asymptotically large $E \gg E_0$, $F \to 0$. In this case, unitarity is manifestly not violated, since the $B+L$ violating processes have a cross section which is comparable to the perturbative $B+L$ conserving processes, where unitarity considerations are irrelevant. Of course physically to have

$$\sigma_{B+L\ viol} \sim \sigma_{B+L\ cons} \sim \frac{1}{\hat{s}} \tag{152}$$

is tremendougly interesting. However, this behaviour for the $B+L$ violating cross section is still characteristic of a weakly coupled theory.

This interesting behaviour for the exponent function F occurs in the recent calculation of Khoze and Ringwald[39] of $B+L$ violating processes, in which they use the valley trajectory of a pure gauge theory. It is, of course, questionable if it is legitimate to completely neglect the Higgs degrees of freedom, except through their appearance in the action function $S[\rho]$ [cf Eq. (65)]. Nevertheless, if one adopts this approximation then one can construct the whole action $S[\Phi^{II}]$ for all separations by solving the valley trajectory constraints[39],[28] Of course, for large separations $S[\phi^{II}]$ is just the sum of the actions for widely separated instantons and anti-instantons, but one has also an explicit form at short distances. The resulting function F, calculated by Khoze and Ringwald[39] is plotted in Fig. 14 and one sees that it smoothly approaches zero for large E, becoming negligible for $E \gtrsim 2E_0$

In view of the smooth behaviour of Fig. 14, it is tempting to suppose that unitarity effects are unimportant, and that at sufficiently high energy, indeed $B+L$ violating processes become a dominant part of electroweak interactions. §

---

§If one were to trust Fig. 14, the energy where this happens is around 40 TeV, which is barely accessible to Eloisatron and completely beyond the range of the SSC!

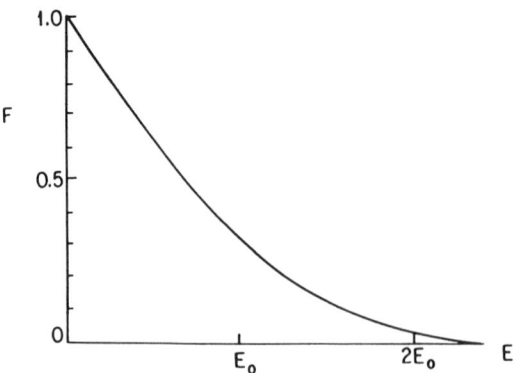

Figure 14. Plot of the function F calculated by Khoze and Ringwald[39] using the Valley method for a pure gauge theory.

However, even with this smooth behaviour for F one can show[40] that multi-instanton effects again strongly modify this behaviour much before F approaches zero. Hence, even in this most favored case, it is not clear whether sizable $B + L$ violating processes really survive at high energy.

6. Envoi

I hope that through these lectures I have been able to convey both the excitement and the tentativeness associated with the phenomena of $B + L$ violation in the electroweak theory. On the one hand, I have shown that there are clear circumstances where a semiclassical approximation is a good starting point for calculation. This is the case for the computation of $B + L$ violating amplitudes in the standard model and for estimating high order contributions for $\lambda\phi^4$ theory. These semiclassical methods, however, invariably lead to growing cross sections at high energies, since the resulting amplitudes are point-like and not suppressed as a function of any of the subenergies.

It is an open question, however, whether these weak coupling theories really become strong at high energies. This requires understanding the nature of the corrections to the semiclassical calculations. Although I have indicated and exemplified the various ways devised to compute these corrections, it is clear that no crisp answer has yet emerged for the phenomena of $B + L$ violation. However, for $\lambda\phi^4$ theory, because of its Borel summability, one can with confidence state that for $\lambda \ll 1$ the theory never becomes strong. For the electroweak theory, on the other hand, the situation is considerably more fluid and can be summarized by these 3 points:

i) The exponent function F could be positive for all energies, leading to an exponentially suppressed $B + L$ violating cross section. However, there are really no convincing arguments why this should be so.

ii) Although multi-instanton effects appear to become important much before the single instanton-antiinstanton cross section approaches the unitarity limit, it is an open question whether this really implies that the $B+L$ violating cross section is bounded by the square root of the 't Hooft tunnelling factor. If this were the case, although the $B+L$ violating cross section would have a spectacular growth from its threshold value, it still would be totally negligible at high energies.

iii) Even though it would be of great interest if the exponent factor F had the smooth behaviour shown in Fig. 14, there are no strong arguments in favor of keeping purely the gauge contributions in solving for the valley trajectory. Furthermore, even such a smooth behaviour for F does not appear to preclude the dominance of multi-instanton effects, effectively cutting off F at values not that differing from those obtained from other more rapidly growing bond functions.

Although it is somewhat discouraging to end a set of lectures with 3 different alternatives for the same phenomena, this state of affairs just illustrates the difficulties associated with trying to make nonperturbative estimates. Furthermore, by not closing this subject in a neat and definitive way it clearly challenges the students of this school to provide their own answers.

7. Acknowledgements

I am grateful to Nino Zichichi and the students in Erice for providing a very lively and stimulating atmosphere. These lectures owe much to conversations and communications I had with numerous colleagues. I would like to thank particularly J. M. Cornwall, T. Gould, S. Khlebnikov, A. Ringwald, M. Shifman and V. Zakharov for their insights. This work was supported in part by the Department of Energy, under Contract DE- AT03-88ER 40384 Mod A006 Task C.

References

1. A. Ringwald, Nucl. Phys. B330 (1990) 1
2. 0. Espinosa, Nucl. Phys. B 343 (1990) 310
3. L. Mc Lerran, A. Vainshtein and M. Voloshin, Phys. Rev. D 42 (1990) 171, 180
4. R. D. Peccei, in **TAUP 89**, ed. A. Bottino and P. Monacelli (Editions Frontieres, Gif-sur-Yvette, 1989)
5. S. Khlebnikov, V. Rubakov and P. Tyniakov, Nucl. Phys. B350 (1991) 441
6. M. Maggiore and M. Shifman, Minnesota preprint, TPI-MINN - 91/24-T 91/27-T
7. C. G. Callan, R. Dashen and D. Gross, Phys. Lett. 638 (1976) 334; R. Jackiw and C. Rebbi, Phys. Rev. Lett. 37 (1976) 172
8. See, for example, R. J. Crewther, Acta. Phys. Austria Suppl. 19 (1978) 46
9. W. A. Bardeen, Nucl. Phys. B75 (1974) 246
10. G. 't Hooft, Phys. Rev. Lett 37 (1976) 8; Phys. Rev. D24 (1976) 3432
11. S. L. Adler, Phys. Rev. 177 (1969) 2426; J. S. Bell and R. Jackiw, Nuovo Cim. 60 (1969) 47
12. M. F. Atiyah and I. M. Singer, Ann. Math. 87 (1968) 484, 546
13. A. A. Belavin, A. M. Polyakov, A. S. Schwartz and Y. S. Tyupkin, Phys. Lett. 59B (1976) 85
14. R. D. Peccei and H. R. Quinn, Nuovo Cim A41 (1977) 1309
15. I. Affleck, Nucl. Phys. B 191 (1981) 445
16. N. Krasnikov, V. Rubakov and V. Tokarev, Phys. Lett 79B (1978) 423
17. V. Zakharov, Minnesota preprint, TPI-MINN-90/7-T
18. M. Porrati, Nucl. Phys. B347 (1990) 371
19. See, for example, J. Zinn Justin **Quantum Field Theory and Critical Phenomena** (Clarendon Press, Oxford 1989)
20. J. M. Cornwall, Phys. Lett 243 B (1990) 271; J. M. Cornwall and G. Tiktopolous, UCLA/90/TEP/56
21. M. Voloshin, Phys. Rev. D43 (1991) 1726
22. H. Goldberg, Phys. Lett. 246B (1990) 445; 257B (1991) 346; H. Goldberg and M. T. Vaughn, Phys. Rev. Lett. 66 (1991) 1267

23. L. Lipatov, JETP Letters 24 (1976) 157; Sov. Phys. JETP 44 (1976) 1055; JETP Letters 25 (1977) 104; Sov. Phys. JETP 45 (1977) 216; For further discussion and a complete set of references, see also[19]
24. M. Maggiore and M. Shifman, Minnesota preprint, TPI-MINN-91/17-T
25. V. V. Khoze and A. Ringwald, Nucl. Phys. B 355 (1991) 351
26. P. Arnold and M. Mattis, Phys. Rev. Lett 66 (1991) 13
27. V. V. Khoze and A. Ringwald, Nucl. Phys. B355 (1991) 351; A. Mueller, Columbia preprint CU-TP-512 (1991); P. Arnold and M. Mattis, Mod. Phys. Lett 6A (1991) 2059; D. Dikonav and V. Petrov, in the Proceedings of the 26th Winter School of the Leningrad Institute of Nuclear Physics, Leningrad (1991)
28. J. Balitsky and A. Yung, Phys. Lett. 168B (1986) 113; A. Yung, Nucl. Phys. B297 (1988) 47
29. V. V. Khoze and A. Ringwald, CERN preprint, CERN-TH-6082/91
30. M. Mattis, Los Alamos preprint, LA-UR-91-2926
31. R. E. Cutkosky, J. Math. Phys. 1 (1960) 429
32. P. Arnold and M. Mattis, Los Alamos preprint, LA-UR-91-1858
33. A. Mueller, Columbia preprint, CU-TP-512 (1991)
34. V. Zakharov, Nucl. Phys. B353 (1991) 683; Max Planck preprint MPI-PAE/PTh 91-11
35. T. Banks, G. Farrar, M. Dine, D. Karabali and B. Sakita, Nucl. Phys. B347 (1990) 581; R. Singleton, L. Susskind and L. Thorlacius, Nucl. Phys. B 343 (1990) 541; A. Casher and S. Nussinov, Tel Aviv preprint 1990
36. A. Mueller, Nucl. Phys. B348 (1991) 310; B353 (1991) 44
37. S. Khlebnikov and P. Tinyakov, CERN preprint, CERN-TH 6146/91
38. H. Aoyama and H. Kikuchi, Phys. Lett. 247B (1990) 75, Phys. Rev. D43 (1991) 1999
39. V. V. Khoze and A. Ringwald, Phys. Lett. 259B (1991) 106
40. T. Gould, UCLA Preprint UCLA/91/TEP/39

CHAIRMAN: R. Peccei

Scientific Secretaries: P. Hernandez, M. Neubert, I. Rothstein

DISCUSSION I

– *Danielsson:*

Could you comment on the status of baryon number processes at high temperatures, as opposed to high energies?

– *Peccei:*

Well, it is a different problem at high energies than it is at higher temperatures, because in the latter case it is an N body to an N body interaction in a thermal bath. In this case there is a Boltzmann factor as opposed to the e^{-1/g^2} suppression. As such, it becomes likely at high temperatures to have thermal fluctuations over the barrier which separates different vacua. These calculations are in fact on better footing than the high energy case.

– *Khoze:*

It is true that at high energies it is more advantageous (in the study of B + L violating processes) to consider pp collisions as opposed to $p\bar{p}$ or e^+e^-?

– *Peccei:*

As far as $p\bar{p}$ is concerned there is essentially no difference apart from small differences in the structure functions, as opposed to e^+e^- where the essential B + L violating vertex has to be generated.

– *Sivaram:*

Is it possible to have B-L violation in some theories?

– *Peccei:*

Yes, you just look for a theory with such an anomaly.

– *Gourdin:*

Will it be difficult to detect the baryon number violation in such processes?

– *Peccei:*

If this process really does occur because of the large number of final particles, it would be almost impossible to pick out the B + L violation. However, you will surely know it is happening because of the multiparticle production.

– *Neubert:*

Is it possible to have non-integer winding number solutions which could enhance the cross-section?

– *Peccei:*

There are classical static solutions with "effective winding number" 1/2 which mediate the high temperature transitions.

– *Pluciennik:*

Could the strong (QCD) instantons be a source of some strong effect in QCD multiparton final states?

– *Peccei:*

I do not know the answer to this question because in the strong instanton case there is an infrared problem which is not completely understood.

– *Ellis:*

I propose a crazy idea. If you look at QCD in perturbation theory, the picture you have of the pomeron in high energy scattering is two gluon exchange. How can one test this idea? One could in principle imagine experiments which do deep inelastic scattering off a pomeron, (such experiments have actually been done) and so one can measure the structure function of a pomeron. So, if these non-perturbative effects are important in generating large hadronic cross-sections, presumably the wave function of the pomeron could contain many gluons, and this would show up in the wave function.

– *Peccei:*

That may be an interesting point.

– *Jain:*

Is it possible that in the high temperature case, there will be fluctuations which will cause transitions with larger changes in winding number?

– *Peccei:*

Yes, that will occur as the temperature gets large, because near the phase transition the Boltzmann factor is not as important, and you can have multiple transitions.

– *Goobar:*

Is it possible to set limits on the "strong" weak interactions from available cosmic ray data?

– *Peccei:*

Two of my colleagues at UCLA, Duncan Morris and Rogerio Rosenfeld, have analysed what you would expect to see in underground detectors. They found that in the most optimistic scenario one would expect to see a multimuon event coming from the bottom, per year per 100 m^2, caused by a neutrino activated B + L violating event near the detector.

– *Junk:*

Is the "too large cross-section" problem just a copy of the infrared divergence problem, but an even nicer example because one has a mass cutoff built into the problem.

– *Peccei:*

No. It is not a similar problem, because this is an s-channel process.

– *Junk:*

Do these classical solutions include the non-linear terms in the equations of motions?

– *Peccei:*

Yes, all the terms are included.

– *Chizhov:*

At present there are already known in physics "pure" quantum states (squeezed, sub-poissonian) which have no classical interpretation. My question is on the completeness and validity of your approach.

– *Peccei:*

This is a very sophisticated question. Given the physical uncertainties in the present calculations, we are not ready to tackle such problems.

CHAIRMAN: R. Peccei

Scientific Secretaries: P. Hernandez, M. Neubert, I. Rothstein

DISCUSSION II

– *Rothstein:*

I am a little concerned about the saddle point approximation you used for the ρ integration. My reason being that it would seem that the integral is dominated in a regime where the constrained instaton solution is no longer valid.

– *Peccei:*

In the "valley approximation" when F gets near zero, the size of the classical solution is at or above $\frac{1}{v}$, and you indeed must worry about this point.

– *U. Vikas:*

What if the field talks back?

– *Peccei:*

This is partially taken into account by computing the fluctuations about the instanton due to the effect of the external legs. This is seen in the calculations done by Khoze, Ringwald, Arnold and Mattis, Mueller, and Khlebnikov et al. of the $\left(\frac{\rho}{t}\right)^6$ term, which will contribute a damping factor of:

$$\exp\left(\frac{-4\pi}{\alpha_W}\frac{9}{16}\left(E/E_0\right)^2\right)$$

which will damp the high energy behaviour.

– *Rothstein:*

If you go to high energies it seems that the instanton may no longer be the relevant solution to expand around. The reason being that the instanton is a process which tunnels from "floor to floor" of the potential, whereas at high energies the initial state has already "climbed" the wall.

– *Peccei:*

These corrections can be taken into account through the prefactor: $R^*(p_1)R^*(p_2)$ in the KRT functional integral. Mueller estimated the effects of the initial state configuration, and found that it damps the cross-section by contributing a factor of:

$$\exp\left(\frac{(-4\pi)}{\alpha_W}C(E/E_0)^{10/3}\right)$$

– *Danielsson:*

The key issue seems to be how to make sense out of a non Borel-summable series. To do that you need extra information beyond perturbation theory. Is it clear that this information is to be found within the electroweak theory? Maybe the theory has to be understood in a larger framework, and then it could be well-defined for high energies.

– *Peccei:*

It is alway problematic when a theory is defined perturbatively, but cannot be summed in perturbation theory. The "valley method" provides a non-perturbative approach trying to solve the equations of motions without having to rely on an expansion in some "illegal" parameter. At the moment, we do not have better techniques to solve non-abelian gauge theories. On the other hand, expanding in "illegal" parameters can, sometimes, at least indicate where interesting physics might happen.

– *Danielsson:*

Could it be that the theory itself is ill-defined?

– *Peccei:*

This can, in principle, happen beyond certain ranges of energies. It seems unlikely, however, that a weakly-coupled theory really becomes ill-defined at high energies. Rather than being an intrinsic problem of the theory itself, I think that the problem here is that you are making an ill-defined expansion.

– *Neubert:*

In many respects, the ultimate method for studying strongly interacting field theories is lattice gauge theory. Is it possible to apply this method to the calculation of σ by computing the function $F(E/E_0)$? For instance, one could think of a numerical solution combining the "valley method" with Monte Carlo-techniques.

– *Peccei:*

The "valley approach" tries to solve integro-differential equations, which can certainly be attacked by numerical methods. To combine analytical and numerical techniques might be the best thing to do in order to solve this problem. On the other hand, I do not believe that it is feasible to use lattice gauge theory for computing something as a function of energy, since even the calculation of static quantities requires enormous effort.

– *Potters:*

In Lipatov's calculation for $\lambda\phi^4$, you calculate an N-loop amplitude and get a perfectly finite answer. I think such amplitudes should diverge unless you regularize your theory. Can you explain this?

– *Peccei:*

In fact, the $2 \rightarrow 2$ amplitude needs subtractions in order to be well-defined, but these are irrelevant for this discussion. One is only interested in the analytical behaviour in the coupling parameter λ and, more precisely, the N-th derivative of the dispersion integral (N being large). Therefore, this is already subtracted.

– *Gourdin:*

Concerning the dispersion relation in the Lipatov approach, you said that one needs to know ImA(λ) near the cut $\lambda \sim 0$. Could it be that subtractions required in the dispersion relation modify the small-λ behaviour?

– *Peccei:*

Making n subtractions in the dispersion relation amounts to adding a polynominal of degree n, on the rhs of the dispersion relation. Since in Lipatov's approach one is only interested in the N-th derivative of the dispersion relation (N being large), such a polynomial is of no relevance.

– *Gourdin:*

What is the origin of the $(E/E_0)^{8/3}$ term? Higher order fluctuations or something else? Why is it so difficult to compute?

– *Peccei:*

In general the difficulty in going to higher order corrections is that the background field propagators are singular and a subtraction of zero modes must be done. As you go to higher orders, the constraints you must solve to determine the orthogonal space in which you are safe are more and more difficult to solve.

– *Wadhwa:*

Using the "valley method" for B + L violation theory, the energy limit you calculated is 40 TeV for F $\to$ 0. What happens in the case of $\lambda\phi^4$ theory?

– *Peccei:*

For $\lambda\phi^4$ theory one can show that σ is always exponentially suppressed and the theory remains weak at high energy. In the case of B + L violation, the Khoze-Ringwald (KR) calculation only includes the pure gauge sector. However, for values of E $\sim$ E_0 (40 TeV) the size of the classical solution is roughly of the same order of 1/v. However, we know that, in order to keep the Higgs sector under control, the size of this solution must be smaller than this scale. Thus, one must take the KR results with caution as Higgs effects, which they do not consider, may be important. From a physical point of view the KR curve is very reasonable. It nicely describes the behaviour of the cross-section at high energies in the case of a strong theory (no unitarity problem) eliminating the unphysical growth of the Ringwald-Espinosa approximation.

– *Wadhwa:*

As an experimentalist, I wonder what kind of physics you are looking for in the TeV range. Is it the weak interaction becoming strong at high energy that you consider most probable?

– *Peccei:*

There could well be many things beyond the Standard Model (SUSY, Technicolor, strings, etc.) but maybe not. Thus, given that one has a model that works very well, what one should do is also to explore to the best of one's ability what the theory is predicting. And one of the things that the SM is telling you is that there is a possible source of B + L violation at high energies, so you must pursue it.

In general, it is necessary to look at those phenomena that in a perturbative treatment of the theory would never appear. It is clear now that a theory is more than its perturbative expansion and that there are also collective non-perturbative effects which could be important. Furthermore, this B + L violation may have crucial cosmological consequences.

– *Sivaram:*

In case you have a different power of ϕ^n, would your series still be Borel summable?

– *Peccei:*

The treatment I have described is only valid for perturbatively well defined theories, that is renormalizable theories. For d = 4, you have only n=3 or 4 to choose if you want your theory to be renormalizable. Furthermore, even for $\lambda\phi^4$ the proof of Borel's summability is a difficult task, although the alternation of signs in the expansion is clearly a clue that this series can be Borel summable.

GEOMETRY AND QUANTUM SYMMETRIES OF SUPERSTRING VACUA

S. Ferrara

CERN, Geneva, Switzerland

Abstract

We report on some recent progress which has been made in the understanding of general features of superstring compactifications. The following aspects are discussed: special geometry of the moduli space of (2,2) vacua, target-space duality symmetries, differential equations for the periods of holomorphic three-forms, and higher loop (genus) string corrections to the tree-level effective action through target-space modular anomaly cancellations.

I. Introduction

The topics covered in this lecture deal with several aspects of superstring theories compactified on (2,2) vacua[1]. N=2 internal superconformal field theories[2] are the quantum (stringy) version of Calabi-Yau (C-Y) manifolds[3], giving N=1 space-time supersymmetry in D=4 dimensions[4]. In particular we will give in section II some general formulae for the geometry of the moduli space of Calabi-Yau manifolds which encode all relevant information for the effective action. In section III the relation of special geometry with the Kähler class and complex structure deformations, the mirror symmetry hypothesis and target-space duality will be considered. In section IV we will show how, at least in certain cases, equations known as Picard-Fuchs equations, allow the determination of the periods of holomorphic three-forms (for C-Y threefolds). For any given C-Y manifold this is equivalent to the dynamical determination of the four-dimensional effective action. In section V target-space modular anomalies will be discussed and their relation, because of supersymmetry, with higher loop corrections to gauge couplings.

II. Special geometry and effective actions

In the present section we will summarize some general formulae which encode all relevant informations for the low-energy dynamics of light particles. By this we mean all effective interactions which may occur in an effective Lagrangian containing up to two-derivative terms.

The striking fact is that the couplings are simple enough to be discussed on a general footing for strings compactified on Calabi-Yau manifolds whose worldsheet interpretation is a c=9, (2,2) superconformal field theory with a quantization condition for the U(1) charge[2].

Much progress has recently been achieved in the study of this particular class of string vacua in three different but related areas; N=2 Landau-Ginzburg models[5], topological N=2 SCFT's[6]-[12] and the C-Y geometry due to its connection with N=2 space-time supersymmetry[13]-[20].

Moreover the study of C-Y compactifications for large volume, compared to the string scale which relies on standard Kaluza-Klein compactification of D=10, N=1 supergravity, has recently been extended to include string corrections through the notion of mirror symmetry[21], which is a property of N=2 superconformal field theories under the exchange of the chiral rings of primary fields. In the target-space interpretation this symmetry allows one to define the notion of C-Y quantum geometry relating pairs of C-Y manifolds with isomorphic even and odd cohomology classes i.e.:

$$\sum_{p+q=3} H^{(p,q)}(C) = \sum_{p=0}^{3} H^{(p,p)}(C')$$

(and the same for $C \to C'$).

This means that Kähler class (complex structure) deformations in C have the opposite meaning in C'.

The relevance of the mirror symmetries comes about because the moduli space for complex structure deformations is purely classical and does not receive α' corrections[22],[23]. Therefore the α' corrected moduli space for the Kähler class deformations in (C) must coincide with the classical moduli space for complex structure deformations in (C') and viceversa. It is well known from string theory that quantum corrections in α' come from world-sheet instanton effects[23],[24]. In the context of algebraic geometry instantons are related to rational curves, so the transition from a classical to a quantum manifold is achieved by properly taking into account all rational curves of any degree in the Kähler class moduli geometry[25].

As a consequence of this mirror duality for (2,2) vacua it is sufficient to discuss the moduli space for the complex structure. In this case, as we will show with some illustrative examples in section IV, it is possible essentially to resolve the problem by some methods of algebraic geometry[26]-[30]. In fact the periods of the holomorphic 3-form satisfy certain linear differential equations, known as Picard-Fuchs equations, whose solutions encode all relevant quantities of the special geometry of the moduli space and hence give all the couplings in the effective action.

From the string point of view these results are equivalent in solving the superconformal field theory, i.e. in giving all perturbed correlation functions in coupling constant space[31],[32]. The equivalence of these methods has recently been shown in some examples[26]-[32].

Let us briefly recall the basic elements of special geometry, as derived from N=2 supergravity in D=4 dimensions[34],[35],[20]. In the next section we will recall its connection to C-Y manifolds and deformation theory. The coordinate free description of special Kähler manifolds was given in Refs. [17],[20]. In primis a special Kähler manifold is a restricted (or Hodge) manifold[36],[37],[38], i.e. the cohomology class of the Kähler form:

$$J = iG_{i\bar{j}} dz^i \wedge d\bar{z}^{\bar{j}} \quad , \quad G_{i\bar{j}} = \partial_i \partial_{\bar{j}} K \qquad \text{II}-1$$

coincides with the first Chern class of a line bundle $\mathcal{L}$: $[J] = [c_1]$. J is the exterior derivative of the U(1) connection A of $\mathcal{L}$:

$$J = dA \quad , \quad A = -\frac{i}{2}(K_{,i} dz^i - K_{,\bar{i}} d\bar{z}^{\bar{i}}) \qquad \text{II}-2$$

which under a Kähler transormation:

$$K \to K - \Lambda - \bar{\Lambda} \qquad \text{II}-3$$

transforms as:

$$A \to A - d\,(\text{Im}\Lambda) \qquad \text{II}-4$$

The U(1) covariant differential of a field $\Phi(z,\bar{z})$ of weight p is[20]:

$$\nabla \Phi = (d + ipA)\Phi \qquad \text{II}-5$$

or in components:

$$\nabla_i \Phi = \left(\partial_i + \frac{1}{2} p K_{,i}\right)\Phi \quad , \quad \nabla_{\bar{i}} \Phi = \left(\partial_{\bar{i}} - \frac{1}{2} p K_{,\bar{i}}\right)\Phi \qquad \text{II}-6$$

A covariantly holomorphic field of weight p satisfies:

$$\nabla_{\bar{i}} \Phi = 0 \qquad \text{II}-7$$

By a change of trivialization, the real U(1) bundle can be reduced to a holomorphic bundle. Indeed by setting:

$$\tilde{\Phi} = e^{-pK/2}\Phi \qquad \text{II} - 8$$

we have:

$$\nabla_i \tilde{\Phi} = (\partial_i + pK_{,i}) \tilde{\Phi} \quad , \quad \nabla_{\bar{i}} \tilde{\Phi} = \partial_{\bar{i}} \tilde{\Phi} \qquad \text{II} - 9$$

In particular the U(1) transformation on $\tilde{\Phi}$:

$$\Phi \rightarrow e^{p/2(\Lambda - \bar{\Lambda})} \Phi \qquad \text{II} - 10$$

induces a holomorphic transformation on $\tilde{\Phi}$:

$$\tilde{\Phi} \rightarrow e^{p\Lambda} \tilde{\Phi} \qquad \text{II} - 11$$

A restricted Kähler manifold is special if and only if there exists a completely symmetric holomorphic three-index section:

$$\partial_{\bar{m}} W_{ijk} = 0 \qquad \text{of weight p} = 2 \qquad \text{II} - 12$$

such that [34],[35],[36],[19]:

$$R_{i\bar{j}\ell\bar{m}} = +G_{i\bar{j}}G_{\ell\bar{m}} + G_{i\bar{m}}G_{\ell\bar{j}} - W_{i\ell p}\overline{W}_{\bar{j}\bar{m}\bar{p}}G^{p\bar{p}}e^{2K} \qquad \text{II} - 13$$

From the Bianchi identities it follows that the (covariantly holomorphic) tensor $C_{iep} = e^K W_{iep}$ satisfies[34],[19]:

$$C_{i\ell p} = \nabla_i \nabla_\ell \nabla_p S \qquad \text{II} - 14$$

when S is a scalar of weight p = 2. Compatibility of eq. II-13 with N=2 space-time supersymmetry allows one to construct the metric and W in terms of some prepotential function F. In Refs. [17], [20] it was shown that on a special n-dimensional Kähler manifold one can always introduce (n+1) holomorphic sections X^I:

$$\partial_{\bar{i}} X^I(z) = 0 \qquad I = 0 \ldots n \qquad \text{II} - 15$$

and a function F(X), holomorphic and homogeneous of degree two in the X'^s:

$$X^I \frac{\partial F}{\partial X^I} = 2F \qquad \text{II} - 16$$

i.e., F is a holomorphic section in $\mathcal{L} \times \mathcal{L}$. Since X^I are holomorphic sections of $\mathcal{L}$, a covariant derivative is defined as:

$$\nabla_i X^I = (\partial_i + K_{,i}) X^I \qquad \text{II} - 17$$

All relevant quantities can now be expressed through the holomorphic section F(X). Let us define:

$$F_{\Lambda\Sigma...} = \frac{\partial}{\partial X^\Lambda}\frac{\partial}{\partial X^\Sigma}...F \quad , \quad N_{\Lambda\Sigma} = F_{\Lambda\Sigma} + \overline{F}_{\bar\Lambda\bar\Sigma} \qquad \text{II} - 18$$

then:

$$G_{i\bar{j}} = \partial_i X^\Lambda \partial_{\bar{j}} \overline{X}^\Sigma \partial_\Lambda \bar\partial_\Sigma K = -f_i^\Lambda f_{\bar{j}}^\Sigma N_{\Lambda\Sigma} \qquad \text{II} - 19$$

when:

$$K = -\ell n(N_{\Lambda\Sigma} X^\Lambda \overline{X}^\Sigma) = -\ell n(X^\Lambda \overline{F}_\Lambda + \overline{X}^\Lambda F_\Lambda) \qquad \text{II} - 20$$

$$f_i^\Lambda = \nabla_i L^\Lambda = e^{K/2}\left(\delta_\Sigma^\Lambda - \frac{X^\Lambda(N\overline{X})_\Sigma}{XN\overline{X}}\right)\partial_i X^\Sigma \qquad \text{II} - 21$$

with $L^\Lambda = e^{K/2} X^\Lambda$:

$$S = -\frac{1}{4} N_{\Lambda\Sigma} L^\Lambda L^\Sigma = -\frac{1}{4} e^K N_{\Lambda\Sigma} X^\Lambda X^\Sigma \qquad \text{II} - 22$$

$$W_{i\ell m} = e^{-K}\nabla_i \nabla_\ell \nabla_m S = e^{-K} f_i^\Lambda f_\ell^\Sigma f_m^\Gamma F_{\Lambda\Sigma\Gamma} = \partial_i X^\Lambda \partial_\ell X^\Sigma \partial_m X^\Gamma F_{\Lambda\Sigma\Gamma} \qquad \text{II} - 23$$

In Ref. [17] it was shown that $\left(X^\Lambda, \frac{i\partial F}{\partial X^\Sigma}\right)$ can be viewed as sections of a flat holomorphic Sp(2n +2.)-bundle whose existence is in fact equivalent to eq. II-13.

Connection of the previous formulae with the effective actions comes from the fact that heterotic or type II superstrings compactified on C-Y threefolds give D=4 superstrings with N=1 or N=2 space-time supersymmetry respectively[13],[14]. Because of a special mapping[14] between heterotic and type II superstrings compactified on a same (2,2) vacuum, the couplings in the effective actions are actually all given by formulae from the special geometry[17]−[19].

Let us first consider briefly type II superstrings. In type II compactifications on C-Y manifolds the moduli scalar corresponding to Kähler-class deformations ((1,1) moduli) are in N=2 vector multiplets in type IIA theories and in hypermultiplets in type IIB theories. The reverse is true for (2,1) moduli corresponding to complex structure deformations. Also because of mirror symmetry the type IIA theory on C turns out to be identical to the type IIB theory in C'. These arguments can be used to show that the moduli geometry is the direct product of two special geometries, one for (1,1) moduli and the other for (2,1) moduli[14],[22],[36]. Because of the mirror hypothesis and absence of stringy corrections one can then confine the discussion to (2,1) moduli in type IIA and IIB theories.

The Kaluza-Klein analysis was performed in Ref. [18]. In type IIA theories the N=2 partners of the (2,1) moduli are $2(h_{(1,2)} + 1)$ Ramond-Ramond scalars. Together with the dilaton and antisymmetric tensor they form a special quaternionic manifold of dimension $4(h_{(2,1)} + 1)$ studied in Ref. [19] which also gives all the coupling in type II effective actions.

In type IIB theories the (2,1) moduli have $h_{(2,1)}$ vector superpartners, which have a non-trivial coupling to the graviphoton. Again all the couplings are fixed by the special geometry[17],[20]. For example the two-derivative couplings of the $(h_{(2,1)} + 1)$ R-R vectors with field strengths $F^{\Lambda\Sigma}_{\mu\nu}(\mu,\nu = 1\ldots 4)(\Lambda,\Sigma = 0\ldots h_{2,1})$ are given by:

$$-\sqrt{-g}(4\mathrm{Re}\mathcal{N}_{\Lambda\Sigma}(z,\bar{z})F^{\Lambda}_{\mu\nu}F^{\Sigma\mu\nu} - 2\mathrm{iIm}\mathcal{N}_{\Lambda\Sigma}(z,\bar{z})F^{\Lambda}_{\mu\nu}F^{\Sigma}_{\rho\alpha}\epsilon^{\mu\nu\rho\sigma}/\sqrt{-g}) \qquad \text{II} - 24$$

where:

$$\mathcal{N}_{\Lambda\Sigma} = -\overline{F}_{\Lambda\Sigma} + \frac{1}{(XNX)}(NX)_{\Lambda}(NX)_{\Sigma} \qquad \text{II} - 25$$

with $F_{\Lambda\Sigma}, N_{\Lambda\Sigma}$ defined by eq. II-18.

Let us now consider N=1 heterotic superstrings[39]. In this case (1,1) and (2,1) moduli are related to $h_{(1,1)}$ (27) E_6 families and $h_{(2,1)}(\overline{27})E_6$ antifamilies. Denoting by $\Phi^i_{27}, \Phi^\alpha_{\overline{27}}(i = 1\ldots h_{(1,1)}, \alpha = 1\ldots h_{(2,1)})$ the complex scalars of the E_6 charged chiral multiplets, all the effective couplings are given by the moduli metrics (given by eq. II-19), by the superpotential:

$$W(\phi,\psi,\Phi_{27},\Phi_{\overline{27}}) = W^1_{i\ell m}(\phi)\,\Phi^i_{27}\,\Phi^\ell_{27}\,\Phi^m_{27} + \\ + W^2_{\alpha\beta\gamma}(\phi)\,\Phi^\alpha_{\overline{27}}\,\Phi^\beta_{\overline{27}}\,\Phi^\gamma_{\overline{27}} \qquad \text{II} - 26$$

and the family metrics[36]:

$$G^{27}_{i\bar{j}} = e^{-(K_1-K_2)/3}\,G^1_{i\bar{j}} \qquad \text{II} - 27$$

$$G^{\overline{27}}_{\alpha\bar{\beta}} = e^{-(K_2-K_1)/3}\,G^2_{\alpha\bar{\beta}} \qquad \text{II} - 28$$

where $K_1(\phi,\bar{\phi}), K_2(\psi\bar{\psi})$ are the Kähler potentials for the two special geometries of (1,1) and (2,1) moduli and $G^1 = \partial\bar{\partial}K^1, G^2 = \partial\bar{\partial}K^2$ are the related metrics given by eqs. II-19, II-20.

Here ϕ,ψ denote the coordinates for (1,1) and (2,1) moduli respectively.

In section V we will show that also some higher-loop corrections to the heterotic effective action can be expressed in terms of quantities related to the special geometry of the moduli space.

III. Deformation of the complex structure and target-space duality symmetry

The relation of the special geometry with algebraic geometry arises with the interpretation of the Kähler metric $G_{i\bar{j}}$ for (2,1) moduli as the corresponding metric for the moduli space for the complex structure deformations, i.e. the Weil-Petersson metric[40],[41],[16]:

$$g_{i\bar{j}}^{WP} = i \int_C \phi_i \wedge \phi_{\bar{j}}^* \qquad \text{III} - 1$$

when ϕ_i is a (moduli-dependent) basis for the $H^{(2,1)}$ Dolbeault cohomology:

$$\phi_i = \phi_{i,ab\bar{c}} dx^a \wedge dx^b \wedge d\bar{x}^c \qquad \text{III} - 2$$

(x are complex coordinates on C).

The precise relation is:

$$G_{i\bar{j}} = \overset{\bullet}{+} \frac{1}{i\int \Omega \wedge \overline{\Omega}} g_{i\bar{j}}^{WP} = -\partial_i \partial_{\bar{j}} \log i \int \Omega \wedge \overline{\Omega}, \; i \int \Omega \wedge \overline{\Omega} = X^A \overline{F}_A + \overline{X}^A F_A \qquad \text{III} - 3$$

where $\Omega(\psi)$ is the deformed holomorphic three-form and by definition:

$$\int \Omega \wedge \phi_i = \int \Omega \wedge \overline{\phi}_i = 0 \qquad \text{III} - 4$$

From the fact that ϕ is a "holomorphic" deformation of the (3,0)-form Ω, it follows that:

$$\phi_i = \nabla_i \Omega = \frac{\partial \Omega}{\partial \psi_i} - \frac{1}{(\Omega, \overline{\Omega})} \left(\frac{\partial \Omega}{\partial \psi_i}, \overline{\Omega} \right) \Omega \qquad \text{III} - 5$$

The holomorphic sections $X^A(\psi), F_A(\psi)$ of the special geometry are identified[15]-[17],[20] with the periods of Ω along the $b_3 = 2(h_{2,1} + 1)$ homology cycles A^A, B_A:

$$\Omega(\psi) = X^A(\psi) \alpha_A + i \, F_A(\psi) \beta^A \qquad \text{III} - 6$$

where α_A, β^A is a (fixed) cohomology basis in H^3 dual to the homology cycles:

$$\int_M \alpha_A \wedge \beta^B = -\int_M \beta^B \wedge \alpha_A = \delta_A^B$$

$$\int_M \alpha_A \wedge \alpha_B = \int_M \beta^A \wedge \beta^B = 0 \qquad \text{III} - 7$$

From the fact that $\int \Omega \wedge \frac{\partial \Omega}{\partial \psi_i} = 0$, it follows that $F_A = \frac{\partial F}{\partial X^A}$ with $X^A F_A = 2F$ and then eq. II-20 follows from eq. III-3.

The holomorphic three-index section given by eq. II-23 is given by the following expression:

$$W_{i\ell m} = i \int \Omega \wedge \frac{\partial \Omega}{\partial \psi_i \partial \psi_\ell \partial \psi_m} \qquad \text{III} - 8$$

while the matrix N (II-18) and f_i^A (II-21)[20]:

$$N_{AB} = i \int \frac{\partial \Omega}{\partial X^A} \wedge \frac{\partial \overline{\Omega}}{\partial \overline{X}^B} \quad , \quad f_i^A = e^{K/2} \int \nabla_i \Omega \wedge \beta^A \qquad \text{III} - 9$$

If we consider the vector space $H^{(3,0)} + H^{(2,1)} \subset H^3(M)$ which varies holomorphically with ψ, a basis is given by:

$$\frac{\partial \Omega}{\partial X^A} = \alpha_A + i\, F_{AB} \beta^B \qquad \text{III} - 10$$

With respect to this basis the period matrix takes the simple form:

$$\omega = (\mathbb{1}, iF) = (\mathbb{1}, \Omega) \qquad \text{III} - 11$$

It then follows that the matrix N is the imaginary part of the period matrix $N = \text{Im}\Omega$. It also follows that[35]:

$$\det(-N) = \det G\, e^{-(n+1)K}$$

$$-\partial_i \partial_{\bar{j}}\, \lg \det N = W_{ipq} \overline{W}_{\bar{j}\bar{p}\bar{q}} g^{WP_{p\bar{p}}} g^{WP_{q\bar{q}}} \qquad \text{III} - 12$$

which relates, in special geometry, the determinant of ImΩ and the determinant of the moduli metric.

All previous formulae apply both for (2,1) and (1,1) moduli, because of the mirror hypothesis. However the existence of a holomorphic prepotential for (1,1) moduli is also a consequence of the special geometry of N=2 supergravity[34]. This allows one to identify a classical (large values of the moduli) limit of the holomorphic prepotential, like the one obtained from a C-Y compactification of D=10 supergravity.

In this case the holomorphic prepotential[14] is a cubic form, related to the intersection matrices[42]:

$$F_{\text{classical}} = i\, d_{ABC} \frac{X^A X^B X^C}{X^0} = (id_{i\ell m} \psi^i \psi^\ell \psi^m)(X^0)^2 \qquad \text{III} - 13$$

where:

$$d_{ABC} = \int_M V_A \wedge V_B \wedge V_C \qquad \text{III} - 14$$

and V_A is a basis for the $H^{(1,1)}$ cohomology.

Eq. III-13 has been written in the "special coordinates"[35],[19] $\phi_i = X^A/X^0 (A = 1\ldots h)$. These are actually the coordinates in which N=2 supergravity was first formulated[34]. In this coordinate basis the Kähler potential and Yukawa couplings become[15]:

$$K = -\log\left(2\mathcal{F} + 2\bar{\mathcal{F}} - (\psi_i - \bar{\psi}_i)(\mathcal{F}_i - \bar{\mathcal{F}}_i)\right) \qquad \text{III} - 15$$

$$W_{i\ell m} = \partial_i \partial_\ell \partial_m \mathcal{F} \quad , \quad \mathcal{F} = (X^{0^{-2}} F(X)) \qquad \text{III} - 16$$

The relation of special geometry with Dolbeault cohomology of C-Y manifolds exhibits a symplectic structure $Sp(b_3; \mathcal{Z})$ corresponding to a change of basis of the homology cycles, as is evident from eq. III-6. From the point of view of N=2, D=4 supergravity, subgroups of $Sp(b_3; R)$ are related to the so-called duality transformations[43],[44],[14]. They correspond to isometries of special Kähler manifolds. Discrete subgroups Γ of $Sp(b_3; \mathcal{Z})$ for which:

$$K(\Gamma\psi) = K - \Lambda(\psi) - \bar{\Lambda}(\bar{\psi}) \qquad \text{III} - 17$$

correspond to the action of global diffeomorphisms on the homology basis. In string theory Γ is the target-space modular group (target-space duality) and it must be modded out to define the moduli space of the corresponding superconformal field theory[45].

The identification of points in the moduli space connected by Γ means actually that the moduli space of string theory is not a manifold M but an orbifold M/Γ with fixed points. In the effective action, Γ appears as an exact symmetry[46],[47] and there are arguments that this symmetry survives in (higher genus) string perturbation theory[48],[49]. In section V we will show how target-space modular anomalies are cancelled in string theory and how this mechanism determines the running of gauge couplings induced by dynamical (moduli) scales. To prepare the ground for a discussion on target-space modular anomalies it is important to realize that a duality transformation Γ on the moduli space $\psi \to \Gamma\psi$ acts with a Kähler phase on the Yukawa couplings and with rescaling and U(1) phases on all the fermions of the theory[50]-[53].

Since fermions are coupled to gauge and gravitational fields, these (field-dependent) phases can induce mixed σ-model-gauge (or gravitational) anomalies through ordinary (one-loop) triangular graphs in which a σ-model U(1) connection couples to gauge or gravitational gauge fields[50]-[53]. We will give here the transformation properties, under a generic discrete isometry Γ, of all relevant quantities appearing in the effective Lagrangian for an N=1 heterotic string compactified on a C-Y manifold with gauge group $E_8 \times E_6$.

These transformations, since they act on sections of line bundle, will act accompanied by Kähler phases.

Let us ψ, ϕ be the coordinates on $M_1 \times M_2$ of a generic moduli space of C-Y manifolds. The Kähler potential, under the action of Γ, will transform as:

$$K \to K - (\Lambda_1 + \overline{\Lambda}_1) - (\Lambda_2 + \overline{\Lambda}_2)$$

where $\Lambda_1(\phi), \Lambda_2(\psi)$ are Kähler phases. Under the same action the Yukawa couplings for 27 and $\overline{27}$ families will change as follows:

$$W^1(\Gamma\psi) \to e^{2\Lambda_1 - 3\gamma_1 \Sigma} W^1(\psi)$$
$$W^2(\Gamma\psi) \to e^{2\Lambda_2 - 3\gamma_2 \Sigma} W^2(\psi)$$

III – 18

and accordingly, the scalar fields of chiral E_6 families:

$$\phi_{27} \to e^{\gamma_1 \Sigma_1 + (\Lambda_2 - \Lambda_1)/3} \phi_{27}$$
$$\phi_{\overline{27}} \to e^{\gamma_2 \Sigma_2 + (\Lambda_1 - \Lambda_2)/3} \phi_{\overline{27}}$$

III – 19

which ensure that the total superpotential W defined by eq. II-26 just cancel the variation of K in the "norm" over the line-bundle:

$$\|W\|^2 = e^G = WW^* e^K$$

III – 20

The variation of the moduli and family metrics is given by:

$$G^1 \to G^1 \, e^{-\gamma_1 + (\Sigma_1 + \overline{\Sigma}_1)}$$

$$G^2 \to G^2 \, e^{-\gamma_2 + (\Sigma_2 + \overline{\Sigma}_2)}$$

$$G^1_{27} \to G^1_{27} \, e^{-\gamma_1(\Sigma_1 + \overline{\Sigma}_1) + (\Lambda_1 + \overline{\Lambda}_1)/3 - (\Lambda_2 + \overline{\Lambda}_2)/3}$$

$$G^2_{\overline{27}} \to G^2_{\overline{27}} \, e^{-\gamma_2(\Sigma_2 + \overline{\Sigma}_2) - (\Lambda_2 + \overline{\Lambda}_2)/3 - (\Lambda_1 + \overline{\Lambda}_1)/3}$$

III – 21

Eq. III-21 can easily be derived from eqs. II-13 and II-27, II-28. Therefore the extra (holomorphic) phase $\gamma\Sigma$ (γ is a real normalization constant) is due to the discrete isometry Γ of the modular group which acts on the fermions as a U(1) gauge transformation with gauge connection:

$$B_\mu = i(B_i, \partial_\mu z_i - B_{\bar{\imath}} \partial_\mu \bar{z}_{\bar{\imath}})$$
$$B_i = \partial_i (\log \det G)$$

III – 22

The holomorphic phases Σ, Λ are actually related to one another since Σ is given by $\log \det \partial f_i^\Gamma / \partial z_j (z_i \to f_i^\Gamma(z))$ and charged matter fields are required to be tensor densities under Γ.

The transformation laws of the fermions are now determined by the bilinear fermionic terms in the effective action which are not chiral invariant (bilinear with an even number of γ matrices) namely:

$$e^{K/2}(W^1 \chi_{27}\chi_{27}\Phi_{27} + W^2 \chi_{\overline{27}}\chi_{\overline{27}}\Phi_{\overline{27}}) \qquad \text{III} - 23$$

$$G^2_{27}\chi_{27}\lambda\Phi_{27} \quad , \quad G^2_{\overline{27}}\chi_{27}\lambda\Phi_{\overline{27}} \qquad \text{III} - 24$$

$$e^{K/2}\phi_\mu \sigma^{\mu\nu}\psi_\nu \quad , \quad \chi^i G_{i\bar{j}}\partial_\mu \psi^{\bar{j}}\gamma^\nu \gamma^\mu \psi_\nu \qquad \text{III} - 25$$

$$\lambda \sigma^{\mu\nu} F_{\mu\nu} \chi_s \qquad \text{III} - 26$$

Eqs. III-13 and III-14, i.e. the Yukawa couplings, fix the phases of the changed E_6 fermions and of the gauginos. Eqs. III-25, III-26 fix the phases of the gravitino, moduli fermions and dilatino. The result is:

$$\chi_{27} \to \chi_{27}\, e^{\frac{\gamma_1}{2}(\Sigma_1 - \overline{\Sigma}_1) - \frac{5}{12}(\Lambda_1 - \overline{\Lambda}_1) - \frac{1}{12}(\Lambda_2 - \overline{\Lambda}_2)}$$
$$e^{\frac{\gamma_1}{2}(\Sigma_1 + \overline{\Sigma}_1) - \frac{1}{6}(\Lambda_1 + \overline{\Lambda}_1) + \frac{1}{6}(\Lambda_2 + \overline{\Lambda}_2)} \qquad \text{III} - 27$$

(the same for $\chi_{\overline{27}}$ with $1 \leftarrow 2$).

$$\lambda \to \lambda\, e^{\frac{1}{4}(\Lambda_1 - \overline{\Lambda}_1) + \frac{1}{4}(\Lambda_2 - \overline{\Lambda}_2)}$$

$$\psi_\mu \to \psi_\mu\, e^{\frac{1}{4}(\Lambda_1 - \overline{\Lambda}_1) + \frac{1}{4}(\Lambda_2 - \overline{\Lambda}_2)} \qquad \text{III} - 28$$

$$\chi_s \to \chi_s\, e^{-\frac{1}{4}(\Lambda_1 - \overline{\Lambda}_1) - \frac{1}{4}(\Lambda_2 - \overline{\Lambda}_2)}$$

$$\chi^i \to \chi^i\, e^{\gamma_1 \Sigma_1}\, e^{-\frac{1}{4}(\Lambda_1 - \overline{\Lambda}_1) - \frac{1}{4}(\Lambda_2 - \overline{\Lambda}_2)} \qquad \text{III} - 29$$

Note that $\lambda, \psi_\mu \chi_s$ undergo a pure U(1) Kähler phase while the matter and moduli fermions acquire extra terms due to the coordinate charge. Also the pure Kähler phase of charged fermions is not standard because the charged chiral scalars transform as holomorphic sections under a Kähler transformation on $M_1 \times M_2$.

Target-space duality transformations can be discussed, in great detail, in Z_n orbifold models which are limiting cases of C-Y manifolds[50]-[53]. In that case the moduli spaces of orbifolds[54] correspond to the so-called untwisted moduli and the duality groups are explicitly known[55]-[57]. We will not report here this analysis, which has been discussed in great detail in recent papers, but rather try to extract from this analysis general conclusions.

The topic we would like to discuss in the last part of this section is the concept of automorphic functions for general C-Y compactification by a generalization of automorphic functions for the modular group SL(2, Z) and their connection to special geometry. In this respect the key ingredient is that, for untwisted moduli of (2,2) orbifolds, the moduli space has the structure M/Γ where M is a symmetric space compatible with special geometry and Γ is a discrete subgroup of M determined by the string quantum symmetries[54]−[57]. The symmetric spaces M are actually all given by particular truncation of the homogeneous space:

$$S = \frac{SO(6,6)}{SO(6) \times SO(6)} \quad \text{III} - 30$$

which is related to the Narain-Sarmadi-Witten[58] classification of the moduli space of toroidal compactifications. Let us call M_1 and M_2 the manifolds for untwisted (1,1) and (2,1) moduli such that $M_1 \times M_2$ is contained in the toroidal six-dimensional moduli space S. Then all Z_n orbifolds correspond to particular choices of the following special symmetric spaces[14],[54]−[57]:

$$M_1 = \left(\frac{SU(1,1)}{U(1)}\right)^3 \quad , \quad \frac{SU(1,1)}{U(1)} \times \frac{SU(2,2)}{SU(2) \times SU(2) \times U(1)} \quad ,$$

$$\frac{SU(3,3)}{SU(3) \times SU(3) \times S(1)} \quad \text{III} - 31$$

$$M_2 = 1 \text{ or } \frac{SU(1,1)}{U(1)}$$

The M_1 spaces correspond to cubic prepotentials of the form:

$$\mathcal{F}^i(\phi_a) = i\, d_{abc}\phi^a\phi^b\phi^c \quad \text{III} - 32$$

while the M_2 space, if not empty, corresponds to a quadratic prepotential:

$$\mathcal{F}^2(\psi) = c + \psi^2 \quad \text{III} - 33$$

The previous equations are written in a special coordinate system as defined by eqs. III-15, III-16. The corresponding Yukawa couplings are therefore given by:

$$W^1_{abc} = i\, d_{abc} \quad , \quad W^2 = 0 \quad \text{III} - 34$$

All (2,2) orbifolds have a complex modulus which corresponds to the breathing mode[54]. This corresponds to a submanifold of M^1 given by the prepotential[14]:

$$\mathcal{F}(\phi) = i\,\lambda\phi^3 \quad \text{III} - 35$$

where λ is a normalization factor. The modular group Γ associated to ϕ is SL(2, Z)[46]. As discussed in Ref. [46] a superpotential term for ϕ, which may arise from non-perturbative stringy effects, must be an automorphic function of Γ. Then the only choice for W(ϕ) which is finite and non-zero inside the fundamental domain is[59]:

$$W(\phi) = \frac{1}{\eta(\phi)^6} = \left[\prod_{m,n}(m + in\,\phi)^{-3}\right]_{\text{reg.}} \qquad \text{III} - 36$$

where $\eta(\phi)$ is the Dedekind function. By interpreting W as a holomorphic section on M[56], its norm is:

$$||W(\phi)|| = |W(\phi)|\,e^{K/2} = \frac{1}{|\eta(\phi)|^6|\phi + \overline{\phi}|^{3/2}} \qquad \text{III} - 37$$

Eq. III-37 defines a modular invariant function, which, in the context of orbifold compactifications, may be interpreted as the free energy of the (fermionic excitations of) the orbifold lattice, as a function of $\phi = R^2 + ib$:

$$\mathcal{F}_{\text{reg.}} = -\log ||W(\phi)||^2 \qquad \text{III} - 38$$

More generally, this connection leads to the suggestive interpretation of a holomorphic superpotential as the (regularized) determinant of the holomorphic mass matrix $\mathcal{M}(\phi)$ of the (infinitely many) fermionic excitations of the lattice[56]:

$$\log \det \mathcal{M}(\phi)_{\text{reg.}} = \log W(\phi) \qquad \text{III} - 39$$

A more interesting case is that of three moduli with $M = (SU(1,1)/U(1))^3$. This is actually a subspace of all spaces in III-31.

The space M/Γ with $\Gamma = SL(2,Z)^3 \times P(3)$ coincides with the (full) moduli space of the Z_3/Z_3 orbifold, as discussed in Ref.[55]. In this case the regularized free-energy is:

$$\mathcal{F}_{\text{reg.}} = \sum_{i=1}^{3}\sum_{m_i,n_i} \left.\frac{|(m_i + n_i\psi_i)|^2}{(\phi_i + \overline{\phi}_i)}\right|_{\text{reg.}}$$

$$= \sum_{i=1}^{3} \log |\eta(\phi_i)|^4(\phi_i + \overline{\phi}) = \qquad \text{III} - 40$$

$$= -\log ||W||^2$$

where:

$$W^{-1}(\phi_i) = \prod_{i=1}^{3}\eta(\phi_i)^2 \quad,\quad K = -\sum_{i=1}^{3}\log(\phi_i + \overline{\phi}_i) \qquad \text{III} - 41$$

Eqs. III-37 - III-41 suggest the following generalization for the holomorphic eingenvalues of the fermionic mass-matrix on an arbitrary (2,2) internal conformal field theory:

$$M(\phi) = (M_I X^I + iN^I F_I) \qquad \text{III} - 42$$

where $X^I(\phi), iF_I(\phi)$ are the periods defined by eq. III-5. Then a modular invariant expression is (formally) obtained as follows[56]:

$$-\log ||W||^2 = -\log (\det MM^+ e^K)$$

$$= \sum_{M_I, N_I} \log \left. \frac{|M_I X^I + iN^I F_I|^2}{\overline{X}^I F_I + X^I \overline{F}_I} \right|_{\text{reg.}} \qquad \text{III} - 43$$

When M_I, N^I transform by the linear action of $Sp(8; Z)$ induced by Γ, the sum over the integers must be restricted to a "modular orbit" of Γ. In the previous example eqs. III-40 - III-41 are reproduced by making the following choices:

$$F(X) = iX^1 X^2 X^3 / X^0 \quad , \quad \phi_i = -iX^i / X^0 \qquad \text{III} - 44$$

$$\begin{aligned}
M_0 &= m_1 m_2 m_3 & N_0 &= n_1 n_2 n_3 \\
M_1 &= n_1 m_2 m_3 & N_1 &= -m_1 n_2 n_3 \\
M_2 &= n_2 m_1 m_3 & N_2 &= -m_2 n_1 n_3 \\
M_3 &= n_3 m_1 m_2 & N_3 &= -m_3 n_1 n_2
\end{aligned} \qquad \text{III} - 45$$

Eq. III-45 specifies the modular orbit of Γ. It is easy to show that the action of the three $SL(2,Z)$ factors on the three sets of (unrestricted) integers (m_I, n_i) corresponds to the action of Γ, embedded in $Sp(8; Z)$, on the eight-dimensional integral vectors (M_I, N^I). It is interesting to stress the fact that existence of the regularized form of eq. III-43, as a norm of a holomorphic section of the line bundle $\mathcal{L}$ is the main assumption of this construction. This is guaranteed provided the objects $M_I X^I + iN^I F_I$ are the eigenvalues of a "good operator" defined over the manifold[60]. We note that for M^I manifolds which are not simply products of $\frac{SU(1,1)}{U(1)}$ spaces, the holomorphic section W defined through eq. III-43 is no longer given in terms of Dedekind functions and it defines a non-trivial holomorphic modular form over the moduli space[56],[57].

Let us make some final comments in this section. As far as eq. III-43 is concerned, it should be possible to compute this formula exactly in a pure Kaluza-Klein context by confining oneself to the "lattice" corresponding to deformations of the complex structure. In fact, based on the considerations of the previous section, one would expect this lattice to be purely classical, in contrast with the case of Kähler class deformations. A good example of this phenomenon is given by the two-dimensional torus[61] in which the duality group Γ is given by the product of two SL(2,Z) groups but one of them is not connected to winding modes and it is purely "classical".

Another point that it is worth mentioning is that, as for SL(2,Z), one expects to find many other holomorphic modular forms for C-Y moduli space M/Γ.

Non-trivial examples of Γ, related to smooth C-Y manifolds, have recently been given[25] together with procedures[25]–[32] to explicitly construct their related special geometry. In the next section we will show how it is possible to determine the periods $X^I(\phi), F_I(\phi)$ in the case of smooth manifolds, using methods of algebraic geometry.

IV. Picard-Fuchs equations and determination of the periods

The determination of the periods $\omega^a(\psi)(a = 1\ldots b_3)$ of the holomorphic 3-form $\Omega(\psi)$ enables us to explore the structure of the moduli space M of complex structure deformations including its global properties, encoded in the (target-space) modular group Γ. Indeed from formulae III-3, III-6, III-8 the metric over M, the three index tensor W, the period matrix Ω are all given in terms of the periods. For instance, the equation for the Kähler potential is given by a bilinear sympletic form over the periods[15]–[17],[20]:

$$e^{-K} = i \int \Omega \wedge \overline{\Omega} = X^A \overline{F}_A + \overline{X}^A F_A = \omega^a \Omega_{ab} \overline{\omega}^b \qquad IV-1$$

with:

$$\omega^a(\psi) = (X^A(\psi), iF_A(\psi)) \qquad a = 1\ldots b_3$$

$$\Omega = \begin{pmatrix} 0 & \mathbb{1} \\ -\mathbb{1} & 0 \end{pmatrix} \qquad IV-2$$

The periods of holomorphic p-forms on certain p-dimensional complex manifolds can be obtained as solutions of linear differential equations, known as Picard-Fuchs equations, which we are going to illustrate.

The periods, at least in the case of manifolds obtained by some defining polynomial in a CP^n space[25]–[27]:

$$F(x) = 0 \qquad IV-3$$

can in principle be computed by explicitly integrating the holomorphic p-form $\Omega(\phi)$ along the homology cycles. This was done in Ref. [25] for the mirror manifold of $P_4(5)$. In a more general setting it is possible to obtain[26]–[30] linear differential equations whose independent solutions give exactly the periods $\omega^a(\phi)$.

These equations turn out to connect methods of algebraic geometry with N=2 superconformal field theories and their topological version[6]−[12] since the very same equations can be derived (at least in some cases) from the associativity of the fusion rule coefficients of the chiral ring of primary fields in TSCFT's, together with the conservation of the U(1) charge and the relation expressing the three-point function has third-derivative (with respect to the deformation parameters) of the topological free-energy[10]. The technique of writing differential equations for the periods is particularly simple in the case of one deformation, with parameter ψ, of the defining polynomial[26]. This applies equally well to a c=3 and c=9, N=2 SCFT namely the (Z_3 orbifoldized) torus[32] and the C-Y threefold which is the mirror of $P_4(5)$. For the torus ($p = 1, b_1 = 2$) the defining polynomial is a cubic in CP^2 while for the C-Y threefold ($p = 3, b_3 = 4$) the defining polynomial is a quintic in CP^4.

In both cases the periods are given by:

$$\omega^a(\psi) = \int_{A,B} \Omega_p(\psi) \qquad \text{IV} - 4$$

when A, B are the homology cycles. It is now possible, by considering the deformed polynomial:

$$F(x, \psi) = F(x) + \psi h(x) \quad |\psi| < 1 \qquad \text{IV} - 5$$

to find a linear differential equation for the ($b_p \times b_p$) matrix C:

$$\frac{dC(\psi)}{d\psi} = C(\psi)B(\psi) \qquad \text{IV} - 6$$

which is equivalent to a b_p-order differential equation for the periods $\omega^a(\psi)$. Actually ω^a are the b_p linearly independent solutions of this equation.

The ($b_p \times b_p$) matrix B is obtained by taking the matrix-elements of an "evolution operator": $x_0 \frac{dF}{d\psi} = x_0 h(x)$ in a certain basis $X^{(\nu)}$ of polynomials with scalar product defined by:

$$< X^{*(\mu)} X^{(\nu)} > = \delta^{\mu\nu} \qquad \text{IV} - 7$$

where $X^{*(\mu)}$ is a dual basis (and in the scalar product only constant terms are kept).

The vector space with basis vector $X^{(\mu)}$ is obtained as follows: let us start with a polynomial of degree d in CP^n, then take the sets:

$$I = \{\omega = (\omega_0, \ldots \omega_{n+1}) | \omega_i \epsilon Z_+^{h+2}, d\omega_0 = \omega_1 + \ldots \omega_{n+1}\} \qquad \text{IV} - 8$$
$$A = \omega \epsilon I, 0 < \omega_i < d, \forall i = 1 \ldots n+1$$

The basic vectors are given by:

$$X^{(\omega)} = x_0^{\omega_0} x^{\omega_1} \ldots x_{n+1}^{\omega_{n+1}} \quad \omega \epsilon A \qquad \text{IV} - 9$$

Let us now define the following covariant derivatives:

$$D_i = x_i \frac{\partial}{\partial x_i} + x_0 x_i \frac{\partial F}{\partial x_i} \quad i = 1\ldots n+1 \qquad \text{IV} - 10$$

Then it can be shown that the quantity:

$$x_0 h(x) X^{(\nu)} \qquad \text{IV} - 11$$

can be expanded in terms of $X^{(\omega)}$ ($\omega \epsilon A$) and covariant derivatives $D_i X^{(\omega)}$ with $\omega \epsilon I$ and $X^{(\omega)}$ having x_0 degree ω_0 less than the polynomial given by eq. IV-11.

Let us generally assume that:

$$x_0 h(x) X^{(\nu)} = B_{\nu\mu}(\psi) X^{(\mu)} + (DX)^{(\nu)}(\psi) \qquad \text{IV} - 12$$

Then since:

$$< X^{*(\mu)}, (DX)^{(\nu)} >= 0 \qquad \text{IV} - 13$$

$B_{\nu\mu}$ is the desired matrix element.

The matrix C is the transpose of the period matrix and eq. IV-6 gives the desired differential equation for $\omega^a(\psi)$.

We would like now to apply these techniques to the two examples in question, the cubic and the quintic, to show the usefulness of the procedure. For the torus we have:

$$F_{\psi=0}(x) = \sum_{i-1}^{3} x_i^3 \quad , \quad h(x) = -3x_1 x_2 x_3 \qquad \text{IV} - 14$$

since d=3, n=2, $0 < \omega_i < 3, \omega_i = (1,2)$ and we have only two basis polynominals:

$$X^{(1)} = x_0 x_1 x_2 x_3 \quad , \quad X^{(2)} = x_0^2 x_1^2 x_2^2 x_3^2 \qquad \text{IV} - 15$$

and three derivatives $D_i X^{(\mu)}$.

It follows that:

$$x_0 h(x) X^{(1)} = -3 X^{(2)}$$

$$x_0 h(x) X^{(2)} = \sum_{i=1}^{2} a_\nu X^{(\nu)} + D(\ldots) \qquad \text{IV} - 16$$

with $a_1 = \dfrac{-\psi}{3(1-\psi^3)}$, $a_2 = \dfrac{3\psi^2}{1-\psi^3}$.

So we get for the B matrix:

$$B_{11} = 0, \ B_{21} = -3, \ B_{12} = \dfrac{-\psi}{3(1-\psi^3)}, \ B_{21} = \dfrac{3\psi^2}{1-\psi^3} \qquad \text{IV} - 17$$

Eq. IV-6 is solved by:

$$C = \begin{pmatrix} \omega_1 & -1/3\omega_1' \\ \omega_2 & -1/3\omega_2' \end{pmatrix} \qquad \text{IV} - 18$$

where $\omega_{1,2}$ are the solutions of the following second-order differential equation:

$$(1-\psi^3)\omega'' - 3\psi^2\omega' - \psi\omega = 0 \qquad \text{IV} - 19$$

If we now define $t = \omega_1/\omega_2$, then the equation for $t(\psi)$ can be re-expressed as an equation for $\psi(t)$ which is related to a ratio of two 3-point functions of the perturbed TSCFT.

This equation was obtained as an equation for the fusion rule coefficients in Ref. [32].

Let us now move to the more interesting case of the C-Y threefold defined by the quintic polynomial:

$$\sum_{i=1}^{5} x_i^5 - 5\psi \, x_1 x_2 x_3 x_4 x_5 = 0 \qquad \text{IV} - 20$$

In this case $h(x) = -5x_1 x_2 x_3 x_4 x_5$ and there are four basis polynominals:

$$\begin{aligned} X^{(1)} &= x_0 x_1 x_2 x_3 x_4 x_5 \\ X^{(2)} &= x_0^2 x_1^2 x_2^2 x_3^2 x_4^2 x_5^2 \\ X^{(3)} &= x_0^3 x_1^3 x_2^3 x_3^3 x_4^3 x_5^3 \\ X^{(4)} &= x_0^4 x_1^4 x_2^4 x_3^4 x_4^4 x_5^4 \end{aligned} \qquad \text{IV} - 21$$

As in the previous case we compute:

$$B_{\mu\nu} = <X^{*(\mu)}, x_0 h(x) X^{(\nu)}> \qquad \text{IV} - 22$$

by expanding $x_0 h(x) X^{(\nu)}$.
We obtain:

$$x_0 h(x) X^{(\nu)} = -5 X^{(\nu+1)}, \quad \nu = 1,2,3, \qquad \text{IV} - 23$$

$$x_0 h(x) X^{(4)} = \sum_{\nu=1}^{4} a_\nu X^{(\nu)} + D(\ldots) \qquad \text{IV} - 24$$

with $a_1 = -\dfrac{1}{125} \dfrac{\psi}{1-\psi^5}$, $a_2 = \dfrac{3}{5} \dfrac{\psi^2}{(1-\psi^5)}$, $a_3 = \dfrac{-5\psi^3}{1-\psi^5}$, $a_4 = \dfrac{10\psi^4}{1-\psi^5}$.

As before, from eq. IV-6 we get for each row a fourth-order differential equation:

$$(1-\psi^5)\omega^{IV} - 10\psi^4 \omega^{III} - 25\psi^3 \omega^{II} - 15\psi^2 \omega^{I} - \psi\omega = 0 \qquad \text{IV} - 25$$

Performing the change of variable $z = \psi^5$ we obtain the solutions in terms of generalized hypergeometric functions:

$$z^{\frac{k}{5}}\, {}_4F_3\left(\frac{k}{5}, \frac{k}{5}, \frac{k}{5}, \frac{k}{5}; \overbrace{\frac{k+1}{5}, \frac{k+2}{5}, \frac{k+3}{5}, \frac{k+4}{5}}; z\right) \qquad \text{IV} - 26$$

with $k = 1,2,3,4$ which reproduces the result of Ref. [25].

In the above method there is no real difference between the cubic and the quintic, with the exception that the matrices C, B are 2×2 in one case and 4×4 in the other case. This is so in spite of the fact that as target-space manifolds they are completely different and the second corresponds, from the world-sheet point of view, to an interacting σ-model which is therefore solved exactly by algebraic methods.

It is also clear that the techniques displayed here can in principle be generalized to other spaces and to the case of multidimensional varieties ($h > 1$ for C-Y threefolds). In that case one generalizes the present equation to partial differential equations giving rise to $b_3 = 2(h+1)$ linearly independent solutions. One interesting point is that these equations together with their monodromy will also determine the "quantum" target-space duality group Γ, which is a subgroup of $Sp(b_3; Z)$ and has a linear action on the periods $\omega^a(\psi_i)$. For the torus this group trivializes to $SL(2, \mathcal{Z})$ and for the mirror of $P_4(5)$ it was recently discussed in Ref. [25].

In the next section we will see that the duality group Γ plays an important role in discussing the effect of higher orders in the effective action. In particular it relates the moduli dependence of gauge couplings to mixed modular-gauge anomalies because of space-time supersymmetry.

The Picard-Fuchs equations, for the c=9 N=2 superconformal theories, corresponding to the C-Y threefolds, can be derived[33] on general grounds, as "holomorphic identities" following from the special geometry of the moduli space or equivalently as N=2 supergravity identities.

Indeed, as explained in sect. III, the deformed holomorphic three-form $\Omega(\psi)$

(eq. III-6) and the (2,1) forms ϕ_i (eq. III-5) are entirely equivalent to the two symplectic vectors:

$$V = (X^\Lambda, iF_\Lambda) \quad , \quad U_i = (\nabla_i X^\Lambda, i\nabla_i X^\Sigma F_{\Sigma\Lambda}) \qquad\qquad \text{IV} - 27$$

according to the formulae given in sect. II. These vectors are the basis elements for the flat $Sp(2h+2;R)$ bundle defined by Strominger in Ref. [17]. In particular, from eqs. IV-21, IV-22, V-15 (second reference of Ref. [20]) it follows that:

$$\mathcal{D}_i V = U_i \qquad\qquad \text{IV} - 28$$

$$\mathcal{D}_i \overline{\mathcal{D}}_{\bar{j}} V = G_{i\bar{j}} \overline{V} \qquad\qquad \text{IV} - 29$$

$$\mathcal{D}_i \mathcal{D}_j V = C_{ijk} G^{k\bar{k}} \overline{U}_{\bar{k}} \qquad\qquad \text{IV} - 30$$

where C_{ijk} are the covariantly holomorphic Yukawa couplings defined by eqs. II-13, II-14. These equations are of course the same as the ones derived from deformation theory of C-Y threefolds[16].

Differentiating once more eq. IV-29 one obtains:

$$\mathcal{D}_i \mathcal{D}_j \overline{U}_{\bar{k}} = 0 \qquad\qquad \text{IV} - 31$$

By regarding C_{ijk} as a square matrix (with jk entries and by assuming it is invertible, for fixed i) one finally gets the identity:

$$\mathcal{D}_i \mathcal{D}_j (C_k^{-1})^{\ell m} \mathcal{D}_k \mathcal{D}_\ell V = 0 \qquad\qquad \text{IV} - 32$$

Since V is holomorphic, eq. IV-32 implies an holomorphic differential relation for V[25],[33]:

$$\left(\sum_{s=0}^{4} a^{(s)}_{ijkm, i_1 \ldots i_s} \partial_{i_1} \ldots \partial_{i_s} \right) V = 0 \qquad\qquad \text{IV} - 33$$

where the $a^{(s)}$ are holomorphic functions of the moduli, computable from eq. IV-32. Note that if the matrix C vanishes then the differential equation degenerates to a second order relation $\mathcal{D}_i \mathcal{D}_j V = 0$ with solution $F(X) = a_{\Lambda\Sigma} X^\Lambda X^\Sigma$ ($a_{\Lambda\Sigma}$ is a constant matrix). In eq. IV-33 the holomorphic coefficients are functions of the Yukawa couplings W_{ijk} and of the Kähler connections $\Gamma^i_{\ell m}, K_i$. These equations are actually holomorphic identities if we plug for W and K their expression given by eqs. II-20 and II-23 in terms of the holomorphic prepotential defined by eq. II-16. Of course eq. IV-32 are much more general than what comes from the deformation of the Hodge structure of C-Y threefolds since they are just a consequence of N=2 supergravity, special geometry and not all N=2 supergravities come from superstring C-Y compactifications.

The same identities can be interpreted[33] as differential equations if we imagine to give, as data, the holomorphic coefficients $a^{(s)}(\psi)$ rather than the prepotential $F(X^\Lambda(\psi))$. This is the case in deformation theory of C-Y threefolds where the Picard-Fuchs equations[26]–[30] for the periods (the holomorphic vector V) allow to directly compute the $a^{(s)}$ coefficients and then to find V (and therefore F) through the fourth order (partial) differential equations given by eq. IV-32. These equations are equivalent to the Picard-Fuchs equations in "arbitrary coordinate systems" and an explicit holomorphic expression of the coefficients $a^{(s)}$ can be given in terms of quantities defined in the special geometry.

For example the $a^{(3)}$ coefficient is always related to the Yukawa coupling (we consider for simplicity the one parameter case) through the equation:

$$W' = -\frac{1}{2} a^{(3)} W \quad \text{(where } a^{(4)} \text{ is normalized to 1)} \qquad \text{IV} - 34$$

given in Ref. [25]. More complicated equations are found for the other coefficients $a^{(s)}(s = 2, 1, 0)$. This approach to Picard-Fuchs systems gives new insight on the relation which occurs between the special geometry of the moduli space and perturbed N=2 topological superconformal field theories. Indeed one can show[33] that a holomorphically flat connection can be obtained within the framework of special geometry acting on a flat holomorphic bundle (which vanishes in the supergravity special coordinates) in complete analogy with N=2 topological field theories[5]–[12].

V. Moduli dependence of gauge couplings and target-space duality anomaly cancellation

Recently, loop-corrections to gauge couplings have been computed[61]–[67] in heterotic string compactification (in four dimensions) with an unbroken gauge group G.

The analysis has been mainly confined to (2,2) orbifold vacua with $G = E_8 \times G'$ by a number of authors[62],[50]–[53],[63] with string and field theory calculations, exploiting the relation between renormalization of gauge couplings and anomalies in N=1 supersymmetric gauge theories[51],[52],[65]–[67].

The peculiar outcome of these investigations is that moduli corrections to gauge couplings are not holomorphic (in the moduli chiral fields) so the question whether these results, which are manifestly target-space duality invariants, compatible with standard supergravity arguments[37],[38], was raised[50]. The apparent paradox came from the fact that a gauge invariant coupling to matter has the general form:

$$g_{ab}(\phi, \overline{\phi}) F^a_{\mu\nu} F^b_{\mu\nu} + \theta_{ab}(\phi, \overline{\phi}) F^a_{\mu\nu} F^b_{\rho\sigma} \epsilon^{\mu\nu\rho\sigma} \qquad \text{V} - 1$$

where, by supersymmetry, the field dependent gauge couplings and θ-angles are required to be harmonic and in fact related to a "complex" holomorphic coupling:

$$f_{ab}(\phi,\overline{\phi}) = f_{ab}(\phi) = g_{ab}(\phi,\overline{\phi}) + \theta_{ab}(\phi\overline{\phi}) \qquad \text{V}-2$$

Consistency with (target-space) modular invariance would seem to require that f is a modular invariant holomorphic function[46] while explicit calculations[62],[63],[50]–[53] show that f is a modular invariant non-holomorphic function. The apparent paradox finds a solution by noticing that target-space duality relates massless to massive modes and a modular invariant result is obtained when all string states circulate in a loop. On the other hand the contribution of massless states does not satisfy the decoupling and gives rise to infra-red divergences which generate non-local terms[50]–[53] in the effective action S_{eff}.

The overall result is summarized by a single (superfield) formula which is manifestly modular invariant and contains both the local and non-local term[70],[51]–[53].

$$S_\Gamma^{1-\text{loop}} = \text{Re}\int d^2\theta W_A^2 \frac{\overline{D}^2 D^2}{\Box} \sum_a C_A^a(H_a + \Lambda_a + \overline{\Lambda}_a) \qquad \text{V}-3$$

where H_a is a real (non-chiral) superfield $H_a(\phi,\overline{\phi})$ and $\Lambda_a(\phi)$ is a chiral superfield function of the (chiral) moduli superfields ϕ, i.e.: $\overline{D}\Lambda_a = 0$.

When added to the three-level term:

$$\text{Re}\int S\, W_A^2 \qquad \text{V}-4$$

eqs. V-3 and V-4 define an effective coupling of the form:

$$\frac{1}{g_{A\text{eff}}^2}(\phi,\overline{\phi}) = S + \overline{S} + \sum_s C_A^a(H_a + \Lambda_a + \overline{\Lambda}_a) \qquad \text{V}-5$$

(a refer to different types of moduli) and we see that the non-holomorficity of $g_{A\text{eff}}^2(\phi,\overline{\phi})$ is related to the non-local superfield in V-3 coming from the triangular graph when massless particles circulate. The term containing Λ_a is local since $\frac{\overline{D}^2 D^2}{\Box}\Lambda_a = \Lambda_a$ and in fact can be interpreted as a local counterterm[51]–[53]:

$$\text{Re}\sum\int d^2\theta\, W_A^2 C_A^a \Lambda_a(\phi) \qquad \text{V}-6$$

needed to cancel the anomaly. In string theory eq. V-6 is just the result of the integration over the heavy string modes for which the decoupling theorem applies. Finally there is an extra source of anomaly cancellation mechanism, i.e. the Green-Schwarz mechanism[71]–[73]. In the S field formation this means that S acquires, through one-loop effects, an anomalous transformation under target-space duality transformations[51]:

$$S \to S + \delta(\phi) \qquad \text{V}-7$$

Eq. V-7 gives a universal contribution to the cancellation of all anomaly diagrams, equal for all gauge group factors, but model dependent. Actually in certain orbifolds, like the Z_3 and Z_7 orbifolds, all anomalies coincide and the contribution of the massive states $\Lambda_a(\phi)$ is completely absent[51]-[53]. When a G.S. term is present the S-Kähler potential gets modified with a non-trivial mixing to the moduli[63],[66]:

$$-\log (S+\bar{S}) \to -\log (S+\bar{S}+\Delta(\phi,\bar{\phi})) \qquad \text{V}-8$$

where, in order to preserve modular invariance, Δ transforms as:

$$\Delta \to \Delta - \delta - \bar{\delta} \qquad \text{V}-9$$

Δ is the universal term in eq. V-5, which is cancelled by the G.S. mechanism. The non-holomorphic term $\sum_a C_A^a H_a(\phi,\bar{\phi})$ comes from the computation of a mixed σ-model gauge anomaly and depends only on the effective couplings of the massless states and their U(1) charges under modular transformations, as defined by eqs. III-27 – III-29. For smooth C-Y manifolds $A = E_8$ or E_6 and "a" refer to (1,1) or (2,1) moduli so we have only four terms to determine. For $A = E_6$:

$$C_A^1 H_1(\psi,\bar{\psi}) = \left[T(G = E_6) - \frac{5}{3} T(R = 27)h_{1,1} - \frac{1}{3} T(\overline{R} = \overline{27})h_{1,2} \right] K_1(\psi,\bar{\psi})$$

$$+ 2T(R = 27)\log \det G^1(\psi,\bar{\psi})$$

$$\text{V}-10$$

$C_A^2 H_2(\psi\bar{\psi})$ is the same with $h_{(1,1)} \leftrightarrow h_{(2,2)}, 2, R \leftrightarrow \overline{R}$.

These formulae can be simplified by using the equation III-12, i.e.= $\log \det N - \log \det G - (h+1)K$.
Then one obtains:

$$\sum_{a=1}^{2} C_{A=E_6}^a H_a(\psi,\bar{\psi}) = [T(E_6) + 2T(27)](K_1 + K_2) + \frac{1}{6}T(27)\chi(K_1 - K_2)$$

$$+ 2T(27)(\log \det N^1 + \log \det N^2) \qquad \text{V}-11$$

and $\chi = 2(h_{1,1} - h_{2,2})$ is the Euler characteristic of C. Also $T(E_8) = 30, T(E_6) = 12, T(27) = T(\overline{27}) = 3$.

For the E_8 anomaly one obtains:

$$T(E_8)(K_1 + K_2) \qquad \text{V}-12$$

Eqs. V-11 and V-12 show that on smooth C-Y manifolds threshold effects of massive states are always present (Λ_a cannot vanish) since for the difference of effective gauge couplings $\frac{1}{g_{E_6}^2} - \frac{1}{g_{E_8}^2}$ we have a non-trivial non-holomorphic function. Also the anomaly cancellation seems to require at least two different automorphic functions for each type of moduli since in general K and log det N have different transformation properties under Γ. Candidate holomorphic functions for threshold corrections were studied in Ref. [56] and discussed in section III.

We would like to add few comments on eq. V-5 as derived from the superfield expression V-3 and formula V-5. The resolution of the puzzle regarding holomorphicity is that the chiral non-local superfield[51],[52]:

$$f_A = \frac{\overline{D}^2 D^2}{\Box} H_A \left(H_A = \sum_a C_A^a H_a \right) \qquad \text{V} - 13$$

has a $\theta = 0$ component given by:

$$f_{A_{\theta=0}} = \left(H_{A_{\theta=0}} + \frac{D}{\Box} \right) + i \frac{\partial_\mu A^\mu}{\Box} \qquad \text{V} - 14$$

For the constant moduli fields $\phi \to <\phi>$ the second and third terms are just absent. As a consequence the non-local chiral superfield f_A at $\theta = 0$ and $\phi = <\phi>$ has a local non-holomorphic component $H_a C_A^a H_a(\phi, \overline{\phi})$. Therefore we conclude that non-holomorphic effective gauge couplings are perfectly compatible with the standard supergravity form, which will produce all desired couplings provided the correct expression for f_A (given by eq. V-13) is used.

For those orbifold models in which the target-space modular group is a product of SL(2, Z) factors, the automorphic functions obtained by integrating out the massive string modes (eq. V-5) are logs of products of Dedekind functions, according to the discussion of section III. Also in this case, since the moduli space M is a symmetric space, the Kähler potential K_a and the curvature term det log G_a are just proportional, so all holomorphic terms are expressible in terms of the same automorphic function[50]-[53].

As a final topic of this section we would like to report on higher loop corrections to gauge couplings in string theories which may also be understood from the anomaly cancellation mechanism[73].

The argument will be made for the E_8 gauge group for which only gauginos contribute to the anomalies. The other gauge groups can be discussed as well but will not be treated here. The idea is to obtain the all-loop β function for a pure N=1 super Yang-Mills theory from purely string arguments.

That this must be so, comes from the fact that the all-loop result[66]-[68]:

$$\frac{\beta(g)}{g^3} = \frac{-3T(G)/16\pi^2}{1 - [T(G)/8\pi^2]g^2} \qquad \text{V} - 15$$

comes from the multiplet structure of the anomaly multiplet of the supercurrent containing the U(1) R-current and the dilatation current[65]-[67]. In string theory all couplings and scales are dynamical variables, so one expects a relation between g^2_{eff} and g^2_{bare} as coming from field equations. This can be seen, in string theory, as a relation between the gauge couplings in the linear multiplet and in the chiral multiplet formulation.

Let us consider the anomaly cancellation mechanism for the E_8 gauge group in superfield notations and in the linear multiplet representation. We also consider, for definiteness, the $\mathcal{Z}_3$ and $\mathcal{Z}_7$ orbifolds for which no counterterms other than the Green-Schwarz mechanism are present[51],[52].

The local effective Lagrangian which cancels the E_8^2-U(1) σ-model anomaly is:

$$S^{1-\text{loop}}_\Gamma = \int d^4\theta [\hat{L}^{-1/2} e^{-1/2K} (S_0 \bar{S}_0)^{3/2} + \lambda \hat{L} K] \qquad \text{V} - 16$$

(here S_0 is a compensating multiplet which sets up the choice of a dynamical scale i.e. gives the normalization of the Einstein term). The combination:

$$\hat{L} = L - \Omega \qquad \text{V} - 17$$

is a gauge-invariant multiplet containing a real linear multiplet L with the Chern-Simons multiplet Ω. Note that $\hat{L}$ is a modular invariant expression (K is the moduli Kähler potential) so a modular transformation:

$$K \to K - \Lambda - \bar{\Lambda} \qquad \text{V} - 18$$

acts on the compensating multiplet S_0 as:

$$S_0 \to S_0 e^{-1/3\Lambda} \qquad \text{V} - 19$$

To go to the S field formulation for the dilaton we have to replace $\hat{L}$ by an unconstrained real superfield V and enforce the linear constraint through a Lagrange multiplier, i.e., a chiral field S:

$$\mathcal{L} \to \mathcal{L}(\hat{L} \to V) - (S + \bar{S})(V + \Omega) \qquad \text{V} - 20$$

By defining $U = \frac{1}{2}\left(\frac{S_0 \bar{S}_0}{2V_e^{k/3}}\right)^{3/2}$, the equation of motion for U becomes:

$$U = \frac{S+\bar{S}}{2} - \frac{T(E_8)}{16\pi^2} K \qquad V-21$$

which gives the one-loop effective coupling in the presence of the Green-Schwarz mechanism by identifying:

$$U = \frac{1}{g_{eH}^2}, \quad \frac{S+\bar{S}}{2} = \frac{1}{g_0^2} \qquad V-22$$

The Green-Schwarz term is not the most general term compatible with the superconformal dimensions and with the anomaly. It can be further completed by replacing:

$$\hat{L}K \to 3\hat{L}\log e^{K/3}\hat{L} \qquad V-23$$

With this modification eq. V-21 gets replaced by:

$$\frac{1}{g_{eH}^2} = \frac{1}{g_0^2} \frac{T(E_8)}{8\pi^2}\left(-\log g_{eff}^2 + c - \frac{3}{2}\log S_0\bar{S}_0\right) \qquad V-24$$

where, to get the one-loop result, we have to substitute $S_0\bar{S}_0 = e^{K/3}$ and neglect the second term on the right-hand side of eq. V-24.

Remembering that $S_0\bar{S}_0$ sets the unit of measure of (mass)^2, eq. V-15 is obtained as follows:

$$\beta(g) = -\frac{1}{g}\frac{dg^2}{d(\log S_0\bar{S}_0)} \qquad V-25$$

A modular-invariant g_{eff}^2 is obtained by setting $S_0\bar{S}_0 \propto e^{K/3}$. Then we get:

$$\frac{1}{g_{eff}^2} = \left(\frac{S+\bar{S}}{2} - \frac{T(E_8)}{16\pi^2}K + c\right) \frac{T(E_8)}{8\pi^2}\log g_{eff}^2 \qquad V-26$$

and the modified S-Kähler potential becomes:

$$\log\left\{g_{eff}^2\left[1 + \frac{T(E_8)}{16\pi^2}g_{eff}^2\right]^{-3}\right\} \qquad V-27$$

where g_{eff}^2 is solved in terms of $S+\bar{S}$ by eq. V-26.

To connect the modular invariant effective gauge coupling V-26 with the running gauge coupling defined in a renormalizable low-energy theory in a given renormalization scheme one should impose on the two couplings appropriate boundary conditions. Note that in eq. V-26 the constant c is undetermined by modular invariant arguments.

Acknowledgements

I would like to thank the following colleagues for the fruitful and stimulating collaboration, on which this review is based:

M. Bodner, C. Cadavid, L. Castellani, A. Ceresole, R. D'Auria, J.P. Derendinger, P. Fré, C. Kounnas, J. Louis, D. Lüst, S. Sabharwal, A. Shapere, A. Strominger, S. Theisen and F. Zwirner.

REFERENCES

[1] For a review, see for instance S. Ferrara, *in* "Strings '90", eds. R. Arnowitt, R. Bigan, M.J. Duff, D.V. Nanopoulos, C. Pope and E. Sezgin (World Scientific, Singapore, 1990), p. 387; *in* "The challenging questions", Subnuclear Series, vol. 27, ed. A. Zichichi (Plenum Press, New York, 1989), p. 103; *in* Proc. 1990 Summer School in High-Energy Physics and Cosmology, ICTP Series in Theoretical Physics, vol. 7, eds. J.C. Pati, S. Randjbar-Daemi, E. Sezgin and Q. Shafi (World Scientific, Singapore, 1990), p. 111; in Mod. Phys. Lett. **A6** No. 24 (1991) (Brief Reviews).

[2] M. Ademollo, L. Brink, A. D'Adda, R. D'Auria, E. Napolitano, S. Sciuto, E. del Giudice, P. di Vecchia, S. Ferrara, F. Gliozzi, R. Musto, and R. Pettorino, Phys. Lett. **62B** (1976) 105;
W. Boucher, D. Friedan, and A. Kent, Phys. Lett. **B172** (1986) 316;
A. Sen, Nucl. Phys. **B278** (1986) 289 and **B172** (1986) 316;
L. Dixon, D. Friedan, E. Martinec, and S.H. Shenker, Nucl. Phys. **B282** (1987) 13;
T. Banks, and L. Dixon, Nucl. Phys. **B307** (1988) 93;
D. Gepner, Nucl. Phys. **B296** (1988) 757 and Phys. Lett. **B189** (1987) 380;
W. Lerche, D. Lüst, and A.N. Schellekens, Nucl. Phys. **B298** (1988) 477.

[3] E. Calabi, *in* "Algebraic geometry and topology", a symposium in honour of S. Lefschetz (Princeton University Press) (1957) 38;
S.T. Yau, Proc. Nat. Ac. Sci. USA **74** (1977) 5798.

[4] P. Candelas, G. Horowitz, A. Strominger and E. Witten, Nucl. Phys. **B258** (1985) 46.

[5] E. Martinec, Phys. Lett. **B217** (1989) 431;
C. Vafa, and N.P. Warner, Phys. Lett. **B218** (1989) 51;
W. Lerche, C. Vafa, and N. Warner, Nucl. Phys. **B324** (1989) 427;
D. Gepner, Phys. Lett. **B222** (1989) 207;
P. Howe, and P. West, Phys. Lett. **B223** (1989) 377;
S. Cecotti, L. Girardello, and A. Pasquinucci, Nucl. Phys. **B328** (1989) 701, Int. J. Mod. Phys. **A6** (1991) 2427;
C. Vafa, Int. J. Mod. Phys. **A6** (1991) 2829;

K. Intrilligator, and C. Vafa, Nucl. Phys. **B339** (1990) 95;

C. Vafa, Mod. Phys. Lett. **A4** (1989) 1615 and Mod. Phys. Lett. **A4** (1989) 1169;

S. Cecotti, Int. J. Mod. Phys. **A6** (1991) 1749 and Nucl. Phys. **B355** (1991) 755;

A. Giveon, and D.J. Smit, Mod. Phys. Lett. **A6**, N. 24 (1991) 2211.

[6] E. Witten, Commun. Math. Phys. **117** (1988) 353; **118** (1988) 411, and Nucl. Phys. **B340** (1990) 281.

[7] T. Eguchi and S.K. Yang, Mod. Phys. Lett. **A5** (1990) 1693.

[8] C. Vafa, Mod. Phys. Lett. **A6** (1991) 337.

[9] K. Li, Caltech preprints CALT-68-1662 and CALT-68-1670 (1990);

B. Blok and A. Varchenko, Princeton preprint IASSNS-HEP-91/5 (1991).

[10] R. Dijkgraaf, E. Verlinde and H. Verlinde, Nucl. Phys. **B348** (1991) 435 and B352 (1991) 59.

[11] A. Giveon and D.J. Smit, preprints LBL-30342, UCB-PTH-91/8; LBL-30388, UCB-PTH-91/10; and LBL-30831, UCB-PTH-91/27 (1991).

[12] S. Cecotti and C. Vafa, preprint HUTP-91/A031, SISSA-69/91/EP.

[13] N. Seiberg, Nucl. Phys. **B303** (1988) 731.

[14] S. Cecotti, S. Ferrara and L. Girardello, Int. J. Mod. Phys. **4** (1989) 2475; Phys. Lett. **B213** (1988) 443.

[15] S. Ferrara and A. Strominger, in "Strings '89", eds. R. Arnowitt, R. Bryan, M.J. Duff, D.V. Nanopoulos and C.N. Pope (World Scientific, Singapore, 1989), p. 245.

[16] P. Candelas, P. Green and T. Hübsch, Nucl. Phys. **B330** (1990) 49;

P. Candelas and X.C. de la Ossa, Nucl. Phys. **B342** (1990) 246.

[17] A. Strominger, Commun. Math. Phys. **133** (1990) 163;

V. Periwal and A. Strominger, Phys. Lett. **B235** (1990) 261.

[18] S. Cecotti, Comm. Math. Phys. **131** (1990) 517.

[19] M. Bodner, A.C. Cadavid and S. Ferrara, Class. Quant. Grav. **8** (1991) 789;

S. Ferrara and S. Sabharwal, Class. Quant. Grav. **6** (1989) L77 and Nucl. Phys. **B332** (1990) 317.

[20] L. Castellani, R. D'Auria and S. Ferrara, Phys. Lett. **B241** (1990) 57 and Class. Quant. Grav. **1** (1990) 1767;

R. D'Auria, S. Ferrara and P. Fré, Nucl. Phys. **B359** (1991) 705.

[21] B.R. Greene and M.R. Plesser, Nucl. Phys. **B338** (1990) 15;

P. Candelas, M. Linker and R. Schimmrigk, Nucl. Phys. **B341** (1990) 383;

P. Aspinwall, C.A. Lütken and G.G. Ross, Phys. Lett. **B241** (1990) 373;

P. Aspinwall and C.A. Lütken, Nucl. Phys. **B353** (1991) 427 and **B355** (1991) 482.

[22] J. Distler, and B. Greene, Nucl. Phys. **B304** (1988) 1; **B309** (1989) 295.

[23] M. Dine, P. Huet, and N. Seiberg, Nucl. Phys. **B322** (1989) 301.

[24] J. Lauer, J. Mas, and H.P. Nilles, Phys. Lett. **B226** (1989) 251.

[25] P. Candelas, X.C. de la Ossa, P.S. Green and L. Parkes, Phys. Lett. **B258** (1991) 118 and Nucl. Phys. **B359** (1991) 21.

[26] A.C. Cadavid and S. Ferrara, Phys. Lett. **B267** (1991) 193.

[27] W. Lerche, D.J. Smit and N.P. Warner, preprints LBL-31104, UCB-PTH-91/39, USC-91/022, CALT-68-1738 (1991).

[28] Z. Maassarani, USC preprint USC-91/023 (1991).

[29] A. Klemm, S. Theisen, and M.G. Schmidt, Karlsruhe preprint TUM-TP-129/91, KA-THEP-91-09, HD-THEP-91-32 (1991).

[30] D.R. Morrison, DUK-M-91-14 (Duke Univ., Durham, 91).

[31] R. Dijkgraaf, E. Verlinde, and H. Verlinde, Nucl. Phys. **B352** (1991) 59.

[32] E. Verlinde and N.P. Warner, Phys. Lett **269** (1991) 96.

[33] S. Ferrara and J. Louis, CERN-TH.6334/91 (November 1991).

[34] B. de Wit and A. Van Proeyen, Nucl. Phys. **B245** (1984) 89;
B. de Wit, P.G. Lauwers and A. Van Proeyen, Nucl. Phys. **B255** (1985) 569.

[35] E. Cremmer, C. Kounnas, A. Van Proeyen, J.P. Derendinger, S. Ferrara, B. de Wit and L. Girardello, Nucl. Phys. **B250** (1985) 385.

[36] L. Dixon, V.S. Kaplunovsky and J. Louis, Nucl. Phys. **B239** (1990) 27.

[37] E. Cremmer, B. Julia, J. Scherk, S. Ferrara, L. Girardello and P. van Nieuwenhuizen, Nucl. Phys. **B147** (1979) 105;
E. Cremmer, S. Ferrara, L. Girardello and A. Van Proeyen, Nucl. Phys. **B212** (1983) 413.

[38] E. Witten and J. Bagger, Phys. Lett. **115B** (1982) 202;
J. Bagger, Nucl. Phys. **B211** (19S3) 302.

[39] D. Gross, J. Harvey, E. Martinec and R. Rohm, Phys. Rev. Lett. **54** (1985) 502; Nucl. Phys. **B256** (1985) 253 and **B267** (1985) 75.

[40] G. Tian, *in* "Mathematical aspects of string theory", ed. C.S.T. Yan (World Scient.) (1987) 629.

[41] R. Briant, and P. Griffith, *in* "Progress in Mathematics" (Birkhäuser) Vol. 36 (1983) 77.

[42] A. Strominger, Phys. Rev. Lett. **55** (l985) 2547;
A. Strominger and E. Witten, Commun. Math. Phys. **101** (1985) 341;
P. Candelas, Nucl. Phys. **B298** (1988) 458.

[43] S. Ferrara, J. Scherk, and B. Zumino, Nucl. Phys. **B121** (1977) 393.

[44] M.K. Gaillard, and B. Zumino, Nucl. Phys. **B193** (1981) 221.

[45] K. Kikkawa and M. Yamasaki, Phys. Lett. **149B** (1984) 357;
N. Sakai and L. Senda, Progr. Theor. Phys. **75** (1986) 692;

V.P. Nair, A. Shapere, A. Strominger and F. Wilczek, Nucl. Phys. **B287** (1987) 402;

A. Giveon, E. Rabinovici and G. Veneziano, Nucl. Phys. **B322** (1989) 167;

A. Shapere and F. Wilczek, Nucl. Phys. **B320** (1989) 167;

M. Dine, P. Huet and N. Seiberg, Nucl. Phys. **B322** (1989) 301;

J. Molera and B. Ovrut, Phys. Rev.**D40** (1989) 1150; J. Lauer, J. Mas and H.P. Nilles, Phys. Lett. **B226** (1989) 251 and Nucl. Phys. **B351** (1991) 353;

W. Lerche, D. Lüst and N.P. Warner, Phys. Lett. **B231** (1989) 417;

M. Duff, Nucl. Phys. **B335** (1990) 610;

A. Giveon and M. Porrati, Phys. Lett. **B246** (1990) 54 and Nucl. Phys.**B355** (1991) 422;

A. Giveon, N. Malkin and E. Rabinovici, Phys. Lett. **B238** (1990) 57;

J. Erler, D. Jungnickel and H.P. Nilles, MPI-Ph/91-90.

[46] S. Ferrara, D. Lüst, A. Shapere and S. Theisen, Phys. Lett. **B233** (1989) 147;

S. Ferrara, D. Lüst and S. Theisen, Phys. Lett. **B233** (1989) 147 and **B242** (1990) 39.

[47] J. Schwarz, Caltech preprints CALT-6S-1581 (1990), CALT-68-1728 (1991), and CALT-68-1740 (1991);

J. Erler, D. Jungnickel and H.P. Nilles, MPI-Ph/91-81.

[48] E. Alvarez and M.A.R. Osorio, Phys. Rev. **D40** (1989) 1150.

[49] D. Gross and I. Klebanov, Nucl. Phys. **B344** (1990) 475.

[50] L.J. Dixon, V.S. Kaplunovsky and J. Louis, Nucl. Phys. **B355** (1991) 649.

[51] J.P. Derendinger, S. Ferrara, C. Kounnas and F. Zwirner, preprint CERN-TH-6004/91, to appear in Nucl. Phys. B.

[52] J. Louis, Stanford preprint SLAC-PUB-5527 (1991).

[53] G.L. Cardoso and B. Ovrut, Pennsylvania preprints UPR-0464T and UPR-0481T (1991).

[54] E. Witten, Phys. Lett. **155B** (1985) 151;

S. Ferrara, C. Kounnas and M. Porrati, Phys. Lett. **B181** (1986) 263;

M. Cvetic, J. Louis and B. Ovrut, Phys. Lett. **B206** (1988) 227;

S. Ferrara and M. Porrati, Phys. Lett. **B216** (1989) 289;

M. Cvetic, J. Molera and B. Ovrut, Phys. Rev. **D40** (1989) 1140.

[55] D. Shevitz, Nucl. Phys. **B338** (1990) 283.

[56] S. Ferrara, C. Kounnas, D. Lüst and F. Zwirner, Nucl. Phys. **B365** (1991) 431.

[57] P. Fré and P. Soriani, Trieste preprint SISSA 90/91/EP (1991).

[58] K.S. Narain, Phys. Lett. **B169** (1986) 369;
K.S. Narain, M.H. Sarmadi and E. Witten, Nucl. Phys. **B289** (1987) 414.

[59] H. Oeguri, and C. Vafa, Nucl. Phys. **B361** (1991) 469.

[60] D. Quillen, Funkt. An. E. Pril, **19** (1985) 37.

[61] R. Dijkgraaf, E. Verlinde, and H. Verlinde, Comm. Math. Phys. **115** (1988) 6491.

[62] V.S. Kaplunovsky, Nucl. Phys. **B307** (1958) 145.

[63] I. Antoniadis, K.S. Narain and T.R. Taylor, Phys. Lett. **B267** (1991) 37.

[64] I. Antoniadis, J. Ellis, R. Lacaze and D.V. Nanopoulos, preprint CERN-TH-6136/91, CPTH-A061-0691-ACT-39-CTP-TAMU-39/91, SPhT/91-093 (1991);
S. Kalara, J. Lopez and D.V. Nanopoulos, Phys. Lett. **269** (1991) 84;
J.P. Derendinger, talk given at the HEP-Conference, Geneva, July 1991;
L. Ibanez, D. Lüst and G. Ross, CERN TH-6241 (1991), to be published in Phys. Lett. B.

[65] S. Ferrara and B. Zumino, Nucl. Phys. **B87** (1975) 207.

[66] D.R.T. Jones, Phys. Lett. **B123** (1983) 45.

[67] M.T. Grisaru and P.C. West, Nucl. Phys. **B354** (1985) 249.

[68] V.A. Novikov, M.A. Shifman, A.L. Vainshtein and V.I. Zakharov, Phys. Lett. **166B** (1986) 29; Nucl. Phys. **B229** (1983) 381 and 407;
M.A. Shifman and A.I. Vainshtein, Minnesota preprint TPI-Minn-91/4-T (1991).

[69] D. Amati, K. Konishi, Y. Meurice, G.C. Rossi and G. Veneziano, Phys. Rep. **162** (1988) 169.

[70] S.J. Gates, M.T. Grisaru, M. Roček and N. Siegel, "Superspace" (Benjamin-Cummings Publ. Co., Reading, Mass., 1983).

[71] M.B. Green and J.H. Schwarz, Phys. Lett **149B** (1984) 117.

[72] S. Ferrara and M. Villasante, Phys. Lett. **B186** (1986) 85;
S. Cecotti, S. Ferrara and M. Villasante, Int. J. Mod. Phys. **A2** (1987) 1839.

[73] M. Dine, N. Seiberg and E. Witten, Nucl. Phys. **B289** (1987) 589.

[74] J.P. Derendinger, S. Ferrara, C. Kounnas and F. Zwirner, Phys. Lett. **B271** (1991) 307.

CHAIRMAN: S. Ferrara

Scientific Secretaries: U. Danielsson, M. Johnson, S. Lau, B. Lindholm, U. Vikas

DISCUSSION

– *Sanchez:*

I have a comment to make about the significance of duality in curved spacetimes such as cosmological backgrounds or two dimensional black holes. Recent results in string theory, in connection with gravity, have shown that duality holds even in the case when the target space has a time or space varying vacuum metric.

– *Ferrara:*

Yes, I am aware of such results.

– *Sanchez:*

What do you have to say about four dimensional string phenomenology, and in particular the flipped SU(5) × U(1) presented here by John Ellis which you claimed was appealing? John Ellis' model made no reference to compactification (it was associated with the so-called four dimensional models), so what can you say about duality for this case?

– *Ferrara:*

There is a more general way to looking at string compactification than just the Lagrangian with the extra dimensions curling up. One takes the first four string coordinates to represent spacetime and the others to represent a two dimensional conformal field theory. The cancellation of the conformal anomaly requires the superstring to exist in ten dimensions. The manifold of compactification is what, in general, is called an internal superconformal field theory. A Calabi-Yau manifold is not just a geometric object; in string theory it has the interpretation as an N=2 superconformal field theory. Hence the results presented here hold for such more general models including 4D-strings constructed out of free fermions.

– *Danielsson:*

For strings moving on compact manifolds there is an instability against the formation of vortices. For a certain value of the radius there is a phase transition: the Kosterlitz-Thouless transition. This breaks duality. Do you have a physical reason why one should exclude vortices?

– *Ferrara:*

I do not know the implication of vortex formation in Calabi-Yau manifolds.

– *Zichichi:*

Does this mean that string theories have a disease?

– *Ferrara:*

I think that the effect of vortex formation is not well understood. However, I do not think it is of any relevance for superstrings moving in 4D-space-time.

– *Vikas:*

I don't think this holds for critical strings. However, this argument probably holds for the non-critical strings in 1 or 2 dimensions.

– *Ferrara:*

I think it would be difficult to show this for strings propagating on Calabi-Yau manifolds.

– *Sivaram:*

I have a comment to make on dimensionality. As you said the electromagnetic coupling is dimensionless in four dimensions, whereas Newton's constant is dimensionless in two dimensions. For gravity we also have the Gauss-Bonnet term, and since in four dimensions this term is a total divergence it is important for the absence of one loop divergences in pure gravity.

– *Ferrara:*

Yes, I agree, but this is peculiar to four dimensions and it explains why pure gravity is one-loop finite at $D=4$.

– *Sivaram:*

Is there any significance of the B-field similar to that of the θ-parameter in QCD?

– *Ferrara:*

Yes, the B-field in four dimensions replaces the θ-parameter and becomes an axion. This field plays an important role in the strong CP problem.

– *Sivaram:*

Can one have an effective running coupling constant related inversely to the size of the manifold?

– *Ferrara:*

Yes, if you compute the one loop corrections to the gauge couplings then you get a dependence on the radius of the manifold.

– *Vikas:*

What are the implications of the Yukawa couplings becoming functions of the radius?

– *Ferrara:*

The Yukawa couplings become like "Clebsch-Gordan" coefficients and have the transformation properties that I have discussed. If one can fix the vacuum expectation values, like the orbifold modulus, one can in principle compute the Yukawa couplings for the different generations. The structure of the Yukawa couplings are determined by this precise symmetry transformation of $R \to 1/R$.

– *Lau:*

While discussing Yukawa couplings you introduced the automorphic tensor λ_{abc}. What is the basis for which the λ_{abc} are the components?

– *Ferrara:*

The Yukawa couplings depend on spacetime scalar fields of complex variables. The scalar fields appear in the effective Lagrangian with a kinetic term of this form:

$$G_{a\bar{b}} \partial_\mu t^a \partial^\mu \bar{t}^b$$

Because of spacetime supersymmetry this metric which multiplies the kinetic term of the scalar fields has to be a Kähler metric:

$$G_{a\bar{b}} = \partial_a \bar{\partial}_b K(t\bar{t})$$

Where the Kähler potential is a function of the scalar fields. In Calabi-Yau compactification the Kähler metric is related to the Yukawa couplings by the intrinsic geometry of the space of the scalar fields. The Riemann tensor of the scalar fields, which is the Riemann tensor of the Calabi-Yau space and describes deformations of the Calabi-Yau space, satisfies the following relation:

$$R_{a\bar{b}c\bar{d}} = G_{a\bar{b}} G_{c\bar{d}} + G_{a\bar{d}} G_{c\bar{b}} - e^{2K} \lambda_{acp} \bar{\lambda}_{\bar{b}\bar{d}\bar{p}} G^{p\bar{p}}$$

The Yukawa couplings are related to Kähler metric by this equation. This shows that the λ_{abc} are tensors in the moduli space. They are tensors in the space of possible radial deformations of the Calabi-Yau space and depend on the moduli coordinates t^i. This relation implies that in some coordinate system these Yukawa couplings are completely integrable: they can be written as the third derivative of some function $F(t^i)$.

$$\lambda_{abc} = \partial_a \partial_b \partial_c F(t^i)$$

This $F(t^i)$ acts like a pre-potential of the moduli space, through which the Yukawa couplings and the metric can be expressed. So one finds that these Yukawa couplings, which are the couplings among different generations, can be viewed as a tensor which is related to the geometry of the moduli space of the Calabi-Yau manifold.

– Jain:

This morning you argued that in searching for a model of quantum gravity one may have to go to spacetime dimensions other than four. If one believes in Kaluza-Klein type models, what reason do you have, independent of specific theoretical considerations, that the compactification scale cannot be much lower than the Planck mass, say 50 TeV?

– Ferrara:

In string theory there are two scales: one is the fundamental string length related to Newton's gravitational constant, the other is related to the string compactification manifold which is arbitrary. This is not fixed at the tree level, and one can in principle set the scale by a dynamical method. The only point to keep in mind is that at that energy scale there is a decompactification, because these radial modes are excited, which should have spectroscopic consequences. This could give new physics.

– Zichichi:

Could you tell us what would be the scale of this phenomenon?

– Ferrara:

The only thing I know is that it cannot be the weak scale. However, nothing dictates the scale. It could in principle be at 200 TeV, in which case these winding modes would be excited at this energy giving new particles.

– Jain:

Are the string loop or threshold corrections to the effective low energy couplings holomorphic in the moduli?

– Ferrara:

No. The duality symmetry relates terms which come from massive excitations to those that come from massless excitations. In fact the contribution from the massless excitations are non-holomorphic while the contributions from the massive excitations are. One considers loop diagrams in string theory with a composite U(1) connection and two gauge bosons in the external lines.

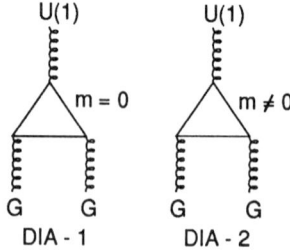

These diagrams are like those in field theory which Altarelli was describing. The intermediate states given in the diagrams can be massless and massive. The diagram with massless particles (DIA-1) gives rise to a non-local term in the effective action. This term has the following form:

$$\frac{\partial_\mu A^\mu}{\Box} F_{\mu\nu}\tilde{F}_{\mu\nu} \tag{1}$$

In string theory DIA-1 is related to gauge coupling by supersymmetry. The condition for this diagram to give a potential anomaly is given by:

$$\sum q T_q(R) \neq 0 \tag{2}$$

(q is a fermion charge and $T_q(R)$ is a group factor). This anomaly is dangerous because it can violate the duality symmetry which we have been discussing. This anomaly diagram is then cancelled in string theory by massive states which give rise to an harmonic term. This harmonic term gives rise to a composite Peccei-Quinn mechanism:

$$\theta(t) F\tilde{F}; \quad \partial_t \partial_{\bar{t}} \theta = 0 \tag{3}$$

where θ is a function of the moduli. The θ above is the imaginary part of an automorphic function (a "complex" holomorphic coupling) and its variation cancels the anomaly diagram. In the gauge sector this mechanism generates another harmonic term (the real part of the mentioned holomorphic function) which combined with the supersymmetric partner of the anomaly term (a local one!) gives rise to:

$$\frac{1}{g_{\text{eff}}^2} F_{\mu\nu}^2 \tag{4}$$

This term is non-holomorphic and seems to break supersymmetry. In local supersymmetric field theories g_{eff}, the gauge coupling must be a harmonic function of the fields. But the term breaking the harmonicity is the consequence of a non-local superfield term coming from a loop integration. This can be written as:

$$\int d^2\theta \omega^2 \frac{\bar{D}^2 D^2}{\Box} \log K$$

On computing the above term one finds that the real part is local while the imaginary part is non-local. The sum of this term with the harmonic one is really modular invariant. Hence, the moduli corrections to the gauge couplings are non-holomorphic but are modular invariant. This can be understood as due to the infra-red divergence coming from the states which arise by integrating out the massless states which give rise to the non-holomorphic term.

A DUALITY BETWEEN STRINGS AND FIVEBRANES

M. J. Duff[†]
Center for Theoretical Physics
Texas A&M University
College Station, Texas 77843

ABSTRACT

Not only does the heterotic string admit a heterotic fivebrane as a soliton but the heterotic fivebrane admits as a soliton a heterotic string. This provides further evidence that the two theories may be dual descriptions of the same physics, with the strong coupling regime of one theory described by the weak coupling regime of the other. To illustrate this, we show how the energy-momentum tensor of the *quadratic* Yang-Mills action associated with the string reduces to that of an elementary fivebrane, and how the energy-momentum tensor of the *quartic* Yang-Mills action associated with the fivebrane reduces to that of an elementary string.

1. Introduction

At the start of the superstring revolution of 1984, many physicists believed that the heterotic string [1] might provide that Holy Grail of theoretical physics: a unified theory of all the forces and particles of nature, including gravity. Six years later this initial euphoria has been dampened by the realization that many of the really basic questions of string theory (e.g. How do strings break supersymmetry? How do strings choose a vacuum state?) cannot be answered within the framework of a weak-coupling perturbation expansion.

In the meantime, a small but dedicated group of theorists were posing a seemingly very different question: If we are going to replace 0-dimensional point

---

[†] Work supported in part by NSF grant PHY-9045132

particles by 1-dimensional strings, why not 2-dimensional membranes or, in general, p-dimensional objects or "p-branes"? [2] This work has revealed that many of the original objections to membranes ("membranes must have ghosts", "membranes have no massless particles in their spectrum", "supermembranes cannot exhibit the κ-symmetry crucial for superstring consistency", "membranes cannot describe chiral fermions", etc) were without foundation.

In this paper, we wish to describe some recent advances in supermembrane theory concerning the "heterotic fivebrane", an object whose existence was conjectured in 1987 [3] but which has recently been rescued from obscurity by Strominger [4], who showed that the heterotic string admits the heterotic fivebrane as a soliton solution. Our new result is that the converse is also true [5]. This lends support to the idea that the two theories are dual descriptions of the same physics, with the strong coupling regime of one theory described by the weak coupling regime of the other. Specifically, we shall see that the dimensionless loop expansion parameters g(string) and g(fivebrane), are related by

$$g \text{ (fivebrane)} = \frac{1}{\sqrt[3]{g \text{ (string)}}} \qquad (1.1)$$

and that the string tension T_2 and fivebrane tension T_6 obey a Dirac quantization rule [6,7]

$$\kappa^2 \, T_2 \, T_6 = n\pi, \qquad n = \text{integer} \qquad (1.2)$$

where κ^2 is the D=10 gravitational constant.

2. Extended objects

Consider some extended object of dimension p moving through spacetime. It sweeps out a worldvolume of dimension $d = p+1$. Thus a $p = 0$ point particle sweeps out a $d = 1$ worldline, a $p = 1$ string sweeps out a $d = 2$ worldsheet, a $p = 2$ membrane sweeps out a $d = 3$ worldvolume and so on. Clearly the dimension of the object cannot exceed the dimension of the spacetime so we must have $d \leq D$ where D is the dimension of spacetime, sometimes called the "target space". Its trajectory is described by the functions $X^M(\xi)$ where X^M is the spacetime coordinate ($M = 0, 1, \ldots, D-1$) and ξ^i are the worldvolume coordinates ($i = 0, 1, \ldots, d-1$). In what follows, we shall frequently make the so-called "static gauge choice" by making the $D = d + (D - d)$ split

$$X^M(\xi) = \left(X^\mu(\xi), \, Y^m(\xi)\right) \qquad (2.1)$$

where $\mu = 0, 1, \ldots d-1$ and $m = d, \ldots D-1$, and then setting

$$X^\mu(\xi) = \xi^\mu \tag{2.2}$$

Thus the only physical degrees of freedom are given by the $(D-d)$ $Y^m(\xi)$.

To describe a super p-brane, we augment the D bosonic coordinates $X^M(\xi)$ with anticommuting fermionic coordianates $\theta^\alpha(\xi)$. Depending on D, this spinor could be a Dirac spinor, a Weyl spinor, a Majorana spinor or a Majorana-Weyl spinor. The action for the super p-brane also has fermionic κ-symmetry which reflects the fact that half of the spinor degrees of freedom are redundant. We may therefore adopt a physical gauge which eliminates the unphysical half. The net result is that the theory exhibits a d-dimensional supersymmetry with equal number of bose and fermi worldvolume degrees of freedom, where the number of supersymmetries is exactly one half of the original spacetime supersymmetries. However, the range of d and D which permit bose-fermi equality (assuming only scalar and spinor fields) is severely limited to the twelve points on the "brane-scan" of Fig. 1. In the next sections we shall focus on the meaning of the horizontal line that connects the $(d=2, D=10)$ superstrings and the $(d=6, D=10)$ superfivebranes.

3. Antisymmetric tensor fields and duality

It is well known that a charged particle couples to an abelian vector potential A_M which displays a gauge invariance

$$A_M \to A_M + \partial_M \Lambda \tag{3.1}$$

and has a gauge invariant field strength

$$F_{MN} = 2\,\partial_{[M} A_{N]} \equiv \partial_M A_N - \partial_N A_M \tag{3.2}$$

Similarly a string couples to a rank-2 antisymmetric tensor potential $A_{MN} = -A_{NM}$ with gauge invariance

$$A_{MN} \to A_{MN} + \partial_{[M} \Lambda_{N]} \tag{3.3}$$

and field strength

$$F_{MNP} = 3\,\partial_{[M} A_{NP]} \tag{3.4}$$

In general, a $(d-1)$-brane couples to a d-form $A_{M_1 M_2 \cdots M_d}$ with

$$A_{M_1 M_2 \cdots M_d} \to A_{M_1 M_2 \cdots M_d} + \partial_{[M_1} \Lambda_{M_2 \cdots M_d]} \tag{3.5}$$

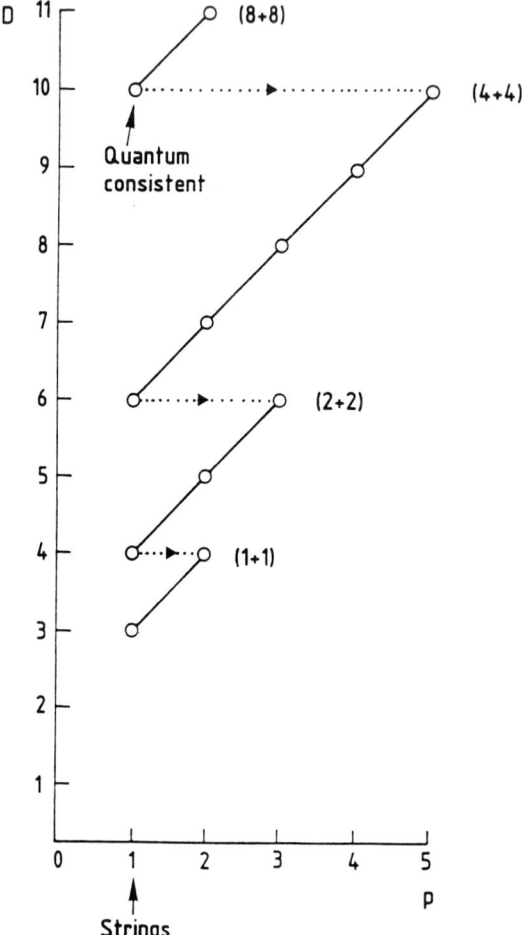

Figure 1. The brane-scan.

and
$$F_{M_1 M_2 \cdots M_{d+1}} = (d+1)\partial_{[M_1} A_{M_2 \cdots M_{d+1}]} \tag{3.6}$$

In the language of differential forms we may write

$$A \to A + d\Lambda \tag{3.7}$$

and

$$F = dA \tag{3.8}$$

from which the Bianchi identity

$$dF \equiv 0 \tag{3.9}$$

follows immediately. In the absence of other interactions, the equation of motion for the d-form potential is

$$d^*F = {}^*J \tag{3.10}$$

where the source J is a d-form. Here we have introduced the Hodge dual operation $*$ which converts a d-form into a $(D-d)$ form e.g.

$$(^*J)^{M_1 M_2 \cdots M_{D-d}} \equiv \frac{1}{d!} \varepsilon^{M_1 M_2 \cdots M_D} J_{M_{D-d+1} \cdots M_D}$$

and $\varepsilon^{M_1 \cdots M_D}$ is the D-dimensional alternating symbol with $\varepsilon^{012 \cdots D-1} = +1$.

A familiar example of the field strength (3.8), Bianchi identity (3.9) and the field equation (3.10) is provided by Maxwell's equations in $D = 4$:

$$F_{MN} = \partial_M A_N - \partial_N A_M \tag{3.11}$$

$$\partial_{[M} F_{NP]} = 0 \tag{3.12}$$

$$\partial_{[M} {}^*F_{NP]} = {}^*J_{MNP} \tag{3.13}$$

The asymmetry between the equation for F and that for $^*F$ corresponds physically to the statement that there are no magnetic monopoles. If we want to restore the symmetry by introducing magnetic monopoles then we must give up equation (3.11) in favour of

$$F_{MN} = \partial_M A_N - \partial_N A_M + \omega_{MN} \tag{3.14}$$

so that

$$\partial_{[M} F_{NP]} = X_{MNP} \tag{3.15}$$

with

$$X_{MNP} = \partial_{[M} \omega_{NP]} \tag{3.16}$$

and hence

$$\partial_{[M} X_{NPQ]} \equiv 0 \tag{3.17}$$

If X is singular,

$$X_{123} = g\delta^3(y) \tag{3.18}$$

we speak of an "elementary monopole" in analogy with the elementary electric charge

$$^*J_{123} = e\delta^3(y) \tag{3.19}$$

Whereas if it is smeared out so as to be regular at the origin we speak of a "solitonic monopole". In both cases, we have

$$e \equiv \int_{S^2} {}^*F = \int_{M^3} {}^*J \tag{3.20}$$

$$g \equiv \int_{S^2} F = \int_{M^3} X \tag{3.21}$$

In this language, the Dirac monopole is elementary whereas the 't Hooft-Polyakov monopole is solitonic. The solitonic monopole is obtained by embedding Maxwell theory in a non-abelian gauge theory with Higgs scalars. It is the presence of the Higgs field term X_{MNP} that smears out the monopole and gives it a finite "size" ρ. From (3.21), however, as we shrink the size to zero we must find

$$\lim_{\rho \to 0} X_{123} = g\delta^3(y) \tag{3.22}$$

The electric charge is conserved by virtue of the field equations and hence corresponds to a Noether charge, whereas the magnetic charge is identically conserved and corresponds to a topological charge. According to the Dirac quantization rule

$$\frac{eg}{4\pi} = \frac{n}{2} \qquad n = \text{integer.} \tag{3.23}$$

The argument generalizes to arbitrary d and D [8]. The usual equations

$$F_{d+1} = dA_d \tag{3.24}$$
$$dF_{d+1} = 0 \tag{3.25}$$
$$d\,{}^*F_{D-d-1} = {}^*J_{D-d} \tag{3.26}$$

imply the presence of an "electric" charge i.e. a $(d-1)$ brane but no "magnetic" charge i.e. no $(D-d-3)$-brane. To restore the symmetry by introducing a $(D-d-3)$-brane we must modify (3.24) to

$$F_{d+1} = dA_d + \omega_{d+1} \tag{3.27}$$

so that the Bianchi identity (3.25) becomes

$$dF_{d+1} = X_{d+2} \tag{3.28}$$

with

$$X_{d+2} = d\omega_{d+1} \tag{3.29}$$

If X is a singular,

$$X_{123\cdots d+2} = g_{D-d-2}\delta^{d+2}(y) \tag{3.30}$$

we speak of an "elementary $(D-d-3)$-brane" whereas if it is smeared out so as to be regular at the origin we speak of a "solitonic $(D-d-3)$-brane". We then have

$$e_d = \int_{S^{D-d-1}} {}^*F_{D-d-1} = \int_{M^{D-d}} {}^*J_{D-d} \tag{3.31}$$

$$g_{D-d-2} = \int_{S^{d+1}} F_{d+1} = \int_{M^{d+2}} X_{d+2} \tag{3.32}$$

From (3.32) we see once again that the solitonic $(D-d-3)$-brane must be such that

$$\lim_{\rho \to 0} X_{12\cdots d+2} = g_{D-d-2}\delta^{d+2}(x) \tag{3.33}$$

The Dirac quantization rule is again

$$\frac{e_d g_{D-d-2}}{4\pi} = \frac{n}{2} \qquad n = \text{integer} \qquad (3.34)$$

Note that, e_d and g_{D-d-2} are not in general dimensionless but rather

$$[e_d] = -\frac{1}{2}(D - 2d - 2)$$
$$[g_{D-d-2}] = \frac{1}{2}(D - 2d - 2) \qquad (3.35)$$

They do become dimensionless when

$$D = 2(d+1) \qquad (3.36)$$

of which the point particle ($d = 1$) in $D = 4$ is the most familiar special case.

4. The elementary string and the elementary fivebrane

The elementary string solution of the three-form version of $D = 10$ supergravity found by Dabholkar et al [9] corresponds to supergravity coupled to the string σ-model. The action for the supergravity fields (g_{MN}, b_{MN}, ϕ) is given by

$$S(\text{string}) = \frac{1}{2\kappa^2} \int d^{10}x \sqrt{-g}\left(R - \frac{1}{2}(\partial\phi)^2 - \frac{1}{2 \cdot 3!} e^{-\phi} H^2\right) \qquad (4.1)$$

where $H = db$, and the string σ-model action is given

$$S_2 = T_2 \int d^2\xi \left(-\frac{1}{2}\sqrt{-\gamma}\,\gamma^{ij}\partial_i X^M \partial_j X^N g_{MN} e^{\phi/2} \right.$$
$$\left. -\frac{1}{2} \epsilon^{ij}\partial_i X^M \partial_j X^N b_{MN}\right) \qquad (4.2)$$

we have denoted the string tension by T_2. The ϕ dependence in (4.2) is fixed by requiring that under the rescalings

$$g_{MN} \to \lambda^{1/2} g_{MN}$$
$$b_{MN} \to \lambda^2 b_{MN} \qquad (4.3)$$
$$e^{\phi} \to \lambda^3 e^{\phi}$$

both $S(\text{string})$ and S_2 scale the same way:

$$S(\text{string}) \to \lambda^2 S(\text{string})$$
$$S_2 \to \lambda^2 S_2 \qquad (4.4)$$

If we make a two-eight split (2.1) and work in the static gauge (2.2), the solution found by Dabholkar et al [9] may be written

$$Y_m = \text{constant} \qquad (4.5)$$

with

$$e^{-2\phi} = e^{-2\phi_0}\left(1 + \frac{k}{r^6}\right)$$
$$g_{\mu\nu} = e^{3(\phi-\phi_0)/2}\eta_{\mu\nu}$$
$$g_{mn} = e^{-(\phi-\phi_0)/2}\delta_{mn} \tag{4.6}$$
$$H_{\mu\nu m} = -\varepsilon_{\mu\nu}\partial_m e^{2\phi-3\phi_0/2}$$

where ϕ_0 is the vev of the dilaton, $k = \frac{\kappa^2 T_2}{3\Omega_7} e^{\phi_0/2}$ and Ω_7 is the volume of the unit seven-sphere, corresponding to an elementary string with "electric" Noether charge

$$e_2 = \int_{S^7} e^{-\phi} \frac{*H}{\sqrt{2}\,\kappa} = \sqrt{2}\,\kappa T_2 \tag{4.7}$$

The elementary fivebrane solution of the seven-form version of $D = 10$ supergravity found by Duff and Lu [7] corresponds to supergravity coupled to the fivebrane σ-model. The action for the supergravity fields $(g_{MN}, a_{MNOPQR}, \phi)$ is given by

$$S(\text{fivebrane}) = \frac{1}{2\kappa^2}\int d^{10}x\,\sqrt{-g}\left(R - \frac{1}{2}(\partial\phi)^2 - \frac{1}{2\cdot 7!}e^{\phi}K^2\right) \tag{4.8}$$

where $K = da$, and the fivebrane σ-model action is given by

$$S_6 = T_6 \int d^6\xi\Bigg(-\frac{1}{2}\sqrt{-\gamma}\,\gamma^{ij}\partial_i X^M \partial_j X^N g_{MN} e^{-\phi/6} + 2\sqrt{-\gamma}$$
$$-\frac{1}{6!}\varepsilon^{ijklmn}\partial_i X^M \partial_j X^N \partial_k X^O \partial_l X^P \partial_m X^Q \partial_n X^R a_{MNOPQR}\Bigg) \tag{4.9}$$

We have denoted the fivebrane tension by T_6. The ϕ dependence in (4.9) is fixed by requiring that under the rescalings

$$g_{MN} \to \sigma^{3/2} g_{MN}$$
$$a_{MNOPQR} \to \sigma^6 a_{MNOPQR}$$
$$e^{\phi} \to \sigma^{-3} e^{\phi} \tag{4.10}$$
$$\gamma_{ij} \to \sigma^2 \gamma_{ij}$$

both $S(\text{fivebrane})$ and S_6 scale the same way

$$S(\text{fivebrane}) \to \sigma^6 S(\text{fivebrane})$$
$$S_6 \to \sigma^6 S_6 \tag{4.11}$$

If we make a six-four split (2.1) and work in the static gauge (2.2), the solution found by Duff and Lu [7] may be written

$$Y_m = \text{constant} \tag{4.12}$$

with

$$\begin{aligned} e^{2\phi} &= e^{2\phi_0}\left(1 + \frac{k}{r^2}\right) \\ g_{\mu\nu} &= e^{-(\phi-\phi_0)/2}\eta_{\mu\nu} \\ g_{mn} &= e^{3(\phi-\phi_0)/2}\delta_{mn} \\ K_{\mu\nu\rho\lambda\tau m} &= \varepsilon_{\mu\nu\rho\lambda\tau}\partial_m e^{-2\phi+3\phi_0/2} \end{aligned} \tag{4.13}$$

where $k = \frac{\kappa^2 T_6}{\Omega_3} e^{-\phi_0/2}$ and Ω_3 is the volume of the unit three-sphere, corresponding to an elementary fivebrane with "electric" Noether charge

$$e_6 = \int_{S^3} e^\phi \frac{*K}{\sqrt{2}\,\kappa} = \sqrt{2}\,\kappa T_6 \tag{4.14}$$

5. The solitonic fivebrane and solitonic string

Consider the equations for the 3-form H arising from the string actions (4.1) and (4.2).

$$H = db \tag{5.1}$$
$$dH = 0 \tag{5.2}$$
$$d(e^{-\phi} *H) = *J_8 \tag{5.3}$$

From the discussion of section 3, they imply the absence of a fivebrane. However, if we modify the Bianchi identity by

$$H = db + \omega_3 \tag{5.4}$$

so that

$$dH = X_4 \tag{5.5}$$

with

$$X_4 = d\omega_3 \tag{5.6}$$

Then this implies, in addition to the string with Noether charge

$$e_2 = \int_{S^7} e^{-\phi} \frac{{}^*H}{\sqrt{2}\,\kappa} = \int_{M^8} \frac{(^*J)_8}{\sqrt{2}\,\kappa} = \sqrt{2}\,\kappa T_2, \qquad (5.7)$$

a fivebrane with topological charge

$$g_6 = \int_{S^3} \frac{H}{\sqrt{2}\,\kappa} = \int_{M^4} \frac{X_4}{\sqrt{2}\,\kappa} = \frac{2n\pi}{\sqrt{2}\,\kappa T_2} \qquad (5.8)$$

obeying

$$\frac{e_2 g_6}{4\pi} = \frac{n}{2} \qquad (5.9)$$

But what is X_4? Just as 't Hooft and Polyakov found a solitonic monopole by embedding Maxwell theory in a larger theory, so Strominger [4] discovered the solitonic fivebrane by embedding the rank-two potential equations into the 2-form version of $D = 10$ supergravity-Yang-Mills theory. And just as the smearing out of the monopole singularity was provided by the Higgs fields, so the smearing out of the fivebrane singularity was provided by the Yang-Mills fields with

$$X_4 = \frac{1}{30\pi T_2} Tr F \wedge F \qquad (5.10)$$

where F is an hermitian matrix in the adjoint representation of either $SO(32)$ or $E_8 \times E_8$. The supergravity equations admit the solutions $F_{\mu\nu} = F_{\mu m} = 0$ where $\mu = 0, 1, \ldots, 5$ and $m = 6, 7, 8, 9$, with

$$F^{mn} = \frac{1}{2} \varepsilon^{mnpq} F_{pq} \qquad (5.11)$$

This is just the instanton solution [10] corresponding to an $SU(2)$ subgroup of $SO(32)$ or $E_8 \times E_8$ for which

$$\frac{1}{480\pi^2} \int_{M^4} Tr F \wedge F = \nu = \text{integer} \qquad (5.12)$$

and hence

$$g = \frac{16\pi\nu}{\sqrt{2}\,\kappa T_2} \qquad (5.13)$$

This is to be compared with (5.8). Hence we find that Strominger's solitonic fivebrane is such that

$$n = 8\nu \qquad (5.14)$$

Now consider the equations for the 7-form K arising from the fivebrane actions (4.8) and (4.9). Then the usual equations

$$K = da \tag{5.15}$$

$$dK = 0 \tag{5.16}$$

$$d(e^\phi *K) = *J_4 \tag{5.17}$$

imply the absence of a string. However, if we modify the Bianchi identity by

$$K = da + \omega_7 \tag{5.18}$$

so that

$$dH = X_8 \tag{5.19}$$

with

$$X_8 = d\omega_7 \tag{5.20}$$

then this implies, in addition to the fivebrane with Noether charge

$$e_6 = \int_{S^3} \frac{e^\phi *K}{\sqrt{2}\,\kappa} = \int_{M^4} \frac{*J_4}{\sqrt{2}\,\kappa} = \sqrt{2}\,\kappa T_6 \tag{5.21}$$

a string with topological charge

$$g_2 = \int_{S^7} \frac{K}{\sqrt{2}\,\kappa} = \int_{M^8} \frac{X_8}{\sqrt{2}\,\kappa} = \frac{2n\pi}{\sqrt{2\kappa T_6}} \tag{5.22}$$

obeying

$$\frac{e_6 g_2}{4\pi} = \frac{n}{2}. \tag{5.23}$$

Now what is X_8? Once again the smearing out of the singularity may be achieved by embedding the rank-six potential equations into the 6-form version of $D = 10$ supergravity-Yang-Mills theory. This time [5]

$$X_8 = \frac{1}{3(2\pi)^3 T_6}\left(\frac{1}{24} TrF_\wedge F_\wedge F_\wedge F - \frac{1}{7200} TrF_\wedge F_\wedge TrF_\wedge F\right) \tag{5.24}$$

The supergravity equations admit the solution $F_{\mu\nu} = F_{\mu m} = 0$ where $\mu = 0, 1$ and $m = 2, \ldots, 9$, with

$$t^{mnpqrstu} F_{pq} F_{rs} F_{tu} = \frac{1}{2} \varepsilon^{mnpqrstu} F_{pq} F_{rs} F_{tu} \qquad (5.25)$$

where $t^{mnpqrstu}$ is given by

$$\begin{aligned}
t^{mnpqrstu} = &-\frac{1}{2} \{(g^{mp}g^{nq} - g^{mq}g^{np})(g^{rt}g^{su} - g^{ru}g^{st}) \\
&+ (g^{pr}g^{qs} - g^{ps}g^{qr})(g^{tm}g^{un} - g^{tn}g^{um}) \\
&+ (g^{mr}g^{ns} - g^{ms}g^{nr})(g^{pt}g^{qu} - g^{pu}g^{qt})\} \\
&+ \frac{1}{2}(g^{np}g^{qr}g^{st}g^{um} + g^{nr}g^{ps}g^{qt}g^{um} \\
&+ g^{nr}g^{ts}g^{pu}g^{qm} + \text{permutations})
\end{aligned} \qquad (5.26)$$

Such a configuration is provided by the instanton solution [11] corresponding to an $SO(8)$ subgroup of $SO(32)$ or $E_8 \times E_8$ for which

$$\frac{T_6}{16\pi} \int_{M^8} X_8 = \nu = \text{integer for } SO(32) \qquad (5.27)$$

$$= 0 \qquad \text{for } E_8 \times E_8 \qquad (5.28)$$

In the case of $SO(32)$, therefore, we find

$$g = \frac{16\pi\nu}{\sqrt{2}\kappa T_6} \qquad (5.29)$$

This is to be compared with (5.22). Hence we find that the Duff-Lu string is such that

$$n = 8\nu \qquad (5.30)$$

also.

We complete this section by giving the formulae for the mass per unit area of the solitonic fivebrane, $\mathcal{M}_6$, and the mass per unit length of the solitonic string, $\mathcal{M}_2$. One finds [4,9]

$$\kappa \mathcal{M}_6 = \frac{1}{\sqrt{2}} g_6 \, e^{-\phi_0/2} \qquad (5.31)$$

$$\kappa \mathcal{M}_2 = \frac{1}{\sqrt{2}} g_2 \, e^{\phi_0/2} \qquad (5.32)$$

Both saturate a Bogonol'nyi bound [9] between the mass and the topological charge. As expected, the fivebrane gets heavier for weak string coupling and the string gets heavier for weak fivebrane coupling.

6. String/fivebrane duality

So far we have discovered that the string equations admit an elementary string solution and a solitonic fivebrane solution and, conversely, that the fivebrane equations admit an elementary fivebrane solution and a solitonic string solution. We now wish to go a step further and assert that the two theories are "dual". In its strongest sense, this means that the heterotic string and heterotic fivebrane are equivalent descriptions of the same physics, just as in quantum mechanics where an electron may be described either by a wave or a particle. In the present context where we consider the limiting field theories, however, we simply make the assumption that S(string) and S(fivebrane) are equivalent i.e. we assume that the metric g_{MN} and dilaton ϕ are the same and that the 7-form field strength K of the fivebrane is dual to the 3-form field strength H of the string. More precisely,

$$K = e^{-\phi} * H \tag{6.1}$$

so that the field equations of the string become the Bianchi identities of the fivebrane and vice-versa. This leads immediately to

$$\begin{aligned} e_2 &= g_2 \\ g_6 &= e_6 \end{aligned} \tag{6.2}$$

and hence, using (5.7), (5.8), (5.21) and (5.22)

$$\kappa^2 T_2 T_6 = n\pi \tag{6.3}$$

which is the result (1.2) quoted in the Introduction.

This assumption also leads to a relation between the dimensionless loop expansion parameters of string and fivebrane. To see this we note that the metrics appearing naturally in the string and fivebrane σ-models (4.2) and (4.9) are related to the canonical metric appearing in (4.1) and (4.8) by

$$g_{MN}(\text{string } \sigma-\text{model}) = e^{\phi/2} g_{MN}(\text{canonical}) \tag{6.4}$$

$$g_{MN}(\text{fivebrane } \sigma-\text{model}) = e^{-\phi/6} g_{MN}(\text{canonical}) \tag{6.5}$$

If we rewrite S(string) and S(fivebrane) in these variables, we find

$$S(\text{string}) = \frac{1}{2\kappa^2} \int d^{10}x \sqrt{-g}\, e^{-2\phi}\left(R + 4(\partial\phi)^2 - \frac{1}{2.3!}H^2\right) \tag{6.6}$$

and

$$S(\text{fivebrane}) = \frac{1}{2\kappa^2} \int d^{10}x \sqrt{-g}\, e^{2\phi/3} \left(R - \frac{1}{2.7!} K^2 \right) \qquad (6.7)$$

This reveals that the string loop coupling constant is given by

$$g(\text{string}) = e^{\phi_0} \qquad (6.8)$$

and the fivebrane loop coupling constant by

$$g(\text{fivebrane}) = e^{-\phi_0/3} \qquad (6.9)$$

Hence we recover the result (1.1) quoted in the Introduction

$$g(\text{fivebrane}) = \frac{1}{\sqrt[3]{g(\text{string})}} \qquad (6.10)$$

Thus strongly coupled strings correspond to weakly coupled fivebranes and vice-versa.

One mystery remains: the elementary string and fivebrane solutions were obtained from the supergravity equations with σ-model sources derived from S_2 and S_6, whereas the solitonic fivebrane and string solutions were obtained from the supergravity equations with Yang-Mills sources, which in canonical variables take the form

$$S(\text{quadratic Yang} - \text{Mills}) = -\int d^{10}x \sqrt{-g}\, \frac{1}{4\pi\kappa^2 T_2} e^{-\phi/2} t^{MNPQ} tr F_{MN} F_{PQ} \qquad (6.11)$$

in the case of the fivebrane, and

$$S(\text{quartic Yang} - \text{Mills}) =$$
$$\int d^{10}x \sqrt{-g}\, \frac{1}{48\kappa^2(2\pi)^3 T_6} e^{\phi/2} t^{IJKLMNOPQ} tr F_{IJ} F_{KL} F_{MN} F_{PQ} \qquad (6.12)$$

in the case of the string. (Here we restrict our attention to $SO(32)$ and F is now in the fundamental representation.) Yet, as discussed in section 3, consistency demands that these sources, denoted generically by J, must be such that

$$\lim_{\rho \to 0} J(\text{quadratic Yang} - \text{Mills}) = J(\text{fivebrane } \sigma - \text{model}) \qquad (6.13)$$

and

$$\lim_{\rho \to 0} J(\text{quartic Yang} - \text{Mills}) = J(\text{string } \sigma - \text{model}) \qquad (6.14)$$

We shall now show that this is indeed the case. First, however, we should spend a little time to justify the choice of Yang-Mills actions (6.11) and (6.12) and corresponding Chern-Simons terms (5.10) and (5.24).

The effective action for the string S(string) + S(quadratic Yang-Mills) and the corresponding quadratic Chern-Simons modification of the Bianchi identity (5.10) may be obtained from demanding conformal invariance of the quantum string (i.e vanishing β functions) or equivalently from scattering amplitude arguments. The result is just the well-known $D = 10$ supergravity couplings [12]. (Actually, there are also gravitational Chern-Simons corrections but these vanish for the solitonic heterotic fivebrane solution [4]). The justification for the effective action S(fivebrane) + S(quartic Yang-Mills) and the corresponding quartic Chern-Simons modification of the Bianchi identity (5.24) will necessarily be more indirect since, although the super fivebrane σ-model is well-known, the heterotic fivebrane σ-model has yet to be constructed. Even if we knew it, the quantization of fivebranes is still in its infancy, and it is doubtful that the fivebrane effective action could yet be derived as rigorously as the string effective action. The point of view we have already adopted is that the fivebrane action is obtained by dualizing the string action, in particular by interchanging field equations and Bianchi identities via $K = e^{-\phi} * H$. However, this process does not respect the loop expansion in the Yang-Mills sector and what is a tree-level effect in string perturbation theory may be a one-loop effect in five-brane perturbation theory, and vice versa. To understand this, we recall the relationships (6.8) and (6.9) between the string loop coupling constant g(string), the five-brane loop coupling constant g(five-brane), and the vacuum expectation value ϕ_0 of the dilaton: g(five-brane) $= (g(\text{string}))^{-1/3} = \exp(-\phi_0/3)$. In string variables each term in the string tree-level action S(string) is proportional to $\exp(-2\phi)$ which reveals that the string loop coupling constant is given by $\exp(\phi_0)$. Similarly, in five-brane variables each term in S(five-brane) is proportional to $\exp(2\phi_0/3)$. Thus the Green-Schwarz anomaly-cancellation term $b \wedge trF \wedge F \wedge F \wedge F$ and also S(quartic Yang-Mills) have no ϕ dependence in string variables and are therefore seen to be one loop in string perturbation theory. The explicit one-loop calculation of this quartic Yang-Mills action has been carried out by Ellis, Jetzer, and Mizrachi [13]. On the other hand, both these terms are *tree level* in five-brane perturbation theory, because they both behave like $\exp(2\phi/3)$ in five-brane variables and must therefore be included in the five-brane tree-level action. By the same token S(quadratic Yang-Mills) and the Chern-Simons term corresponding to $a \wedge trF \wedge F$ are one loop in five-brane perturbation theory since they are independent of ϕ when written in five-brane variables. We therefore omit them from the five-brane tree-level action.

In arriving at (6.12) and (5.24), we have also employed the equation $\kappa^2 T_2 T_6 = \pi$ which relates the two elementary tensions. Note that T_2 has dimension 2 and T_6 has dimension 6, which is another reason for expecting a quartic classical Yang-Mills action for the five-brane. This causes no problems with unitarity. We emphasize that the *exact* string and five-brane actions are equivalent; it is merely the division into "classical" plus "quantum" which is different in the two cases.

7. σ-model sources from Yang-Mills sources

We shall now return to the consistency conditions (6.13) and (6.14). The energy-momentum tensor of the five-brane action (4.9) with as yet undetermined tension $\tilde{T}_6$ is

$$T^{MN}(\text{five-brane}) = -\tilde{T}_6 \int d^6\xi \sqrt{-\gamma}\, \gamma^{ij} \partial_i X^M \partial_j X^N e^{-\phi/6} \frac{\delta^{10}(x-X)}{\sqrt{-g}} \quad (7.1)$$

where

$$\gamma_{ij} = \partial_i X^M \partial_j X^N g_{MN} e^{-\phi/6} \quad (7.2)$$

whereas the energy-momentum tensor of the quadratic Yang-Mills action is

$$T^{MN}(\text{quadratic Yang-Mills}) = \frac{e^{-\phi/2}}{\pi\kappa^2 T_2}\, tr(F^{MP} F^N{}_P - \frac{1}{4} g^{MN} F^{PQ} F_{PQ}) \quad (7.3)$$

On the face of it, they are very different. Under the following circumstances, however, they coincide. Let us pick the static gauge

$$X^\mu(\xi) = \xi^\mu, \qquad \mu = 0,\ldots 5 \quad (7.4)$$

and substitute the solution

$$Y^m = \text{constant}, \qquad m = 6,\ldots 9 \quad (7.5)$$

in (7.1) and the solution of the Yang-Mills equation

$$F_{\mu\nu} = 0, \qquad F_{\mu n} = 0, \qquad F_{mn} = 4D \text{ instanton} \quad (7.6)$$

in (7.3), then T^{mn} and $T^{\mu n}$ vanish for both the five-brane and Yang-Mills. As for $T^{\mu\nu}$, we have since $\gamma_{\mu\nu} = g_{\mu\nu} e^{-\phi/6}$ and $\sqrt{-\gamma} = \sqrt{-^6g}\, e^{-\phi/2}$

$$\sqrt{-g}\, T^{\mu\nu}(\text{five-brane}) = -\tilde{T}_6 \int d^6\, xi\sqrt{-\gamma}\, \gamma^{\mu\nu}\delta^{10}(x-X)e^{-\phi/6}$$

$$= -\tilde{T}_6 \int d^6\xi\, e^{-\phi/2} g^{\mu\nu}\sqrt{-^6g}\,\delta^{10}(x-X)$$

$$= -\tilde{T}_6 e^{-\phi/2} g^{\mu\nu} \delta^4(y)\sqrt{-^6g} \tag{7.7}$$

and

$$\sqrt{-g}\, T^{\mu\nu}(\text{quadratic Yang-Mills}) = -\frac{1}{4\pi\kappa^2 T_2}\, e^{-\phi/2} g^{\mu\nu}\sqrt{-^6g}\,\sqrt{^4g}\, tr F^{mn} F_{mn} \tag{7.8}$$

but from (5.12)

$$\lim_{\rho\to 0}\sqrt{^4g}\, tr F^{mn} F_{mn} = 32\pi^2 \delta^4(y) \tag{7.9}$$

where $\nu =$ instanton number $= 1$. Hence

$$\sqrt{-g}\, T^{\mu\nu}(\text{quadratic Yang-Mills}) = -\frac{8\pi}{\kappa^2 T_2}\, e^{-\phi/2} g^{\mu\nu}\delta^4(y)\sqrt{-^6g} \tag{7.10}$$

which agrees with $T^{\mu\nu}$ (five-brane) of (7.7) provided

$$\kappa^2 T_2 \tilde{T}_6 = 8\pi \tag{7.11}$$

This is just the Dirac quantization rule (6.3) with $n=8$. Thus as we shrink the instanton to zero, the solitonic fivebrane goes over into an elementary fivebrane with 8 times the fundamental tension.

Exactly the same thing happens as we consider the elementary string solution [9] and the solitonic string solution [5]. The energy-momentum tensor of string action (4.2) with as yet undetermined tension $\tilde{T}_2$ is

$$T^{MN}(\text{string}) = -\tilde{T}_2 \int d^2\xi\sqrt{-\gamma}\,\gamma^{ij}\partial_i X^M \partial_j X^N e^{\phi/2}\frac{\delta^{10}(x-X)}{\sqrt{-g}} \tag{7.12}$$

whereas the energy-momentum tensor of the quartic Yang-Mills action is

$$T^{MN}(\text{quartic Yang-Mills}) = -\frac{e^{\phi/2}}{24(2\pi)^3\kappa^2 T_6}\left[\frac{\delta t^{IJKLPQST}}{\delta g_{MN}} tr F_{IJ} F_{KL} F_{PQ} F_{ST}\right.$$
$$\left. -\frac{1}{2} g^{MN} t^{IJKLPQST} tr F_{IJ} F_{KL} F_{PQ} F_{ST}\right] \tag{7.13}$$

Again we pick the static gauge

$$X^\mu(\xi) = \xi^\mu \qquad \mu = 0, 1 \tag{7.14}$$

and substitute the solution

$$Y^m = \text{constant} \qquad m = 2, \ldots 9 \tag{7.15}$$

in (7.12) and the solution of the Yang-Mills equation

$$F_{\mu\nu} = 0, \qquad F_{\mu n} = 0, \qquad F_{mn} = 8D \text{ instanton} \tag{7.16}$$

in (7.13), then $T^{\mu m}$ and T^{mn} vanish for both the string and Yang-Mills. As for $T^{\mu\nu}$, we have since

$$\sqrt{-\gamma}\, \gamma^{\mu\nu} = \sqrt{-g}\, g^{\mu\nu} \tag{7.17}$$

$$\begin{aligned}\sqrt{-g}\, T^{\mu\nu}(\text{string}) &= -\tilde{T}_2 \int d^2\xi \sqrt{-g}\, g^{\mu\nu} e^{\phi/2} \delta^{10}(x - X) \\ &= -\tilde{T}_2 e^{\phi/2} \sqrt{-^2 g}\, g^{\mu\nu} \delta^8(y)\end{aligned} \tag{7.18}$$

and

$$\sqrt{-g}\, T^{\mu\nu}(\text{quartic Yang-Mills}) = \frac{e^{\phi/2}}{48(2\pi)^3 \kappa^2 T_6} \sqrt{-^2 g} \sqrt{^8 g}\, g^{\mu\nu} t^{ijklmnpq} tr F_{ij} F_{kl} F_{mn} F_{pq} \tag{7.19}$$

but from (5.27)

$$\lim_{\rho \to 0} \sqrt{^8 g}\, t^{ijklmnpq} tr F_{ij} F_{kl} F_{mn} F_{pq} = -8 \times 48\pi (2\pi)^3 \nu \delta^8(y) \tag{7.20}$$

where $\nu = $ instanton number $= 1$. Hence

$$\sqrt{-g}\, T^{\mu\nu}(\text{quartic Yang-Mills}) = -\frac{8\pi e^{\phi/2}}{\kappa^2 T_6} g^{\mu\nu} \sqrt{-^2 g}\, \delta^8(y) \tag{7.21}$$

which agrees with $T^{\mu\nu}$(string) of (7.18) provided once again

$$\kappa^2 \tilde{T}_2 T_6 = 8\pi \tag{7.22}$$

So as we shrink the instanton to zero, the solitonic string goes over into elementary string with 8 times the fundamental tension.

We have focussed our attention on the energy-momentum tensors, but the same consistency conditions (6.13) and (6.14) can be shown to hold for the sources appearing on the right hand side of the dilaton equations of motion and those of the antisymmetric tensors. In the latter case, the Yang-Mills Chern-Simons terms reduce, in the limit $\rho \to 0$, to the corresponding Wess-Zumino terms in the σ-model.

8. Recent developments and outstanding problems

In our string soliton solution [5] we employed the SO(8) 8-dimensional instanton of Ref. [11] since this solves our generalized self-duality constraint (5.25). Recently, Harvey and Strominger [14] have employed the SO(7) 8-dimensional instanton of Ref. [15]. This is sometimes called the "octonionic instanton" since it satisfies a self-duality equation similar to (5.11) but where ε^{mnpq} is replaced by the SO(7) invariant tensor c^{mnpq} related to the octononic structure constants. The role of this solution in the grand scheme of things is not at all clear (it breaks N=16 supersymmetry to N=1 as opposed to N=8). Because they employ the quadratic Yang-Mills action (6.11) and Chern-Simons term X_4 of (5.10), the Harvey-Strominger string has infinite mass per unit length in contrast to our solution which has finite mass per unit length given by (5.32) as a consequence of employing the quartic Yang-Mills action (6.12) and Chern-Simons term X_8 of (5.24).

Callan, Harvey and Strominger [16] have observed that our fivebrane solution (4.13), also solves the field equations of D=10 Type IIA and Type IIB supergravity and hence that there exist both Type IIA and Type IIB fivebranes in D=10. These were previously thought not to exist [17]. The no-go theorem is circumvented because the gauge-fixed d=6 worldvolume actions involve higher spin zero modes corresponding to d=6 (2,0) tensor supermultiplets $(B^-_{\mu\bar{\nu}}, \lambda^I, \phi^{IJ})$ and (1,1) vector supermultiplets $(B_\mu, \chi_i, A_i{}^j, \xi)$. They conjecture that these fivebranes are dual to Type IIA and Type IIB superstrings in D=10, in analogy with the heterotic duality discussed in this paper. One might object that the elementary solutions of section 4 are not "solitons" because of the δ-function singularities at the origin. However, when regarded as a fivebrane soliton, the Dabholkar et al string solution (4.6) is non-singular, and when regarded as a string soliton the Duff-Lu fivebrane solution (4.13) is non-singular too [16].

The scattering of these solitonic strings and solitonic fivebranes has recently been investigated by Callan and Khuri [18] who conclude that, at least in the case of strings where we already know the answer, the amplitudes coincide with those of fundamental strings.

Horowitz and Strominger [19] have shown that the solitonic super p-branes may be regarded as the extreme mass = charge "Reissner-Nordstrom" limit of a more general class of solutions with event horizons: "blackbranes".

Campbell et al [20] have speculated that the topological charge associated with the fivebrane may become deconfined above a critical temperature T_H analogous to quarks in QCD.

Kalara and Nanopoulos [21], invoking a strong/weak coupling symmetry conjecture of Font et al [22], have suggested that string/fivebrane duality may even solve those deep problems in the quantum mechanics of black holes associated with the loss of quantum coherence.

The heterotic string involves two kinds of loop expansion: quantum D=10 string loops (L) and classical d=2 σ-model loops (ℓ). Similarly, the heterotic fivebrane presumably requires quantum D=10 fivebrane loops (L') and classical d=6 σ-model loops (ℓ'). We have already seen in section 6 how one loop effects in strings (L=1) can correspond to tre- level effects in fivebranes (L'=0), and vice versa. What happens at higher L and L'? We have recently shown [23] that under string/fivebrane duality the roles of quantum loops and σ-model loops are reversed: $L' = \ell'$-1 and $L' = \ell$-1. Thus, in some sense, the theory may be regarded as purely classical.

All this rapid progress in supermembrane theory has highlighted several outstanding problems: (1) Can we be sure that solutions of the heterotic or Type II supergravities can be extended to exact string solutions? Some progress in this direction has recently been made by Callan, Harvey and Strominger [24] who make use of the exact conformal field theory of Antoniadis et al [25]. (2) Does the recent discovery of Type II p-branes now allow for *open* p-branes too? (3) The fact that $E_8 \times E_8$ differs from SO(32) in having no independent fourth order Casimir means that the topological charge of our string solution (5.28) vanishes for $E_8 \times E_8$. The corresponding string configuration is not smeared out by the Yang- Mills field but looks like the elementary string of section 4. What physical significance should be ascribed to this? (4) Despite the accumulating evidence for string/fivebrane duality, the quantization of fundamental fivebranes must still overcome many obstacles e.g. the non-renormalizability of the d=6 σ-model, the fact that p-branes may exhibit a continuous spectrum for p>1 and last, but not least, the fact that no algorithm exists for classifying all d=6 topologies.

Much, as they say, remains to be done.

Note added

We have recently constructed a self-dual Type IIB superthreebrane [26] which represents a new point ($d = 4, D = 10$) on the brane-scan. Earlier no-go theorems [17] are circumvented because there are spin 1 fields on the worldvolume. In fact, the gauge-fixed theory on the worldvolume is described by a $d = 4$ $N = 4$ Maxwell supermultiplet.

Acknowledgement

This work was carried out in collaboration with my graduate student Jian Xin Lu, a World Laboratory Scholar.

References

1. D. J. Gross, J. A. Harvey, E. Martinec and R. Rohm, Nucl. Phys. **B256**, 253 (1985).
2. M. J. Duff, E. Sezgin and C. N. Pope, *Supermembranes and Physics in 2+1 Dimensions* (World Scientific, 1990).
3. The existence of a heterotic fivebrane was conjectured on the basis of the dual formulations of $D = 10$ Supergravity in M. J. Duff, Class. Quantum Grav. **5**, 189 (1988) and M. J. Duff in *The Superworld II*, edited by A. Zichichi (Plenum, New York, 1990).
4. A. Strominger, Nucl. Phys. **B343**, 167 (1990).
5. M. J. Duff and J. X. Lu, Phys. Rev. Lett. **66**, 1402 (1991).
6. M. J. Duff and J. X. Lu, Nucl. Phys. **B354**, (1991) 129.
7. M. J. Duff and J. X. Lu, Nucl. Phys. **B354**, (1991) 141.
8. R. I. Nepomechie, Phys. Rev. **D31** (1984) 1921; C. Teitelboim, Phys. Lett. **B167**, (1986) 69.
9. A. Dabholkar, G. W. Gibbons, J. A. Harvey, and F. Ruiz-Ruiz, Nucl. Phys. **B340**, 33 (1990).
10. A. A. Belavin, A. M. Polyakov, A. S. Schwartz and Yu. S. Tyupkin, Phys. Lett. **59B**, 85 (1975).
11. B. Grossman, T. W. Kephart and J. D. Stasheff, Commun. Math. Phys. **96**, 431 (1984); D. H. Tchrakian, J. Math. Phys. **21**, 166 (1980).
12. A. H. Chamseddine, Nucl. Phys. **B185**, 403 (1981); E. Bergshoeff, M. de Roo, B. de Wit, and P. Van Nieuwenhuizen, Nucl. Phys. **B195**, 97 (1982); G. F. Chapline and N. S. Manton, Phys. Lett. **120B**, 105 (1983).
13. J. Ellis, P. Jetzer and L. Mizrachi, Nucl. Phys. **B303**, 1 (1988).
14. J. A. Harvey and A. Strominger, Phys. Rev. Lett. **66**, 549 (1991).
15. D. B. Fairlie and J. Nuyts, J. Phys. **A17**, 2807 (1984); S. Fubini and H. Nicolai, Phys. Lett. **155B**, 369 (1985).
16. C. G. Callan, J. A. Harvey and A. Strominger, PUPT-1244, EFL-91-12 (1991).
17. A. Achucarro, J. M. Evans, P. K. Townsend and D. L. Wiltshire, Phys. Lett. **B198**, 441 (1987).
18. C. G. Callan and R. R. Khuri, PUPT-1218 (1990).
19. G. Horowitz and A. Strominger, UCSBTH-91-06.
20. B. Campbell, J. Ellis, S. Kalara, D. V. Nanopoulos and K. Olive, CERN-TH-5833/90.

21. S. Kalara and D. V. Nanopoulos, CTP-TAMU-14/91.
22. A. Font, I. Ibanez, D. Lüst and F. Quevado, CERN-TH-5790/90.
23. M. J. Duff and J. X. Lu, Nucl. Phys. **B357**, 534 (1991).
24. G. Callan, J. A. Harvey ad A. Strominger PUPT-1233 (1991).
25. I. Antoniadis, C. Bachas, J. Ellis and D. V. Nanopoulos, Phys. Lett. **B211**, 393 (1988) and Nucl. Phys. **B328**, 117 (1989).
26. M. J. Duff and J. X. Lu, CTP-TAMU-29/91, to appear in Phys. Lett. B.

CHAIRMAN: M. Duff

Scientific Secretaries: U. Danielsson, M. Johnson, S. Lau, M. Potters

DISCUSSION I

– *Sanchez*:

My first question concerns an important issue raised in your discussion this morning. What about the quantization of these 5-branes, or could you say something about the quantization of extended objects in general?

– *Duff*:

Okay, let me go back to this brane scan (see figure in lecture). You remember I said that at the classical level, at least, supersymmetry allowed for these 12 types of extended objects (I will talk about D=10 threebranes tomorrow) . In particular we recover as a special case a well known result that the Green-Schwarz superstring, which corresponds to p=1, can exist in 3,4,6 and 10 dimensions. I should mention, since it cropped up in John Ellis's discussion, that when I talk about a ten dimensional string, I really mean a string which has the corresponding central charge. It could equally well mean a four dimensional string with all the right kinds of internal symmetries. Now it's known that of the four superstrings that exist classically only one of them is quantum mechanically consistent, and that's the 10 dimensional one. This is so because the others have anomalies which render the theory inconsistent. That's what singles out ten as the critical dimension of the string.

So the natural question then is: "which, if any, of these other extended objects survives quantum consistency?" Now unfortunately that's a much more difficult question to answer for several reasons. The first one is that you can always select a gauge in the case of the string which renders the equations of motion linear, and then you can do the usual mode expansion. Then away you go with your string calculations. For all p>1, there is no clever choice of gauge that renders the equations of motion linear, and so you are stuck with a difficult nonlinear differential equation. The best that you can do so far is some linearized semiclassical approximation, so you have to take those limitations into account. However, you can say a few things, and, in fact, this problem was addressed by Bars, Pope, and Sezgin a year or two ago. You know in the case of strings that if you go into the lightcone gauge, where the theory is manifestly unitary, then the problem of anomalies manifests itself as a lack of Lorentz invariance. So you find that these three strings (D=3,4,6) are not Lorentz invariant, but this one is (D=10). They did a similar thing for p-branes: they went to a lightcone gauge, and they asked about Lorentz invariance. What they found was that not only these three strings, but all the points on the three sequences (labelled by $\mathcal{H}, \mathcal{C}$, and $\mathcal{R}$) were quantum mechanically inconsistent. Also they showed that (D=10, p=1) and (D=11, p=2) were quantum mechanically consistent. On the basis of that analysis, you might conclude that the only candidate for a consistent quantum theory of gravity is the one we already know (and possibly the eleven dimensional super membrane). In particular, it seems to exclude the 5-brane, which is the subject of my lectures.

However, the 5-brane that they were analyzing then was the non-heterotic 5-brane: it didn't have any internal symmetry degrees of freedom. So it's not surprising that they found that it was quantum mechanically inconsistent. If you did the same thing for the superstring, and you didn't include the Yang-Mills field, you would also discover that it was quantum mechanically inconsistent. What should be done now is to repeat their analysis, this time putting in the heterotic 5-brane. Although we know from the soliton analysis (which I'll discuss tomorrow) that the heterotic 5-brane exists, no one has yet succeeded in writing down the Lagrangian for it. As a result we cannot do this kind of consistency check and establish that it is as quantum mechanically consistent as the string. Now all the indications are that it is, and there are two arguments for that. Number one, the original motivation for postulating the thing in the first place was the existence of the two anomaly-free formulations of supergravity. The conjecture is that just as one corresponds to the string, so the other corresponds to the 5-brane. Number two is simply that it arises as a soliton in a quantum mechanically consistent theory, namely, the heterotic string. For all these reasons I believe that the heterotic 5-brane (provided, of course, that the gauge group is $E_8 \times E_8$ or $SO(32)$) will be quantum mechanically consistent. Now it has recently been discovered by Callan, Harvey, and Strominger that in addition to heterotic 5-branes there are so-called type II 5-branes. They claim that the type II 5-brane is dual to the type II superstring in just the same way the heterotic 5-brane is conjectured to be dual to the heterotic string. It may be that the kind of consistency check that we need to do would be much simpler in the type II case, so I think that this is what needs to be done first. The encouraging thing about the type II 5-brane is that it has 8+8 bose/fermi degrees of freedom, which made the $\mathcal{O}$-sequence consistent in the Bars, Pope, Sezgin analysis. The $\mathcal{O}$-sequence has 8+8 degrees of freedom, the $\mathcal{H}$-sequence has 4+4, the $\mathcal{C}$-sequence has 2+2, and the $\mathcal{R}$-sequence has 1+1. As it turns out, you need the 8+8 to overcome the anomaly problem. This is yet another encouraging sign that the type II 5-brane has the right degrees of freedom for consistency.

If you allow me to extrapolate from all this circumstantial evidence, my conclusion would be that for every consistent ten-dimensional closed superstring (type IIA, type IIB, and heterotic) there is also a consistent ten-dimensional closed 5-brane (again type IIA, type IIB, and heterotic). This, then, summarizes what is known about the quantum mechanical consistency.

– *Sanchez:*

Thank you, but I am not happy with these comments about quantization. All your comments are based on the philosophy of considering p-branes from the point of view of an effective field theory, because you are comparing with what is known about superstring theory. Let us take another point of view. In string theory we know how to deal with quantization. For instance we have answers to concrete questions such as: "what does the mass spectrum look like?" However, for the membrane case will the spectrum be discrete or continuous, and which kinds of massless particles are there? Also how is the critical dimension derived in the membrane case. In string theory we have a guide, even if it's not completely proved, and that is the conformal invariance of these two dimensional sigma models. These types of issues should be addressed from the point of view of, say, canonical quantization or path integral quantization. Your picture is a way to approach the problem, and it's interesting. However....

– *Duff:*

Well you have raised several issues so let me try to deal with each of them. The first point you made, and I agree with you that it's important, is that what I said this morning was at the level of field theory. You could ask whether these soliton solutions which were first obtained by looking at the field theory limit are really exact solutions of string theory. There has been some recent progress on precisely that question by Callan, Harvey, and Strominger who have now constructed exact conformal field theories corresponding to the 5-brane solitons, so that should make you happy.

Two related issues you raised, equally important as the one above, were: (a) do they have massless particles in their spectrum, which they must of course if they are to be of physical interest; and (b) is the spectrum discrete or continuous? The first question has been answered affirmatively, by the same group Bars, Pope, and Sezgin using the same semiclassical methods. The more worrisome issue is the second one. There is a qualitative difference between an extended object with two or more spatial dimensions as compared with the string. That is if you write the Hamiltonian in, say, the lightcone gauge, you can see that the energy of the extended object is given by its area. Now if you pull a small piece of string out of an existing string you change its length, and so you cannot perform that operation without expending energy. On the other hand, if you take a membrane you can pull a string-like hair out of it without changing its area. Therefore no energy is expended in that operation. Another way of stating this is that there are valleys or flat directions in the potential in the case of membranes and above, which are not present in the case of strings. Now this has been analysed by de Wit and Nicolai who asked whether this implies a discrete or continuous spectrum. Once again you can't do the calculation in the full-blown theory because it's too complicated.

So they looked at a toy model, which they claimed was a good approximation to the full theory. In that toy model they found, curiously enough, that in the bosonic membrane this degeneracy is lifted by vacuum self energy effects and you get a discrete spectrum. In the supersymmetric membrane supersymmetry does, for once, something bad, and the vacuum effects between the bosons and fermions cancel, so you recover the continuous spectrum. On the basis of that they were led to say that p-branes with p>1 have continuous spectra.

What reaction do we have to that? Since theirs was a toy model, do the results carry over to the full theory? Also, if these are to be realistic models, supersymmetry must be broken, and perhaps the discreteness is restored in this breaking. A third possibility was pointed out to me several weeks ago by Halpern at the Stony Brook string conference. He said to me: "Do these 5-branes which you claim are dual to strings, have a discrete or a continuous spectrum?" I said: "Well I don't know, but if they are dual to strings I would guess they have a discrete spectrum since strings have a discrete spectrum." He replied: "Oh no, if they are dual to strings I would expect them to have a continuous spectrum." I asked him to explain. His logic is worth thinking about. The statement in string theory that you have this infinite tower of stable massive particles is, of course, just a tree level statement. When unitarity corrections are taken into account, the massive particles become unstable and acquire a nonzero width. Now imagine the situation in a string theory where you allow the string coupling to go to infinity. According to him that would manifest itself as a continuous spectrum. Since in this duality picture a strongly coupled string is a weakly coupled 5-brane, if Halpern is right a weakly coupled 5-brane should have a continuous spectrum. That is about all I can say on the subject.

– Sanchez:

Another point which arises from what you have said is whether this effective action for this 5-brane can be constructed from some beta function or scattering amplitude arguments.

– Duff:

That is also a pertinent question. For the case of the string we know how to construct the low energy effective lagrangian in the way you've alluded to. For the heterotic 5-brane no one has yet succeeded in writing down the σ-model action. Point number one. Point number two is, as I mentioned in my talk, that the quantization of 5-branes is still in its infancy. Therefore, we cannot derive the 5-brane effective Lagrangian in the same direct way as for the string.

The point of view I am adopting is that the field theory limit of the 5-brane is obtained by dualizing the field theory limit of the string. Everything I say will be based on that assumption. It leads to what look like consistent conclusions in the framework of this duality picture. I should mention that you would expect that the Lagrangian obtained by dualizing the string Lagrangian would give you the same physics. However, although there is a one-to-one correspondence, this correspondence does not respect the loop expansion and it can mix up different loop orders. Tomorrow we'll see that what is a tree level effect for the string can be a one-loop effect for the 5-brane and vice-versa. Simply by keeping track of powers in our loop expansion parameters when dualizing, some interesting and curious results arise.

– *Sanchez:*

We know that the d=2 sigma model in string theory is renormalizable, whereas a d=6 sigma model for your 5-brane is not conformally invariant and we don't know how to manage its quantum properties.

– *Duff:*

Right. You've managed in your one question to highlight the three outstanding problems in 5-brane theory. Well, actually two. I mentioned the third one this morning: the inability to classify six dimensional topologies. But the other two that I think are the most pressing are the continuous spectrum problem and the apparent nonrenormalizability of the six dimensional sigma model. I'm not sure that the apparent nonrenormalizability is a genuine problem or a pseudo-problem. After all, even in string theory, if you include the massive modes in your sigma model, your sigma model is nonrenormalizable. Nevertheless, string theory makes sense, so maybe we shouldn't be too worried about the nonrenormalizability of the sigma model. And in fact, my guess as to what takes the place of conformal invariance in the case of the string is finiteness. Namely, you write down a six dimensional sigma model with background fields and you find that it's a nonrenormalizable field theory. However, for very specific choices of the background you will get a finite result, which is another way of saying what happens with the string.

Once again, these are statements that I made the first time I talked about membranes in Erice. But I'm much more confident now than I was then, because in the meantime we've discovered that the 5-brane is a soliton of the string and the string is a consistent quantum theory. Somehow, and we've yet to discover how in detail, all these problems of 5-branes must be taken care of, since they're part and parcel of a consistent string theory. Let me reiterate. This discovery of the solitons means that 5-branes are there, whether you like them or not. If you buy strings, and of course you're free not to, you have to buy 5-branes in the same package.

– Danielsson:

You mentioned the problems in the loop expansion and the classification of six dimensional manifolds. Six dimensional manifolds presumably have many kinds of topological invariants, an unknown number I presume. Shouldn't each of these invariants be associated with some coupling constant as with the string coupling constant associated with the genus? What does this mean for the case of the 5-brane and its duality with the string?

– Duff:

Nobody knows, is the short answer, but I can give you a half-baked response. As you've said, for strings, where we have a two dimensional world sheet, the loop expansion looks something like this:

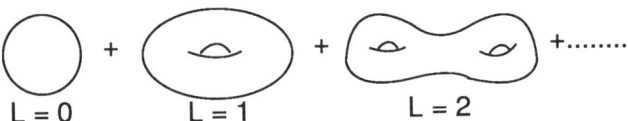

The number of holes has a topological characterization in terms of the Euler number $\chi_2 = \int d^2\xi \sqrt{-\gamma}\, R^{(2)}$, and the Euler number of the world sheet is related to the number of loops by:

$$\chi_2 = 2(1 - L)$$

This is consistent because the Euler number is the alternating sum of Betti numbers. (The pth Betti being the number of closed surfaces of dimension p that cannot be shrunk to a point) $\chi_2 = b_0 - b_1 + b_2$. And there's a theorem that says that the first is equal to the last, and that $b_0 = b_2 = 1$, and so $\chi_2 = 2 - b_1$ and the Euler number is bounded above by 2, which is consistent with the loop expansion. So everything is well understood for strings. However, for 5-branes everything is very far from being well understood because of the classification problem in higher dimensions. It's not even clear what summing over topologies means, and it's even less clear what that has to do with the loop expansion. Now let us speculate a bit. What topological number could represent the number of loops? I don't know, but let's use the Euler number again. A different way of looking at it in strings is that you can have a coupling between the two dimensional curvature and the dilaton.

$$\int d^2\xi \sqrt{-\gamma} R^{(2)} \phi$$

If the dilaton is given a nonvanishing constant vacuum expectation value then this is just the Euler number times the expectation value. This is so since the Euler number is the integral of the two dimensional curvature, and so if you plug this result into your partition function, then you can justify this relationship between the

Euler number and the loop expansion. In terms of the dilaton vacuum expectation value the string coupling constant is given by:

$$g(\text{STRING}) = e^{\phi_0}$$

Now in the case of the 5-brane the coupling constant is:

$$g(5-\text{BRANE}) = e^{-\phi_0/3}$$

and so you might be tempted to write down a coupling of the dilaton to a six dimensional Euler density, instead of a two dimensional Euler density. The six dimension Euler density is a complicated expression containing three Riemann tensors. So we might have the following expression:

$$-\frac{1}{3} \int d^6\xi (d = 6 \text{ Euler density})\phi$$

The -1/3 takes into account the appropriate 5-brane expansion parameter. Perhaps you could in that way justify the relation between Euler number and loop expansion for the 5-brane in the same way that you did for the string. There is a problem with this guess, however. Once again the Euler number corresponds to an alternating sum of Betti numbers.

$$\chi_6 = b_0 - b_1 + b_2 - b_3 + b_4 - b_5 + b_6$$
$$= 2 - 2b_1 + 2b_2 - b_3 = 2(1 - L)$$

But whereas χ_2 was bounded above by two, here the number of 5-brane loops can go negative. This seems like a crazy argument. Alternatively, perhaps you should only sum over topologies for which this formula makes sense. I tend to favour this last proposition, since we have already established that it is impossible to sum over all six dimensional topologies. Perhaps the requirement that the string and 5-brane yield the same physics will provide the answer to which subset of six dimensional topologies should be summed over. Of course this is all conjecture.

– Hasan:

If I understand correctly you want the 5-branes to be solitons because you want to introduce a particle nature in the theory. I want to know if the 5-branes satisfy the following conditions to be a soliton: (1) they preserve their shape while travelling in the medium, (2) their amplitude is a function of the velocity, and (3) if two solitons interact they preserve their shapes but with a change of phase.

– Duff:

Irrespective of what I want, the 5-branes are solitons and they are not particle like, they are extended objects. The particle interpretation would come about from the vibrational modes of the 5-brane in the same way that it arises in strings.

Now with regard to your second question as to whether I mean solitons in the old fashioned sense, like the solitary wave solutions to the sine-Gordon equation, the answer is no. I am using the term in its looser definition - as a classical lump of energy which is stabilized by a topological conservation law, just as a magnetic monopole is a "soliton" of a GUT theory. This clearly does not satisfy your criteria.

– *Hasan:*

Solitons are the solutions of nonlinear PDE's which contain two important terms: (1) a nonlinear term, and (2) a dispersive term. These terms compensate to make the solutions solitary. If your 5-brane is travelling in vacuum, how do you get the dispersive term?

– *Duff:*

I'm not quite sure what you mean by the dispersive term. You are right in asserting that they are nonlinear PDE's. It turns out that these solitons satisfy what's called a no-force condition. This means that from single soliton solutions you can get multi-soliton solutions by linear superposition. This is strange, because you have nonlinear equations. The no force condition implies that if you have two 5-branes with the same orientation, the mutual gravitational attraction that you get from the exchange of the graviton and the dilaton is exactly cancelled by a repulsive term from the antisymmetric tensor. That's why you can superimpose many of these 5-branes and still have a stable configuration.

– *Potters:*

In your answer to the first question you said that the 10 dimensional superstring (not heterotic) is not quantum consistent. Could you elaborate on this?

– *Duff:*

In fact the only quantum consistent D=10 superstrings are the heterotic strings with gauge group $E_8 \times E_8$ or $SO(32)$ and the type IIA and type IIB superstrings. (These are closed; there is also the type I which is open and also needs $SO(32)$). With any other gauge group - in particular no gauge group - the heterotic theory is inconsistent.

– *Potters:*

What do you mean by type II?

– *Duff:*

The type IIA string has N=2 spacetime supersymmetry, whereas the heterotic has N=1. The division into type IIA and type IIB is that in type IIA the two fermions have opposite chirality (so the theory is nonchiral) and in type IIB they have the same chirality (so the theory is chiral).

– *Peccei:*

You've more or less answered my question which concerned this de Wit-Nicolai proof of the continuous spectrum. I hadn't realize that they'd worked with a toy model.

– *Duff:*

Perhaps "toy" is too derisory a word. Let me be more precise. What they did was this. There is an old result due to Goldstone and his student Hoppe. In the case of 2+1, if you write the theory in the lightcone gauge, the Hamiltonian is that of a Yang-Mills SU(N) quantum mechanical model in the limit N goes to infinity. That was the Goldstone-Hoppe result. They were looking at the 2+1 membrane, and nobody knows what happens for more than 2+1. That's another caveat that I should make. What de Wit and Nicolai did was look at the supersymmetric Yang-Mills SU(N) quantum mechanical model for finite N. They established that it had a continuous spectrum, and then took the limit of N going to infinity. That's a very dubious limit because you are going from a finite number of degrees of freedom to an infinite number of degrees of freedom. Anyway, that's what they did.

– *Lisi:*

We have seen this morning that the duality imposes a "seesaw" relation between the coupling constants for strings and the 5-branes. Is there a similar relation between the mass spectra of the string and 5-brane?

– *Duff:*

That's a very good question, but I don't know the answer. Their massless spectra coincide, and that is basis of the claim that one is just the dual version of the other. The massless spectra are just those of the corresponding supergravity theory. Now what about the massive spectrum? I'm not sure what duality would imply for the massive spectrum. I don't know.

– *Sivaram:*

When you wrote down the action for the p-brane in general and for the string, you had a term in the action for the p-brane which was absent in the action for the string. It was a d-2 term which looks like a cosmological constant. Can you clarify as to why it is absent in the string case?

– *Duff:*

Let me say first of all that in the Dirac-Nambu-Goto version of the theory, the action is the same for whatever value of d. It is just the area that the object sweeps out. The difference arises when you want to write the action in the Polyakov or Howe-Tucker form which involves this auxiliary metric γ_{ij}. Then there is a difference between d=2 and d≠2, which is in fact related to conformal invariance. For the string this auxiliary metric introduces a symmetry which is not present in the Nambu-Goto action: the Weyl invariance, in which this metric is scaled by an arbitrary function of ξ. That's absent for d>2. One way to understand that is to look at the equations for arbitrary d, and note that the equation obtained by varying with respect to γ_{ij} (which is the vanishing of the energy momentum tensor) goes like this. If I take the trace of this equation and rearrange it I then get two answers according to whether d=2 or not. In the case when it's not equal to two, I simply learn that γ_{ij} is the induced metric on the world volume, which is the quantity which appears here. If I substitute this back in, I get back to where I started. For the d=2 case, all I can derive is this condition up to an arbitrary conformal factor, which reflects the conformal invariance of the theory. Nevertheless, if I substitute it back in here I regain the Nambu action again. So the purpose of this world volume cosmological term, is to ensure that I get back to the Nambu-Goto version of the theory when I eliminate γ_{ij}.

– *Sivaram*:

Since the couplings of the string and the 5-brane are inversely proportional to the cubed root, does this imply that the tensions are similarly related?

– *Duff*:

Good question. As far as the loop coupling constants are concerned we have the following relation:
$$g_2^2 = \frac{1}{g_6^6}$$
(For an arbitrary (d-1)-brane this generalizes to:
$$g_d^d = \frac{1}{g_{D-d-2}^{D-d-2}}$$

The reason I draw this to your attention is that in the self dual case where d=D-d-2, then g=1. This means you presumably can't trust perturbation theory. For D=10 this is the case for the 3-brane. A 3-brane is dual to itself in D=10). Let me now answer your question. You find that the tensions obey a Dirac quantization rule:

$$\kappa^2 T_2 T_6 = n\pi \qquad n = \text{integer}$$

κ^2 is the D=10 Newton's constant. (More about that tomorrow.)

– *Sivaram:*

Is there an analog of the string duality symmetry, i.e. R goes to 1/R? Is there something similar here?

– *Duff:*

Yes there is, but it's complicated. I can give you the reference to a paper I've written with my graduate student on the topic: M.J. Duff and J.X. Lu, Nucl. Phys. B47 (1990) 394.

– *Sivaram:*

And another point, please. It has been conjectured that a black hole in its final stage of evaporation could make a transition to a superstring. It would now look as if the transition to a 5-brane could prevent a final singularity. Could you comment on this?

– *Duff:*

As a matter of fact there is a recent paper by Horowitz and Strominger on black p-branes. They solve the supergravity equations and find solutions with event horizons. But these are generalizations of a black hole. Instead of having a two sphere surrounding a point-like black object, you can have black string with an event horizon around it. You can have a black membrane or, indeed, a black 5-brane with event horizons. These are two parameter solutions, reminiscent of the Reissner-Nördstrom black hole solutions that carry both mass and charge. Just as in the Reissner-Nördstrom solution, there is an extreme limit you can take when the mass is equal to the charge. So these black-branes, for lack of a better word, have such a limit. In that limit they become precisely the super p-branes that I wrote down on the brane scan. So the super p-branes can be considered as the endpoint of a black-brane collapse.

– *Sanchez:*

Before I go on to my last question, I would like to comment on your answer regarding black p-branes. Again, those solutions refer to the effective field theory limit of the theory.

– *Duff:*

The supersymmetric ones are exact. I agree with you that the solutions with the event horizon are still within the framework of the effective theory. The inability to find an exact solution applies equally well to the black hole of course, not just to the black p-brane.

– *Sanchez:*

Anyway, my last question is much more precise. Let us adopt your point of view about using duality as a tool for understanding. However, I am still unhappy about the lack of a proof of duality. Also let us adopt your point of view about solitons, which is that the 't Hooft-Polyakov monopole is obtained by the embedding of Maxwell's theory in a nonabelian gauge theory with Higgs' fields. Here you are going in the same direction, you find solitonic solutions by embedding or adding Yang-Mills type sources.

– *Duff:*

That's precisely the topic of my first three transparencies tomorrow, so perhaps we can return to it then.

– *Spanos:*

There are strings modelled in more than four spacetime dimensions where boundary conditions are considered. Are there any similar models in the case of membranes.

– *Duff:*

Yes, that's a good question. My feeling is that there must be, but no one has done that yet. In the language of compactification, what's lacking is the Frenkel-Kac mechanism that you have in the string. Thinking in terms of the free fermionic formulations, no one's done that for membranes either. In fact the ability to go between bosons and fermions on the world sheet does not apply for d>2, so it's not even clear what you mean by the fermionic formulation. These are all very good open questions.

– *Junk:*

On heterotic strings left and right travelling excitations decouple. For heterotic 5-brane we have more directions for excitations to propogate, so they may not decouple. How is "heteroticity"' manifested for the 5-branes?

– *Duff:*

Some of the properties of the heterotic string are shared by the heterotic 5-brane, and others which are peculiar to the two dimensionality of the world sheet cannot be shared by the 5-brane. If fact the ability to split your modes into left movers, which depend on (x+t), and right movers, which depend on (x-t), has no analogue in d>2.

The features you do have in common relate to the fact that there are certain properties of the fermions from supersymmetry that are common to all dimensions 2 modulo 4. 2 and 6 are examples. You can define left-handed and right-handed supersymmetry generators, and in 2 mod 4 dimensions you can have unequal numbers of them i.e. you can have (p,q) supersymmetry with p not equal to q. The heterotic string is an example of that. The same applies to the 6 dimensional world sheet of the heterotic 5-brane, so that feature is common to both. The $E_8 \times E_8$ or $SO(32)$ Yang-Mills internal symmetry is also common to both. So that's the reason for using the word heterotic, even though -as you point out- the concept of left and right movers has no analog.

CHAIRMAN: M. Duff

Scientific Secretaries: U. Danielsson, M.Johnson, S. Lau, M. Potters

DISCUSSION II

– *Sanchez:*

What about the compactification of 5-branes from ten dimensions to four dimensions?

– *Duff:*

In the Calabi-Yau approach you could look at the low energy Lagrangian and see if the same solution that works for strings works for 5-branes. There is however an unresolved problem. The relevant Lagrangian is the Ellis-Jetzer-Mizrachi Lagrangian which is quartic in the field strength. At first sight the Calabi-Yau does not solve the corresponding field equations. There are two possibilities: either 5-branes rule out Calabi-Yau or you need to modify the solution. (In string terms you can look at the one-loop Lagrangian and see if the same compactification goes through). We are currently looking into this.

– *Jain:*

Could you explain how string membrane duality restricts terms in the effective Lagrangian?

– *Duff:*

Take for example the cosmological constant. For the string the relevant term in the Lagrangian is given by:

$$\sqrt{-g(\text{string})}e^{2\phi(1-L)}$$

where $g_{MN}(\text{string}) = e^{\phi/2}g_{MN}$ and L is the number of string loops. For the 5-brane we have:

$$\sqrt{-g(\text{fivebrane})}e^{-2\phi(1-L')/3}$$

where $g_{MN}(\text{fivebrane}) = e^{-\phi/6}g_{MN}$ and L' is the number of fivebrane loops. Using string fivebrane duality these should be equal for some non-negative integers L and L'. But no such L and L' exist so there is no cosmological term. In the case of the Einstein term these expressions are replaced by:

$$\sqrt{-g(\text{string})}e^{-2\phi(1-L)}R(\text{string})$$

and

$$\sqrt{-g(\text{fivebrane})}e^{-2\phi(1-L')/3}R(\text{fivebrane})$$

There is now a unique solution, namely $L = L' = 0$. In other words the Einstein term is allowed only at tree level and is not renormalized by higher loops. Furthermore, there are higher powers of the Reimann tensor which you might think are allowed at higher loops but are in fact ruled out. It would be interesting if someone could do the two-loop calculations to see if the terms we claim are ruled out are indeed ruled out.

Is this the answer to the problem of the cosmological constant? At first sight it looks better than any previous attempt because at no stage I have explicitly used supersymmetry of the vacuum. As you all know if you insist on supersymmetry of the vacuum the cosmological constant is zero. The hard part is to explain why it apparently remains zero after you break supersymmetry. My feeling is that although this argument did not explicitly use supersymmetry of the vacuum it is implicitly doing so. Because if supersymmetry breaks it will be a non-perturbative phenomenon. So it is not something you should see in the loop expansion. If you ask if string/fivebrane duality allows mass terms for the gaugino and the gravitino the answer is no by the same argument. We already know that the low energy string Lagrangian has no cosmological term in the absence of supersymmetry breaking. So our results can be seen as a consistency check.

– Kretschmer:

Are there any attempts to generalize the ideas of p-branes to manifolds with non-integer dimensions?

– Duff:

No, not as far as I know.

– *Lau:*

For just the classical version of four 5-brane theory, can you isolate the physics degrees of freedom? In other words can the constraints be solved and the reduced phase-space which admits a sympletic structure be constructed?

– *Duff:*

No, you cannot solve the constraints. For instance if you look at the lightcone formulation you get horrible equations. So, although you can formally set up the constraints you cannot solve them exactly as in string theory.

– *Hasan:*

What kind of low energy physics does your theory predict?

– *Duff:*

We are trying to show that the 5-brane and the string are different formulations of the same theory. Hence the predictions are the same. The problem that superstring phenomenologists face is that you cannot in perturbation theory answer the important questions like how strings break supersymmetry and why they should choose a vacuum state with $SU(3) \times SU(2) \times U(1)$ symmetry and the right fermion families out of all the billions of possibilities. These really basic questions cannot be answered in string theory at weak coupling. So the hope is by studying weakly coupled fivebranes we can get a window on the strong coupling regime of string theory. I think this has to be done if we are to answer these basic phenomenological questions.

– *Sanchez:*

Singular solutions in string theory can appear as regular in 5-brane theory and vice versa. Can this be used to resolve some issues regarding singularities in space-time?

– *Duff:*

Yes indeed, this has to do with the fact that the coupling of the string to the metric and the dilaton is different from the 5-brane case. The moral is that what you mean by a singular space-time is somewhat subjective. It depends on the test object which you use to probe the singularity.

– *Zichichi:*

What you are saying disturbs me very much indeed, because if I have a hole I have a hole. It does not depend on anything besides the fact that the hole is there. Singularity is singularity. I must say that from the physics point of view I am completely confused.

– *Duff*:

I agree with you that it is confusing. The way Callan, Harvey and Strominger explain it, is that the singularity is still there but it takes an infinite time to reach it.

THEORETICAL IMPLICATIONS
OF PRECISION ELECTROWEAK DATA

G. Altarelli

CERN, Geneva, Switzerland

1. Introduction

The main goal of LEP 1 is to perform precision tests of the standard electroweak theory[1] at the Z peak. Theoretical predictions in the Standard Model for all relevant observables have been developed in detail[2]. I refer the reader to my talks[3] at some recent conferences for concise summaries and for many relevant discussions that I will not repeat here. One starts from the Standard Model Lagrangian and a conveniently chosen set of input parameters. The interesting quantities are computed in perturbation theory. The lowest-order formulae plus one-loop radiative corrections[4], often improved by important renormalization group resummations, provide a sufficiently accurate approximation to match the precision of realistic experiments and to allow quite significant tests of the theory. For LEP physics, a self-imposing set of input parameters is given by $\alpha_s, \alpha, G_F, m_Z, m_f$ and m_H. Clearly the Fermi coupling $G_F = 1.16637(22) \times 10^{-5} \text{GeV}^{-2}$ is conceptually more complicated than $\alpha_{\text{weak}} = \frac{g^2}{4\pi}$ (which would more naturally accompany $\alpha = 1/137.036$ and α_s) or $\sin^2 \theta_W$ or m_W, but is preferred for practical reasons because it is known with all the desirable accuracy. Similarly, m_Z has now been measured at LEP with remarkable precision. This preliminary task of LEP in view of precision tests of the Standard Model has already been accomplished to a nearly final degree of accuracy.

The LEP results on m_Z, as summarized at the Geneva Conference[5], are reported in Table 1[†]. The resulting relative precision is impressive: $\delta m_Z/m_Z = 2.3 \times 10^{-4}$.

Among the quark and lepton masses, m_f, the main unknown is the top quark mass. Our ignorance of m_t is at present a serious limitation for precise tests of the electroweak theory because the radiative corrections are relatively large for large m_t and depend quadratically on $m_t^{[3],[4]}$. This fact can be used to put stringent constraints on m_t from the existing electroweak measurements, in particular an upper bound on m_t, to be discussed in detail later. As for lower bounds on m_t the best results arise from the failure to observe the t quark at e^+e^- and hadron colliders. LEP and SLC lead to a model-independent bound $m_t \gtrsim 45$ GeV. From CDF one learns that $m_t \gtrsim 89$ GeV, provided that the t quark semi-leptonic branching ratio is as predicted by the Standard Model.

The Higgs mass m_H is largely unknown. One of the most impressive performances of LEP up to now has been the dwarfing[6] of all previous lower bounds on m_H. For the mass of the minimal Standard Model Higgs boson, ALEPH was

TABLE 1

Experiment	m_Z (GeV)
ALEPH	91.182 ± 0.009
DELPHI	91.177 ± 0.010
L3	91.181 ± 0.010
OPAL	91.161 ± 0.009
AVERAGE	91.175 ± 0.005 (Stat.) ± 0.020 (LEP) $\simeq$ 91.175 ± 0.021

able to establish the lower limit $m_H \gtrsim 51$ GeV. Less stringent but comparable limits were also obtained by the other LEP experiments (OPAL $m_H \gtrsim 47.3$ GeV, L3: $m_H \gtrsim 47.6$ GeV, DELPHI: $m_H \gtrsim 42$ GeV). For the two-doublet Higgs sector of the minimal supersymmetric extension of the Standard Model[7],[8], the corresponding limit is: $m_h \gtrsim 43$ GeV (ALEPH).

The upper limit on m_H is mainly from theoretical arguments of consistency and is not equally sharp. It is well known that for $m_H \gtrsim 0.6$-1 TeV the Standard Model becomes affected by serious problems[9] (e.g., Landau singularities moving down to energies of order 1 TeV), certainly the perturbative framework is no more reliable and weak interactions become strong. For this reason, most computations of radiative corrections are given for $m_H < 1$ TeV. The sensitivity of the radiative

---

† In writing this Proceedings I decided to update the experimental data by also including results not yet available when the talk was given.

corrections to variations of m_H in the range 50 GeV $< m_H <$ 1 TeV is not large. In a sense, this level of accuracy fixes the goal for precision tests of the Standard Model because the clarification of the symmetry breaking sector of the theory is the main target of present-day experiments.

Finally, for electroweak calculations involving hadrons, the value of the QCD coupling α_s, must also be specified. The best value of α_s at the Z mass, obtained from experiments at energies lower than m_Z (in particular from deep inelastic scattering) is given by[10] $\alpha_s(m_Z) = 0.11 \pm 0.01$. At LEP one finds[10] something like $\alpha_s(m_Z) \simeq 0.12 \pm 0.01$. (Note that I am more conservative than usual on errors which are dominated by theoretical uncertainties). The QCD corrections to processes involving quarks are typically of order $\frac{\alpha_s}{\pi}$. As a consequence the stated error on α_s, leads to a few per mille relative uncertainty on the corresponding predictions.

2. Precision Tests of the Standard Electroweak Theory

From the above discussion it is clear that the set of input parameters can be separated into two parts. On the one hand, $\alpha, G_F, m_Z, m_{f_{light}}$ are well known and the ambiguities associated with these quantities on the radiative corrections are quite small. We can add α_s to this class, in that, if it is true that the experimental error on α_s, is relatively large, it only enters as a small correction to electroweak processes involving hadrons and is practically irrelevant for purely leptonic processes. Also, by fitting the electroweak data one cannot obtain a better value for $\alpha_s(m_Z)$ than that derived from QCD tests. On the other hand m_t and m_H are largely unknown. Thus, for each relevant observable, one can only express the prediction of the Standard Model as a function of m_t and m_H, obtained by using the best available calculations of radiative corrections, with $\alpha, \alpha_s, G_F, m_Z$ and $m_{f_{light}}$ fixed at their experimental values with the corresponding errors. By comparing these predictions with experiment one can check their mutual consistency and derive constraints on m_t and m_H.

Actually the sensitivity on m_H is so small that for all the measured quantities the ambiguity due to varying m_H in the range 50 GeV $< m_H <$ 1 TeV is far below the present experimental error, so that for practical purposes, at the present stage of accuracy, the relevant predictions can be plotted as functions of m_t in the form of a band of values determined by $\delta m_H, \delta m_Z, (\delta \alpha_s)$ (see figs. 1,7).

Note that from this point of view $\sin^2 \theta_W$ is not a primary quantity. It is not part of the set of input parameters. It is a derived quantity that one could even decide not to introduce at all. I stress this point in order to make it clear that all disputes over which is the better definition of $\sin^2 \theta_W$ beyond the tree level are completely secondary. First of all it is always true that physical results are independent of definitions. Differences in physical results obtained from a different definition of input parameters (scheme dependence) can at most occur by terms of higher order, due to the truncation of the perturbative series at a given order. But for $\sin^2\theta_W$ its precise definition is only necessary to compute it from the input parameters, but cannot matter for the prediction of observables because, with the choice specified above, $\sin^2\theta_W$ is not taken as an input parameter of the theory.

The widespread use of expressing the experimental result to each given observable in terms of the corresponding value of $\sin^2\theta_W$ (within a specified definition for it) is no longer adequate at the present level of sophistication. For comparing the constraining power of different experiments it would be more appropriate to quote the range of m_t implied by each of them[11] (see Table 4). In fact, $\sin^2\theta_W$ is just one particular observable of the theory. Outside the domain of precision tests, with appropriate definitions of $\sin^2\theta_W$, one can write simple improved Born approximations that include the main contributions of radiative corrections (e.g., large logarithms and terms of order $G_F m_t^2$). While for precision tests the use of as complete as possible radiative corrections is mandatory, these approximate formulae are very useful for our understanding of the pattern of radiative corrections and for everyday-life estimates of rates and experimental sensitivities.

One common definition[12] of $\sin^2\theta_W$ is:

$$\sin^2\theta_W = 1 - \frac{m_W^2}{m_Z^2} \equiv s_W^2 \tag{1}$$

to all orders of perturbation theory. Clearly in this case the observables s_W^2 and m_W are directly equivalent given that m_Z is among the input parameters. In the Standard Model, s_W^2 can be computed from the input parameters by the relation:

$$s_W^2 c_W^2 \equiv \left(1 - \frac{m_W^2}{m_Z^2}\right)\frac{m_W^2}{m_Z^2} = \frac{\pi\alpha}{\sqrt{2}G_F} \frac{1}{m_Z^2} \frac{1}{1 - \Delta r} \tag{2}$$

where $c_W^2 = 1 - s_W^2$ and $\Delta r \equiv \Delta r(\alpha_s, G_F, m_Z, m_f, m_H)$ is the effect of radiative corrections. The quantity Δr as a function of the input parameters has been studied in great detail[13]. The result for s_W^2, obtained starting from the average LEP value for m_Z (see Table 1), as a function of m_t, is plotted in fig. 1, where the uncertainties for 50 GeV $< m_H <$ 1 TeV and $\delta m_Z = \pm$ 21 MeV are also visible. We see that m_t is the main unknown in the calculation of $\frac{m_W}{m_Z}$ from m_Z, followed in importance by the ambiguity from varying the Higgs mass in the above range, while the remaining uncertainty from the experimental error on m_Z is very small.

When the available direct experimental information on $\frac{m_W}{m_Z}$ is added, the sensitivity of s_W^2 to m_t provides a very strong constraint on m_t. $\frac{m_W}{m_Z}$ is directly measured at hadron colliders and can also be obtained (assuming the validity of the Standard Model) from the ratio $R_\nu = \sigma^{NC}/\sigma^{CC}$ of neutral current (NC) to charged current (CC) cross-sections in neutrino-nucleus deep inelastic scattering. The value of $\frac{m_W}{m_Z}$ has been measured at hadron colliders[14]. From CDF and UA2 we have the results reported in Table 2.

TABLE 2

Experiment	$\frac{m_W}{m_Z}$	$s_W^2 = 1 - \frac{m_W^2}{m_Z^2}$
CDF	0.8768 ± 0.0046	0.231 ± 0.008
UA2	0.8841 ± 0.0043	0.2184 ± 0.0077
AVERAGE	0.8807 ± 0.0031	0.2245 ± 0.006

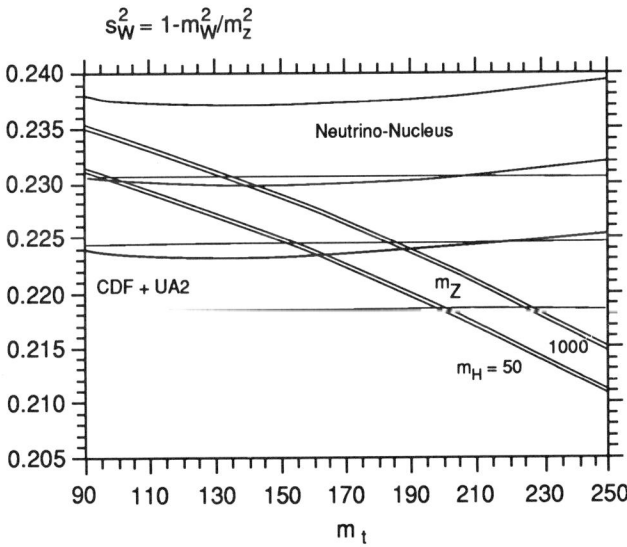

Fig. 1. Labelled by m_Z is the theoretical prediction for s_W^2 obtained from m_Z as a function of m_t for $m_H = 50 - 1000$ GeV (the uncertainty due to $\delta m_Z = 21$ MeV leads to the double boundaries). The error bands implied by the CDF and UA2 measurements of m_W/m_Z and by the data on R_ν are also shown.

By combining $\frac{m_W}{m_Z}$ with the LEP value for m_Z one obtains $m_W = 80.30 \pm 0.28$ GeV. The corresponding average value of s_W^2 is also shown in fig. 1 as a horizontal band, obviously independent of m_t, in the $s_W^2 - m_t$ plane.

As is well known, the value of s_W^2 extracted from R_ν is also nearly independent of m_t in the interesting range of values for the top mass. This fact arises from a largely accidental cancellation[15], specific to this process and to the Standard Model, between two different sources of m_t dependence, as discussed in the following.

In general, at tree level, the four-fermion interaction from Z exchange is given by:

$$M_{if} = \frac{\sqrt{2}G_F m_Z^2}{D(s)} \rho_{tree} \qquad (3)$$

$$(J_3^i - 2\sin^2\theta_W J_{em}^i) \cdot (J_3^f - 2\sin^2\theta_W J_{em}^f)$$

where $D(s)$ is the Z propagator and $J_3^{i,f}$, $J_{em}^{i,f}$ are the weak isospin and electromagnetic currents for the fermion i or f. Excluding pure QED corrections, electroweak radiative corrections[4] modify M_{if} according to:

$$M_{if} = \frac{\sqrt{2}G_F m_Z^2}{D(s)} \rho_{i,f} \qquad (4)$$

$$(J_3^i - 2k_i\sin^2\theta_W J_{em}^i) \cdot (J_3^f - 2k_f\sin^2\theta_W J_{em}^f) + \ldots$$

where $\rho_{i,f} = \rho_{tree}(1+\delta\rho_{i,f})$, $k_a = 1+\delta k_a (a = i, f)$ are different for different fermions and depend on the scheme adopted (for example δk_i depend on the definition of $\sin^2\theta_W$). The ellipses indicate possible additional non-factorizable terms (for example from box diagrams). Let us call "large" radiative corrections those terms containing large logarithms, i.e. $\frac{\alpha}{\pi} \ln \frac{m_Z^2}{m_{flight}^2}$ or quadratic dependences on m_t, i.e., $\sim G_F m_t^2$. For large enough m_t, the bulk of the contribution of electroweak radiative corrections arises from these terms[3],[4]. The "large" contributions to $\delta\rho_{i,f}$ and δk_f in eq. (4) are universal, i.e., they are the same at fixed q^2 for all i and f (except for b quarks). If for $\sin^2\theta_W$ one adopts the definition $s_W^2 = 1 - \frac{m_W^2}{m_Z^2}$ one obtains[4]:

$$1 - \Delta r = (1 - \Delta\alpha)\left(1 + \frac{c_W^2}{s_W^2}\delta\rho + \text{"small"}\right) \qquad (5)$$

$$\rho \cong 1 + \delta\rho = 1 + \frac{3G_F m_t^2}{8\pi^2\sqrt{2}} + \text{"small"} \qquad (6)$$

$$k \cong 1 + \delta k = 1 + \frac{c_W^2}{s_W^2} \delta\rho + \text{"small"} \qquad (7)$$

(for b quarks there are additional large terms) where Δr is defined in eq. 2 and $\Delta\alpha$ arises from the running of the QED coupling:

$$\alpha(m_Z) = \frac{\alpha}{1 - \Delta\alpha} \qquad (8)$$

$\Delta\alpha$ is dominated by large logs and its value is given by[4]:

$$\Delta\alpha \simeq 0.0601 \pm 0.0009 \qquad (9)$$

(or $\alpha(m_Z)^{-1} \simeq 128.8 \pm 0.1$). Note that both δk and Δr contain the large term $\delta\rho$ enhanced by the factor $\frac{c_W^2}{s_W^2}$. Logarithmic scale violations of order $\frac{\alpha}{\pi} \ln q^2/m_Z^2$ are included in the "small" terms (which is only appropriate for $q^2 \gg m_{\text{flight}}^2$). The ratio $R_\nu = \frac{\sigma_{NC}}{\sigma_{CC}}$ for ν-N scattering is given in terms of s_W^2 by:

$$R_\nu = \rho_{\nu N}^2 \left(\frac{1}{2} - k_{\nu N} s_W^2 + \frac{5}{9} (k_{\nu N} s_W^2)^2 (1+r) \right) + \ldots \qquad (10)$$

where $r = (\sigma^{\bar{\nu}}/\sigma^\nu)_{CC} \simeq 0.4$ is also measured. The tree approximation (with $\rho_{\text{tree}} = 1$) is recovered for $\rho_{\nu N} = k_{\nu N} = 1$. Some large logarithms from the radiative corrections to σ^{CC} are also included in $\rho_{\nu N}$. But for the sake of this argument we are only considering the $G_F m_t^2$ terms. For fixed $R_\nu =$ (the experimental value) and $s_W^2 \sim 0.23$ there is a strong cancellation in the Standard Model between the m_t dependence of $\rho_{\nu N} \simeq 1 + \delta\rho$ and of $k_{\nu N} \simeq 1 + \frac{c_W^2}{s_W^2} \delta\rho$, so that as a result $\delta s_W^2 \simeq 0.2 \delta\rho$, where $\delta\rho$ is given in eq. (6). For realistic values of m_t the resulting contribution of the quadratic m_t terms is no more dominant.

The most precise experimental results on R_ν were obtained by the CHARM[16] and CDHS[17] collaborations at CERN. The original results on $\sin_W^2$ were given for fixed m_t and m_H. CHARM obtained $s_W^2 = 0.236 \pm 0.005$ (exp) ± 0.005 (th) for $m_t = 45$ GeV and $m_H = 100$ GeV, while the CDHS result was $s_W^2 = 0.2275 \pm 0.005$ (exp) ± 0.005 (th) for $m_t = 60$ GeV and $m_H = 100$ GeV. The theoretical error arises from hadronic uncertainties and the effect of the charm threshold. An average at $m_t = 60$ GeV and $m_H = 100$ GeV gives $s_W^2 = 0.232 \pm 0.006$ (where the error 6×10^{-3} is obtained as $6 \times 10^{-3} = \sqrt{\left(\frac{5}{\sqrt{2}}\right)^2 + 5^2} \times 10^{-3}$). The corresponding combined result at different values of m_t and m_H can also be obtained from the known form of the radiative corrections. The result is shown[18] in fig. 1.

There are many more less precise experimental results on s_W^2 from low energy neutral current data, most of them being well known[19],[20]. These additional data are all consistent among them and with the results in fig. 1.

We now consider the implications for the standard electroweak theory of the LEP results on the Z partial widths and asymmetries. In figs. 2-4, we compare the data on the Z widths (collected in Table 3) with the predictions of the Standard Model, obtained by the programme ZSHAPE[21] which includes a state of the art set of electroweak radiative corrections.

TABLE 3

	ALEPH	DELPHI	L3	OPAL	Average
Γ_Z (MeV)	2484 ± 17	2465 ± 20	2501 ± 17	2492 ± 16	2487 ± 10
Γ_h (MeV)	1744 ± 15	1726 ± 19	1747 ± 16	1739 ± 17	1740 ± 9
Γ_ℓ (MeV)	83.1 ± 0.7	83.4 ± 0.8	83.5 ± 0.7	83.0 ± 0.7	83.2 ± 0.4
$R = \dfrac{\Gamma_h}{\Gamma_\ell}$	21.00 ± 0.20	20.70 ± 0.29	20.93 ± 0.22	20.95 ± 0.22	20.92 ± 0.11
Γ_{inv} (MeV)	491 ± 13	488 ± 17	501 ± 14	504 ± 15	496.2 ± 8.8
Γ_e (MeV)	83.8 ± 0.9	82.4 ± 1.2	82.5 ± 0.9	82.9 ± 1.0	83.0 ± 0.5
Γ_μ (MeV)	81.4 ± 1.4	86.9 ± 2.1	86.0 ± 1.6	83.2 ± 1.5	83.8 ± 0.8
Γ_τ (MeV)	82.4 ± 1.6	82.7 ± 2.4	85.6 ± 1.9	82.7 ± 1.9	83.3 ± 1.0

Results from LEP. The average also includes systematic errors as given by J. Carter[5] at the Geneva Conference. The average value of Γ_{inv} corresponds to $N_\nu = 2.99 \pm 0.05$ which is the best determination of the number of light neutrinos from LEP.

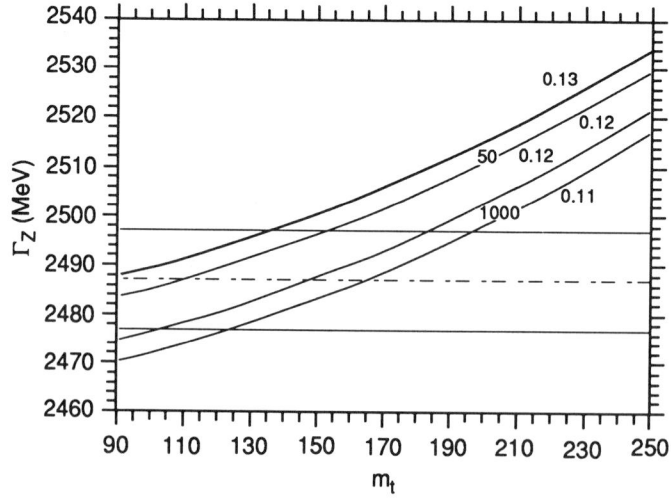

Fig. 2. Γ_Z vs m_t as predicted by the Standard Model for $m_H = 50-1000$ GeV and $\alpha_s(m_Z) = 0.11$-0.13 compared with the LEP result.

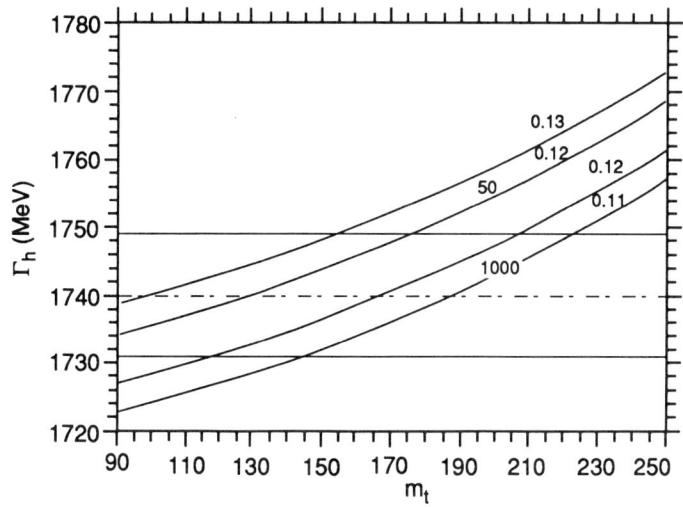

Fig. 3. Γ_h vs m_t as predicted by the Standard Model for $m_H = 50 - 1000$ GeV and $\alpha_s(m_Z) = 0.11$-0.13 compared with the LEP result.

Totally equivalent predictions are obtained by other complete calculations of the line shape[22]–[26]. The predicted widths are plotted as a function of m_t and compared to the data. The ambiguities corresponding to m_H = 50-1000 GeV and $\alpha_s(m_Z)$ = 0.11 - 0.13 are also indicated. Note that the leptonic width (obtained by assuming e – μ – τ universality) gives a particularly strong constraint on m_t, because of its insensitivity to $\alpha_s(m_Z)$, the rather low central value and the small experimental error.

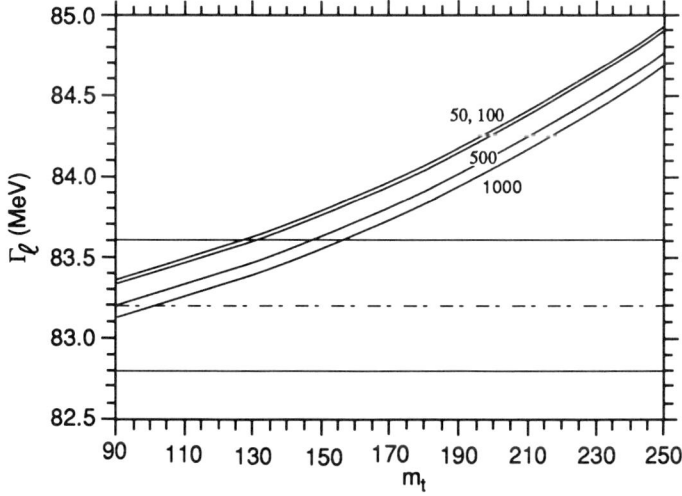

Fig. 4. Γ_ℓ vs m_t as predicted by the Standard Model for $m_H = 50 - 1000$ GeV compared with the LEP result.

217

Additional important information is provided by the measurement of a number of asymmetries. In particular we refer to the forward-backward asymmetries for charged leptons $\left(A_{FB}^{\ell}\right)$ and for the b-quark $\left(A_{FB}^{b}\right)$ and the τ polarization asymmetry $\left(A_{pol}^{\tau}\right)$. The value of those asymmetries, combined over the LEP experiments, are given in the following[5]. For A_{FB}^{ℓ} one has:

$$A_{FB}^{\ell}(\sqrt{s} = m_Z) = 0.0163 \pm 0.0036 \tag{11}$$

obtained from the average value $y = \bar{v}^2/\bar{a}^2 = 0.0048 \pm 0.0012$ quoted by V. Nikolaenko[27] and J. Carter[5], by using the relation:

$$A_{FB}^{\ell} = \frac{3y}{(1+y)^2} + 0.002 \tag{12}$$

which corresponds to the definitions of ZFITTER[24]. For A_{FB}^{b} the result, after correction for the $B - \overline{B}$ mixing effect, is given by:

$$A_{FB}^{b}(\sqrt{s} = m_Z) = 0.126 \pm 0.022 \tag{13}$$

Actually the measured value includes the effect from QCD corrections. We prefer to consider the value of A_{FB}^{b} with the QCD correction[28] removed:

$$\left(A_{FB}^{b}\right)_{measured} \simeq \left(A_{FB}^{b}\right)_{E-W} \left(1 - \frac{0.79\,\alpha_s(m_Z)}{\pi}\right)$$
$$\simeq 0.97 \left(A_{FB}^{b}\right)_{E-W} \tag{14}$$

Dropping the E-W subscript, in the following we shall use:

$$A_{FB}^{b}(\sqrt{s} = m_Z) = 0.130 \pm 0.022 \tag{15}$$

and the corresponding theoretical prediction. Finally, for A_{pol}^{τ} one has:

$$A_{pol}^{\tau}(\sqrt{s} = m_Z) = 0.134 \pm 0.035 \tag{16}$$

The experimental results on the asymmetries are compared with the Standard Model predictions in figs. 5-7.

When all the data are combined I find the following range for the top quark mass:

$$m_t = 140 \pm 35 \text{ GeV} \tag{17}$$

The central value of m_t is smaller, $\simeq 120$ GeV, for light Higgs ($m_H = 50 - 100$ GeV), while, it is larger ($\simeq 160$ GeV) for heavy Higgs ($m_H = 0.5 - 1$ TeV). The above result on m_t is in agreement with other recent analyses of the data[29].

The exact range of m_t depends on all sorts of details, e.g. the set of data which are included, the weight given to the CDF bound on m_t, the range assumed for $\alpha_s(m_Z)$ (here $\alpha_s(m_Z) = 0.12 \pm 0.01$) and so on. The ranges of m_t implied by each experiment are listed in Table 4. We repeat that m_t is a more adequate figure of merit for experiments than $\sin^2\theta_W$ (defined in one or another way) because the contraints on m_t are the obvious goal of precision tests of the electroweak theory at present.

TABLE 4

	$m_H = 50 - 1000$ GeV	$m_H = 0.5 - 1$ TeV
$\frac{m_W}{m_Z}$	150 ± 50 GeV	180 ± 50 GeV
R_ν	100 ± 60	140 ± 50
Γ_Z	90 ± 60	150 ± 50
Γ_h	100 ± 75	170 ± 50
Γ_ℓ	80 ± 50	90 ± 60
A_{FB}^ℓ	140 ± 60	180 ± 60
A_{pol}^τ	130 ± 120	170 ± 120
A_{FB}^b	195 ± 100	330 ± 100
Average	121 ± 22	162 ± 21

3. Toward a Model Independent Analysis of the Data

In the present section we propose a different way[30] of analysing the data which does not necessarily assume the validity of the Standard Model from the start and takes into account recent theoretical studies on the parametrisation of possible effects of new physics on precision experiments[31]–[36]. Given the set of input parameters as specified above, we start from the basic observables m_W/m_Z, Γ_ℓ and A_{FB}^ℓ. We assume charged lepton universality, which is supported by the data at the present level of accuracy, so that Γ_ℓ and A_{FB}^ℓ refer to the corresponding average data. From these three quantities we can isolate the corresponding dynamically significant corrections Δr_W, $\Delta\rho$ and $\Delta k'$, which contain the small effects

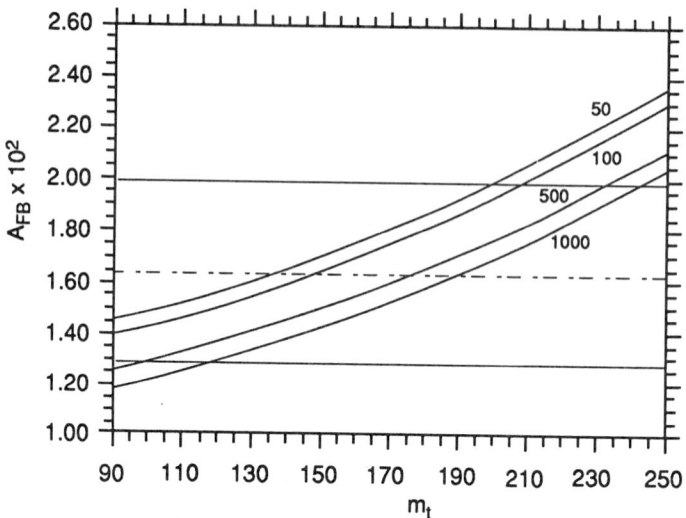

Fig. 5. $A^\ell_{FB} \times 10^2$ vs m_t as predicted by the Standard Model for $m_H = 50-1000$ GeV compared with the LEP result.

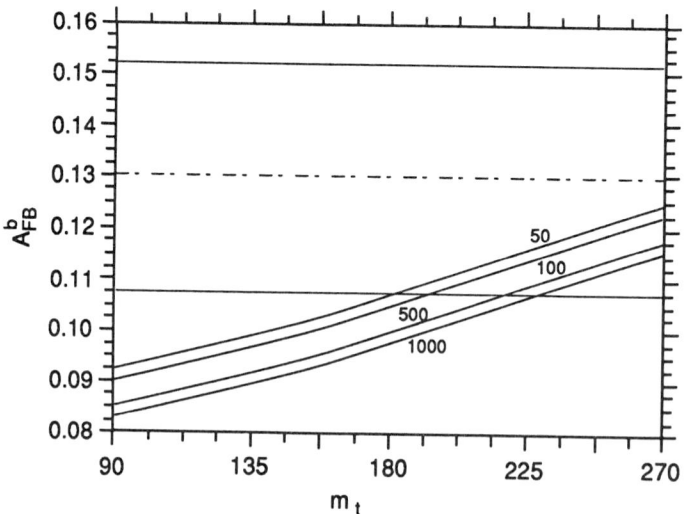

Fig. 6. A^b_{FB} vs m_t as predicted by the Standard Model for $m_H = 50 - 1000$ GeV compared with the LEP result.

one is trying to disentangle, and are defined in the following. First we introduce Δr_W as obtained from m_W/m_Z by the relation (see eq. (2)):

$$\left(1 - \frac{m_W^2}{m_Z^2}\right)\frac{m_W^2}{m_Z^2} = \frac{\pi\alpha(m_Z)}{\sqrt{2}G_F\, m_Z^2(1 - \Delta r_W)} \tag{18}$$

Here $\alpha(m_Z) = \alpha/(1 - \Delta\alpha)$ is fixed to the conventional value $1/128.8$ (recall eq. 9)) so that the effect of the running of α due to known physics is extracted from $(1 - \Delta r) = (1 - \Delta\alpha)(1 - \Delta r_W)$. We take $G_F = 1.16637 \times 10^{-5}$ GeV^{-2}. A possible departure from these values would then be included in Δr_W. In order to define

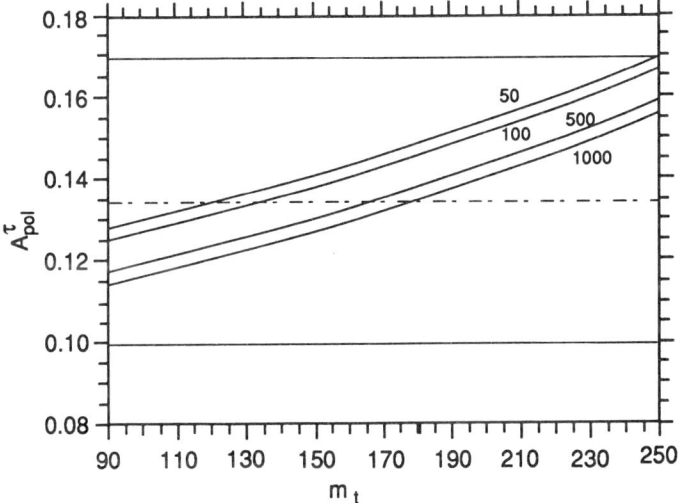

Fig. 7. A_{pol}^τ vs m_t as predicted by the Standard Model for $m_H = 50 - 1000$ GeV compared with the LEP result.

$\Delta\rho$ and $\Delta k'$ we consider effective vector and axial-vector couplings g_V and g_A for the on-shell Z to charged leptons, defined by the formulae:

$$\Gamma_\ell = \frac{G_F m_Z^3 \left(g_V^2 + g_A^2\right)}{6\pi\sqrt{2}} \tag{19}$$

$$A_{FB}^\ell (\sqrt{s} = m_Z) = \frac{3\, g_V^2\, g_A^2}{\left(g_V^2 + g_A^2\right)^2} \tag{20}$$

Note that Γ_ℓ stands for the inclusive partial width ($\Gamma(Z \to \ell\bar{\ell}$ + photons). We could extract from $(g_V^2 + g_A^2)$ the factor $(1 + 3\,\alpha/4\pi + \ldots)$ which is induced in Γ_ℓ from final state radiation, but we prefer the simpler definition of eq. (19). The asymmetry in eq. (20) is obtained from the data after deconvolution of initial state radiation. Contributions from box diagrams and imaginary parts of vertex functions and propagators are absorbed in the definition of g_A and g_V (of course, this can only be possible for a single channel, which we chose to be the charged lepton channel). In terms of g_V and g_A, $\Delta\rho$ and $\Delta k'$ are given by[33]:

$$g_A = -\frac{\sqrt{\rho}}{2} = -\frac{1}{2}\left(1 + \frac{\Delta\rho}{2}\right)$$

$$\frac{g_V}{g_A} = 1 - 4\bar{s}_W^2 = 1 - 4(1 + \Delta k')s_0^2 \qquad (21)$$

In eq. (21) $\bar{s}_W^2$ is a particular effective $\sin^2\theta_W$ for on-shell $Z^{[4],[37]-[39]}$, while s_0^2 is the corresponding quantity before non pure-QED corrections, given by:

$$s_0^2 c_0^2 = \frac{\pi\alpha(m_Z)}{\sqrt{2}G_F m_Z^2} \qquad (22)$$

with $c_0^2 = 1 - s_0^2$ ($s_0^2 = 0.23145$ for $m_Z = 91.175$ GeV).

In the Standard Model Δr_W, $\Delta\rho$ and $\Delta k'$, for sufficiently large m_t, are all dominated by quadratic terms in m_t of order $G_F m_t^2$ [40] and in this limit one has $\Delta r_W \sim -c_0^2/s_0^2 \Delta\rho \sim (c_0^2 - s_0^2) s_0^2 \Delta k'$ 4). As new physics can more easily be disentangled if not masked by large conventional m_t effects, it is convenient to keep $\Delta\rho$ while trading Δr_W and $\Delta k'$ for two quantities with no contributions of order $G_F m_t^2$. We thus introduce the following linear combinations[39]:

$$\epsilon_1 = \Delta\rho$$

$$\epsilon_2 = c_0^2 \Delta\rho + \frac{s_0^2 \Delta r_W}{(c_0^2 - s_0^2)} - 2s_0^2 \Delta k' \qquad (23)$$

$$\epsilon_3 = c_0^2 \Delta\rho + (c_0^2 - s_0^2) \Delta k'$$

Clearly ϵ_2 and ϵ_3 no longer contain terms of order $G_F m_t^2$ but only logarithmic terms in m_t. The leading terms for large Higgs mass, which are logarithmic, are mainly contained in ϵ_1 but are also present in ϵ_3. In the Standard Model one has the following "large" asymptotic contributions[4],[40]-[41]:

$$\epsilon_1 = \frac{3G_F m_t^2}{8\pi^2 \sqrt{2}} - \frac{3G_F m_W^2}{4\pi^2 \sqrt{2}} \mathrm{tg}^2\theta_W \, \ell n\left(\frac{m_H}{m_Z}\right) + \ldots$$

$$\epsilon_2 = -\frac{G_F m_W^2}{2\pi^2\sqrt{2}} \ln\left(\frac{m_t}{m_Z}\right) + \ldots \qquad (24)$$

$$\epsilon_3 = \frac{G_F m_W^2}{12\pi^2\sqrt{2}} \ln\left(\frac{m_H}{m_Z}\right) - \frac{G_F m_W^2}{6\pi^2\sqrt{2}} \ln\left(\frac{m_t}{m_Z}\right) + \ldots$$

In passing, we note the following interesting alternative expression for the leptonic width in terms of $\bar{s}_W^2$ defined in eqs. (19-22) and ϵ_3 (valid in linear approximation in ϵ_1 and ϵ_3):

$$\Gamma_\ell = \frac{\alpha(m_Z) m_Z}{48 \bar{s}_W^2 \bar{c}_W^2} \left(1 + \frac{\epsilon_3}{\bar{c}_W^2}\right) [1 + (1 - 4\bar{s}_W^2)^2] \qquad (25)$$

The definitions in eqs. (18-23) are quite general and do not commit us to any particular model. The epsilons are useful in that they provide a very efficient parametrization of the most important input data with respect to the sensitivity to new physics so that they represent a convenient starting point for a model-independent analysis of the data. One can then formulate a hierarchy of simple and rather general assumptions valid in large classes of models which are needed in order to relate the epsilons to a progressively larger set of observables.

Starting from the hadron collider result on m_W/m_Z given in Table 2, by combining it with the LEP value for the Z mass; $m_Z = 91.175 \pm 0.021$ GeV, and using eq. (1) one obtains:

$$\Delta r_W = (-2.2 \pm 1.7) 10^{-2} \qquad (26)$$

similarly from the LEP results on the charged lepton partial width and the forward-backward asymmetry given in Table 3 and eq. (11) one finds:

$$g_A^2 = 0.2495 \pm 0.0012 \qquad (27)$$

$$g_V/g_A = 0.074 \pm 0.008 \text{ or } \bar{s}_W^2 - 0.2315 + 0.0020 \qquad (28)$$

One can now use eq. (21) to derive the results for $\Delta\rho$ and $\Delta k'$:

$$\epsilon_1 = \Delta\rho = (-0.19 \pm 0.49) 10^{-2} \qquad (29)$$

$$\Delta k' = (0.02 \pm 0.87) 10^{-2} \qquad (30)$$

Finally the corresponding results for ϵ_2 and ϵ_3 are obtained from eq. (23):

$$\epsilon_2 = (-1.10 \pm 0.88) 10^{-2} \qquad (31)$$

$$\epsilon_3 = (-0.14 \pm 0.67) 10^{-2} \qquad (32)$$

We can make contact with the notation of Refs. [31-32] by setting $\Delta\rho = \alpha T$, $\epsilon_2 = -\alpha U/(4s_0^2)$ and $\epsilon_3 = \alpha S/(4s_0^2)$. We then obtain:

$$S = -0.18 \pm 0.85 \;,\; T = -0.26 \pm 0.67 \;,\; U = 1.4 \pm 1.1 \tag{33}$$

Note, however, that in Refs. [31,32] S and T are defined in the specific context of dominance of vacuum polarisation corrections from new physics, while the present definitions are general. Also, in Refs. [31],[32], S and T are defined as deviations from the theoretical predictions of the Standard Model for given values of m_t and m_H, while the values in eq. (33) are unsubtracted.

The predictions of the Standard Model for ϵ_1, ϵ_2 and ϵ_3 have been studied in detail in Ref. [30]. We refer the reader to this paper for a description of the calculations, a comparison of results from several available programmes of radiative corrections and a discussion of the theoretical errors on the epsilons. Here we only quote the main results. In fig. 8 the predicted values of the epsilons are shown together on the same scale. Note that indeed ϵ_2 and ϵ_3 are much flatter than ϵ_1 in m_t (this is especially true for ϵ_3, while some dependence is still visible in ϵ_2 due mainly to logarithmic terms in m_t which are three times larger for ϵ_2 than for ϵ_3, as is seen from eq. (24)). In figs. 9-11 the results are displayed and compared with the present experimental values. Only the comparison of $\Delta\rho$ from theory and experiment implies a strong constraint on m_t, while for ϵ_2 and ϵ_3 one observes consistency with the Standard Model with no important constraints on m_t and m_H. However, ϵ_2 and ϵ_3 impose interesting bounds on new physics, as we shall see.

In the Standard Model, the knowledge of ϵ_1 and ϵ_3 allows one to determine all other observables measured at the Z pole and related to charged leptons, such as A^τ_{pol} and A_{LR}. As mentioned in the introduction, this is also true in a very large class of models where new physics only contributes at $q^2 = m_Z^2$ through either vacuum polarisation amplitudes and/or vertex corrections of the form $\Delta V_\mu(Z \to \ell+\bar{\ell} = \bar{u}(\Delta g_A \gamma_5 + \Delta g_V)\gamma_\mu u$. It is therefore worthwhile to establish this connection in as general terms as possible.

Together with eq. (19), but in place of eq. (20), one could have used A^τ_{pol} to define the effective on-shell vector and axial vector Z couplings to charged leptons, via the relation:

$$A^\tau_{pol} = \frac{2g'_A g'_V}{g'^2_V + g'^2_A} \tag{34}$$

We have adopted different symbols for the effective couplings because g_V and g_A, defined from A_{FB}, certainly cannot be confused with g'_V and g'_A obtained from A^τ_{pol}, to the level of accuracy at which the LEP experiments are aiming. On the

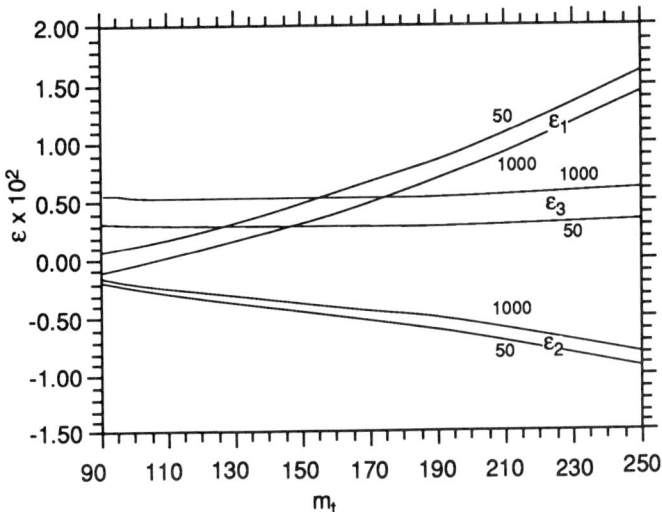

Fig. 8. Theoretical predictions for ϵ_1, ϵ_2 and ϵ_3 in the Standard Model for $m_H = 50 - 1000$ GeV.

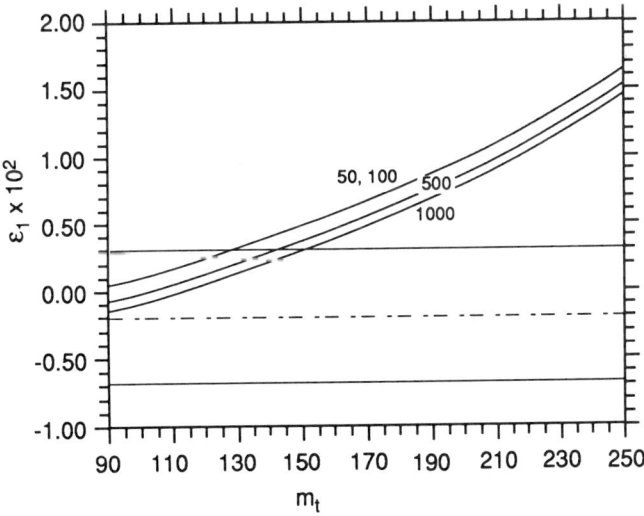

Fig. 9. $\epsilon_1 \times 10^2$ vs m_t as predicted by the Standard Model for $m_H = 50 - 1000$ GeV compared with the LEP result.

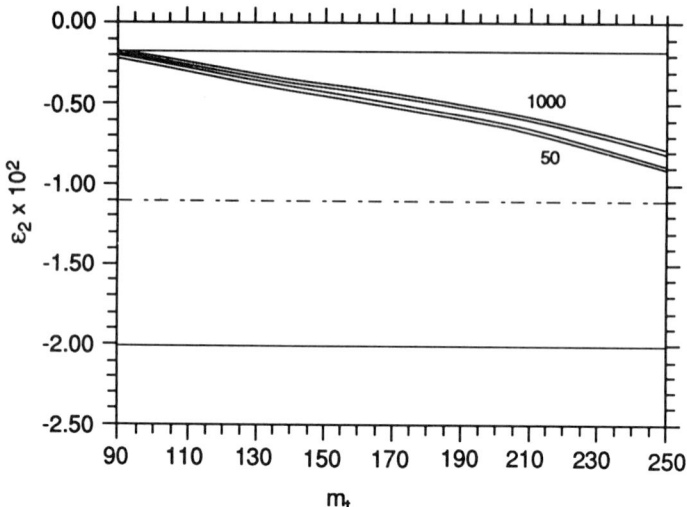

Fig. 10. $\epsilon_2 \times 10^2$ vs m_Z as predicted by the Standard Model for $m_H = 50-1000$ GeV compared with the LEP result.

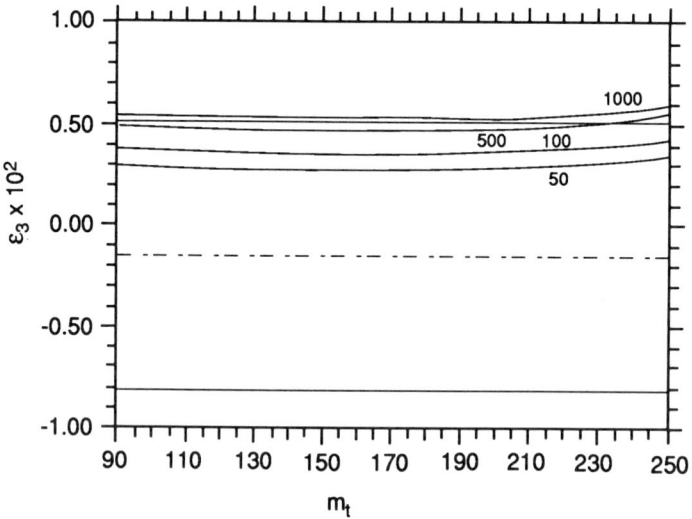

Fig. 11. $\epsilon_3 \times 10^2$ vs m_t as predicted by the Standard Model for $m_H = 50-1000$ GeV compared with the LEP result.

contrary A_{pol}^τ and A_{LR} can be considered as equivalent in this respect. In fact we have checked that the relative differences between the two asymmetries are smaller than 10^{-4} in the Standard Model.

The set of ϵ_i', related to g_A' and g_V' in the same way as the set of ϵ_i to g_A and g_V as in eq. (21), are also sizeably different. One can easily understand the corresponding pattern of corrections. The same effective $\sin^2\theta_W$ would describe A_{FB} and A_{pol}^τ when only quadratic and logarithmic terms in m_t and m_H are included, because such terms only arise from vacuum polarisation diagrams, which affect the asymmetries in the same way. Thus the effective values of $\sin^2\theta_W$ defined from one or the other asymmetry can only differ by a constant term in m_t and m_H. In fact numerically we find:

$$(\bar{s}_W^2)_{pol} = (\bar{s}_W^2)_{FB} + \delta \; : \; \delta = (1.3 \pm 0.2)10^{-3} \quad (35)$$

The main contribution to δ is from the constant $+0.002$ that appears in eq. (12).

Clearly, by the same argument that leads to the independence of δ from m_t and m_H, one obtains that the same shifts also apply to all cases of new physics in oblique or universal vertex corrections. Starting from the experimental value of A_{pol}^τ obtained by LEP (eq. (16)) (see also fig. 7) which corresponds to (from eq. (34) using $g_V'/g_A' = 1 - 4(\bar{s}_W^2)_{pol}$):

$$(\bar{s}_W^2)_{pol} = 0.2332 \pm 0.0045 \quad (36)$$

we can apply the correction in eq. (35) and obtain $(\bar{s}_W^2)_{FB} = 0.2319 \pm 0.0045$ and hence a new value for $\Delta k'$:

$$\Delta k' = (0.2 \pm 1.9)10^{-2} \quad (37)$$

We now proceed to consider quantities that also involve quarks and are still measured at the Z pole. In the Standard Model, to obtain a prediction for these quantities a range of values for $\alpha_s(m_Z)$ has to be assumed. Also, the large quadratic terms in m_t from the $Z \to b\bar{b}$ vertex are to be taken into account. Beyond the Standard Model, it is obvious that a stronger form of universality must be assumed in order to directly transfer into the quark sector the information embodied by the ϵ_i's measured in the charged lepton sector. Quark-lepton universality is automatic in the case of oblique corrections, while it is in general violated if vertex corrections are important.

Particularly important quantities, (nearly) independent of $\alpha_s(m_Z)$, are Γ_b/Γ_h and A_{FB}^b. Similar quantities from charm or light quarks are not equally interesting both in terms of experimental precision and because of the large m_t term [42] in the $Z \to b\bar{b}$ vertex corrections. For example, the data on Γ_b/Γ_h in principle allow us to set a bound on m_t which is independent of assumptions on the absence of exotic contributions to ϵ_1. However, such a limit is not very interesting at the moment because a comparatively large error is introduced by ambiguities associated with the semileptonic branching ratio or by other uncertainties in the case of DELPHI that uses a purely hadronic b-selection criterium. On the contrary, A_{FB}^b is almost unaffected by the presence of large m_t-dependent vertex corrections. Schematically the reason is that $A_{FB}^b = 3\eta_e\eta_b$ with $\eta = g_V g_A/(g_V^2 + g_A^2)$, so that $\delta A_{FB}^b = 3(\eta_e\delta\eta_b + \eta_b\delta\eta_e)$. We see that the sensitivity on η_b, which contains the $Z->b\bar{b}$ vertex correction, is strongly suppressed by the small factor η_e. This is confirmed by an accurate numerical calculation of A_{FB}^b. We define a new quantity $(\bar{s}_W^2)_b$ by the identity:

$$A_{FB}^b = 3 \frac{1 - 4(\bar{s}_W^2)_b}{1 + \left(1 - 4(\bar{s}_W^2)_b\right)^2} \frac{\beta\left(1 - \frac{4}{3}(\bar{s}_W^2)_b\right)}{\beta^2 + \frac{3-\beta^2}{2}\left(1 - \frac{4}{3}(\bar{s}_W^2)_b\right)^2} \quad (38)$$

where $\beta = \sqrt{1 - 4m_b^2/m_Z^2}$. We can explicitly evaluate the relation, as a function of m_t and m_H, between $(\bar{s}_W^2)_b$ and the similar quantity $(\bar{s}_W^2)_{FB}$ previously defined from the charged lepton asymmetry A_{FB}^ℓ. We obtain:

$$(\bar{s}_W^2)_b = (\bar{s}_W^2)_{FB} + \delta_b, \quad \delta_b = (0.9 \pm 0.15)10^{-3} \quad (39)$$

(valid for $m_t = 90\text{-}300$ GeV, $m_H = 50\text{-}1000$ GeV).

As a consequence, if quark-lepton universality is assumed, one can use the present combined LEP result on A_{FB}^b given in eq. (15) which is equivalent, by eq. (38) to:

$$(\bar{s}_W^2)_b = 0.2268 \pm 0.0040 \quad (40)$$

and obtain an independent input on $\Delta k'$. In fact we derive from eqs. (39) and (40) that $(s_W^2)_{FB} = 0.2259 \pm 0.0040$, and then, from eq. (21), the result:

$$\Delta k' = (-2.4 \pm 1.7)10^{-2} \quad (41)$$

While flavour universality of new physics is the crucial assumption that is needed to relate different measured quantities at the Z pole, some hypotheses on the absence of new sources of substantial q^2 dependence have to be formulated in order to add low energy measurements to the picture. Actually, in most of the relevant cases both flavour universality and q^2 independence have to be combined in order to make contact with important experiments, such as ν (or $\bar\nu$) $-$ N deep inelastic scattering and atomic parity violation. In models where oblique corrections, which directly possess flavour universality, are dominant, sizeable q^2-dependent effects from new physics are absent if terms involving second and higher derivatives with respect to q^2 can be neglected in vacuum polarization form factors. This is a good approximation in models with no decoupling, where first-order derivatives lead to effects of order 1 in the limit when the scale Λ of new physics becomes very large. On the other hand, in models with decoupling, like those considered in Ref. [36], first order derivatives are of order v^2/Λ^2, where v is a parameter of the order of the electroweak scale (typically the Higgs vacuum expectation value). In this case the effect of second order derivatives, of order m_Z^2/Λ^2, is not relatively negligible. Note, however, that in these models both effects are quite small unless the scale Λ of the new physics is very close to the domain of energies of present experiments.

If one assumes flavour and lepton-quark universality and no additional q^2-dependence, the available data on neutrino-nucleus deep inelastic scattering and on parity violation in Cs atoms lead to further constraints on ϵ_1 and ϵ_3, while they have no direct effect on ϵ_2. The present data[16]-[17] on R_ν and $R_{\bar\nu}$, the ratios of neutral to charged current processes in deep inelastic neutrinos scattering on nuclei, imply the following constraints[20],[34]:

$$R_\nu : \epsilon_1 - 0.34\epsilon_3 = (-0.07 \pm 0.45)10^{-2}$$
$$R_{\bar\nu} : \epsilon_1 - 0.02\epsilon_3 = (-1.34 \pm 0.95)10^{-2} \qquad (42)$$

with our definition of epsilons (for comparison, we also report the result from the ratio of neutrino to antineutrino scattering on electrons[43], which gives[20],[34] $\epsilon_3 - 0.74\, \epsilon_1 = (0.13 \pm 2.12)10^{-2}$).

Similarly, the results on parity violation in Cs[44],[45] lead to a value of ϵ_3, while the sensitivity to ϵ_1 is accidentally almost exactly cancelled due to the particular ratio of protons to neutrons in Cs[34]. Neglecting the ϵ_1 contribution, one finds in this case the general result:

$$Q_W = -72.84 \pm 0.13 - 102\epsilon_3 \qquad (43)$$

The present experimental value[44],[45]:

$$(Q_W)_{exp} = -71.04 \pm 1.81 \qquad (44)$$

implies the result:

$$\epsilon_3 = (-1.8 \pm 1.8)10^{-2} \qquad (45)$$

Note that in eqs. (42),(45) the quoted values for ϵ_1 and ϵ_3 are inclusive of all standard and possibly non-standard effects.

In a large variety of different models the epsilon parameters are suitable for a discussion of the possible effects of new physics on the various observables. There is an extensive and still rapidly growing literature on the subject. A summary is given in Ref. [30].

We now start from the model-independent determination of the epsilons given in eqs. (29),(31),(32) and progressively make various stages of assumptions that allow us to combine an increasing large set of data. We first assume that there are no new physics effects that can invalidate the connection with A_{pol}^τ (e.g. peculiar four-fermion interactions which could affect A_{FB}^ℓ but not A_{pol}^τ). As a second step we assume flavour and lepton-quark universality and we include hadronic quantities measured at the Z-peak. In particular we consider A_{FB}^b which is precisely measured and relatively unaffected by $\alpha_s(m_Z)$. Then, in addition, we assume the absence of additional sources of q^2 dependence beyond the Standard Model and include neutrino-nucleus deep inelastic scattering and parity violation in Cs atoms.

The inclusion of the data on A_{pol}^τ allows us to combine the values of $\Delta k'$ given in eqs. (30) and (37), thus obtaining:

$$\Delta k' = (0.04 \pm 0.81)10^{-2} \qquad (46)$$

We can then evaluate ϵ_1, ϵ_2 and ϵ_3 by using the known values of Δr_W and Γ_ℓ and the above value of $\Delta k'$. This gives:

$$\epsilon_1 = \Delta\rho = (-0.19 \pm 0.49)10^{-2}$$
$$\epsilon_2 = (-1.11 \pm 0.87)10^{-2} \qquad (47)$$
$$\epsilon_3 = (-0.12 \pm 0.63)10^{-2}$$

This stage is permissible in all models, including the case of a new Z' from an extra U(1) provided the universality of charged leptons is maintained.

We now add hadronic quantities measured at the Z pole. The b-asymmetry A_{FB}^b leads to the determination of $\Delta k'$ given in eq. (41) which can be combined with those leading to eq. (46). The resulting value is:

$$\Delta k' = (-0.40 \pm 0.70)10^{-2} \qquad (48)$$

At this stage one could also include the hadronic width Γ_h and/or the total width Γ_Z, for $\alpha_s(m_Z)$ varying in a specified interval, for example in the range measured at LEP, as given is Section 1. The presence of large vertex corrections in the Z → b$\bar{\text{b}}$ vertex makes the relation with the epsilons strongly m_t dependent (as is also the case for Γ_b/Γ_h). For these reasons we restrict our attention to A_{FB}^b at this stage. For the epsilons we then obtain:

$$\epsilon_1 = \Delta\rho = (-0.25 \pm 0.49)10^{-2}$$
$$\epsilon_2 = (-0.96 \pm 0.86)10^{-2} \qquad (49)$$
$$\epsilon_3 = (-0.41 \pm 0.58)10^{-2}$$

The next step, valid if the effects of a large q^2 difference can be described as in the Standard Model, is to include the low energy data, in particular neutrino-nucleus deep inelastic scattering and parity violation in Cs atoms. By combining the low energy results given in eqs. (42) to (45) with the rest of the data one finds:

$$\epsilon_1 = \Delta\rho = (-0.07 \pm 0.36)10^{-2}$$
$$\epsilon_2 = (-0.80 \pm 0.81)10^{-2} \qquad (50)$$
$$\epsilon_3 = (-0.28 \pm 0.51)10^{-2}$$

At each stage, in the $\epsilon_1 - \epsilon_3$ plane, the experimental result is compared in figs. 12a to 12d with the predictions of the Standard Model for different values of m_t and m_H. It is well known[34] that the data on atomic parity violation in Cs push the value of ϵ_3 on the negative side [see eq. (45)]. We see that this tendency toward negative values of ϵ_3 is also supported by the very recent data on A_{FB}^b (see eq. (49)). The effect of A_{FB}^b can be appreciated from fig. 12e which is a fit to the same data as fig. 12d but with A_{FB}^b removed.

As seen from the overall summary in fig. 12d, the central value of ϵ_3 is negative and about 1σ away from the Standard Model prediction. Clearly, models where an additional positive contribution to ϵ_3 is predicted are a fortiori discouraged, as in the case for the class of technicolour models[30],[31]-[35],[46] leading to the band of values (corresponding to $N_{TC} = 4$ and one technifamily) also shown in fig. 12d. The values of ϵ_2 are consistent with the Standard Model, with a still rather large error (fig. 13).

Summarising, we have shown that it is possible and indeed useful to introduce the parameters ϵ_1, ϵ_2 and ϵ_3 (or equivalently S, T and U) following a general definition independent on assumptions like the dominance of oblique corrections as were made in previous discussions. The presence of three parameters is related to the fact that $m_W/m_Z, \Gamma_\ell$ and A_{FB}^ℓ carry qualitatively different information and are the most precisely measured observables defined at the W/Z mass scale (beyond the input parameter m_Z). ϵ_2 and ϵ_3 are good indicators for the presence of new physics effects, because the uncertainties due to our ignorance of m_t are concentrated in ϵ_1. The epsilons have been studied in the Standard Model as functions of m_t and m_H and the associated theoretical errors were estimated.

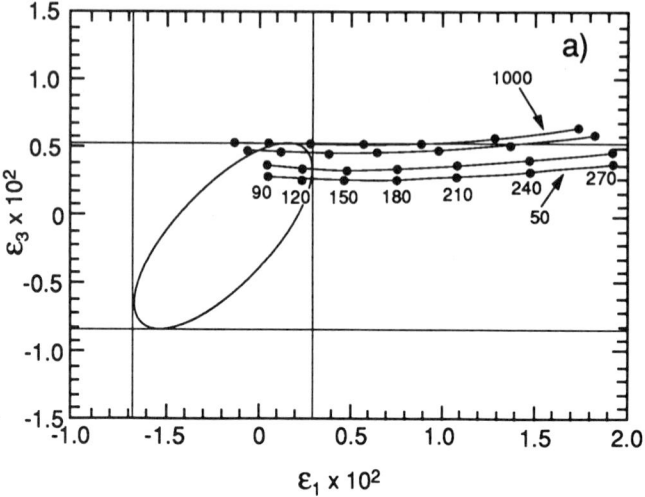

Fig. 12a

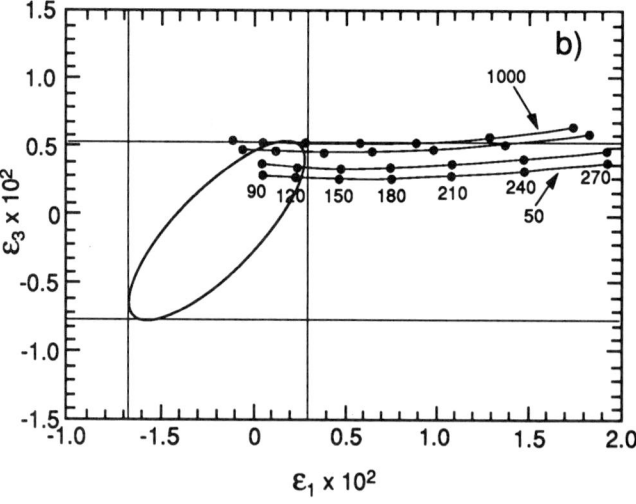

Fig. 12b

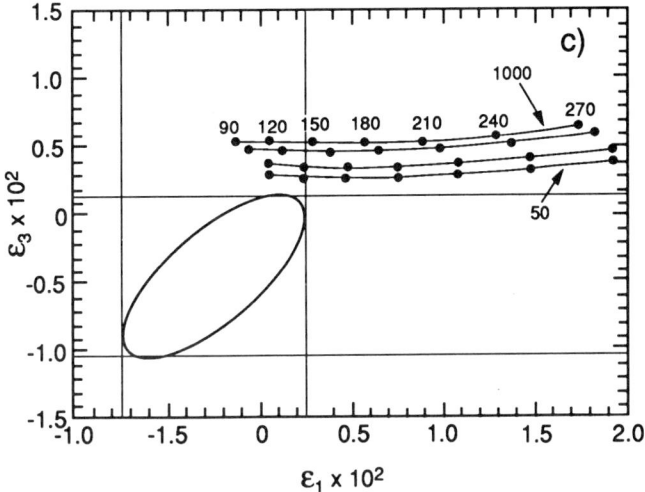

Fig. 12c

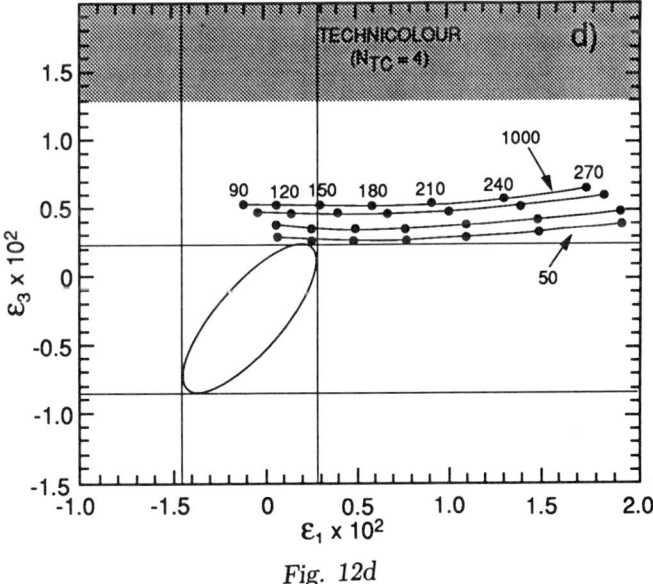

Fig. 12d

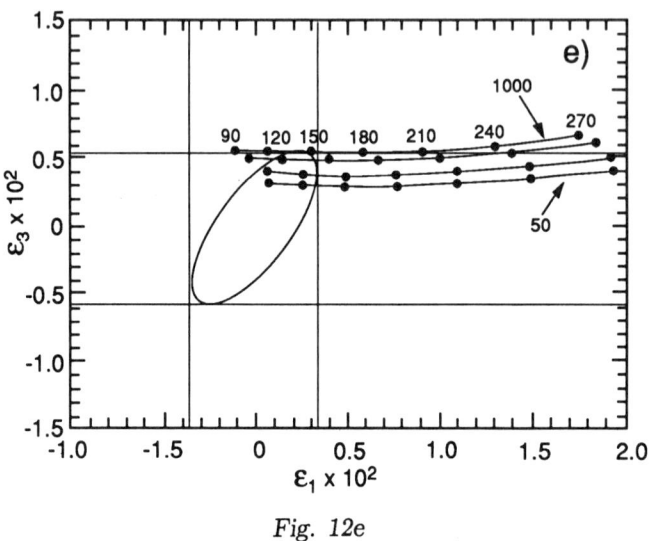

Fig. 12e

Fig. 12 Data on ϵ_1 and ϵ_3. The 1σ ellipses and their projections on the ϵ_1, ϵ_3 axes are shown. The Standard Model prediction is also displayed for reference purposes (the four solid lines are for different values of m_H, m_H = 50, 100, 500, 1000 GeV, and the dots mark values of m_t in the range m_t = 50-270 GeV).

Cases a) to d) correspond to different input data. a) is obtained from Γ_ℓ and A_{FB}^ℓ. In b) the data on A_{pol}^τ have been also taken into account. The result for A_{FB}^b is added in c). All data, including low energy experiments (eq. 50), are included in d). The predictions of QCD-like versions of technicolour (with N_{TC} = 4 and one technifamily) are also shown for comparison. Case e) corresponds to the all data but for A_{FB}^b.

ϵ_3 turns out to be particularly interesting, being independent of m_t with very good accuracy, sensitive to Higgs sector and likely to collect large contributions in models with no decoupling. Present data on A_{FB}^b and on parity violation in atomic Cesium favour values of ϵ_3 smaller than in the Standard Model (although compatible with it).

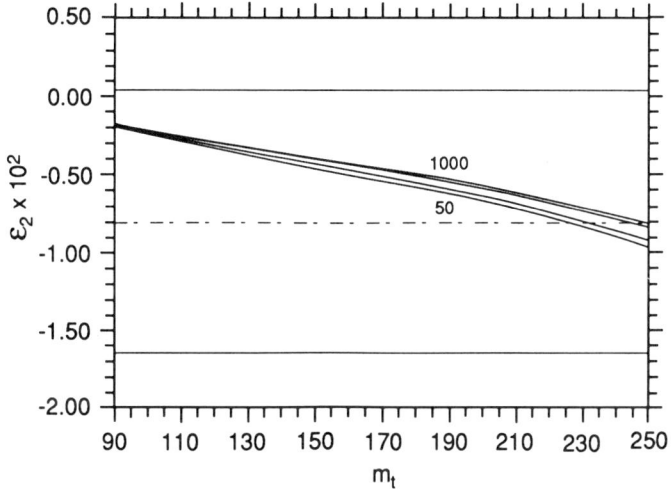

Fig. 13. ϵ_2 vs m_t in the Standard Model for different values of m_H (m_H = 50, 100, 500, 1000 GeV) with the experimental result in eq. (50), obtained from all the data which are available at present.

4. Heavy Flavours at LEP

We have already discussed the importance of A_{pol}^τ and A_{FB}^b for precision tests of the electroweak theory. In this section a number of additional important results of LEP on heavy flavour phenomenology will be discussed. These results include Γ_b, the b semileptonic branching ratio $B_{SL} \equiv B(b \to \ell\nu X)(\ell = e, \mu)$ and the τ-lifetime.

A quantity which is directly measured at LEP is the product $B_{SL}\Gamma_b/\Gamma_h$. The present experimental results are given in Table 5[47],[48].

TABLE 5

$B_{SL}\dfrac{\Gamma_b}{\Gamma_h}$	$= 0.0224 \pm 0.0009 \pm 0.0007$ ALEPH $(e+\mu)$
	$= 0.0230 \pm 0.0012 \pm 0.0011$ DELPHI (μ)
	$= 0.0259 \pm 0.0005 \pm 0.0007$ L3 $(e+\mu)$
	$= 0.0226 \pm 0.0007 \pm 0.0013$ OPAL (μ)
	$= 0.0239 \pm 0.0010$ Average

Two main applications of these results are possible: either one takes Γ_b/Γ_h from the Standard Model and obtains B_{SL}, or one can input B_{SL} from other measurements and derive a value for Γ_b/Γ_h. While both ways are interesting, to me the determination of B_{SL} is more important, because the LEP number is very precise and the situation with B_{SL} is a little mysterious. Inserting the Standard Model value $\Gamma_b/\Gamma_h \simeq 0.217$ one obtains:

$$\begin{aligned}B_{SL} &= (10.3 \pm 0.4 \pm 0.3)\% \text{ ALEPH}\\ &= (10.6 \pm 0.6 \pm 0.5)\% \text{ DELPHI}\\ &= (11.9 \pm 0.2 \pm 0.3)\% \text{ L3}\\ &= (10.4 \pm 0.3 \pm 0.6)\% \text{ OPAL}\end{aligned} \qquad (51)$$

Note that the L3 result is somewhat different from the other ones. The L3 collaboration has also given an independent determination of B_{SL} from the ratio $N\ell\ell/N_\ell$ of dilepton vs single lepton events, which leads to $B_{SL} = (11.3 \pm 1.2)\%$.

These results can be compared with those obtained at PEP and PETRA:[47],[48]

$$B_{SL} = 12.0 \pm 0.7\% \qquad (52)$$

At the $\Upsilon(4S)$ there is a collection of recent, precise results, which, however, in principle do not refer to the same average as measured at higher energies. The $\Upsilon(4S)$ results are[47]:

$$\begin{aligned}B_{SL} &= 10.2 \pm 0.54\% \text{ ARGUS}\\ &= 10.5 \pm 0.45\% \text{ CLEO}\\ &= 10.0 \pm 0.5\% \text{ CUSB}\\ &= 12.2 \pm 0.9\% \text{ X-BALL}\end{aligned} \qquad (53)$$

$$10.4 \pm 0.3\%$$

We see that L3 and PEP/PETRA suggest a value B_{SL} at high energies larger than at the $\Upsilon(4S)$, while the other LEP experiments tend to confirm the low energy value.

From a purely theoretical point of view a value of B_{SL} below 12% is difficult to accommodate in a spectator picture of b decays[49]. This is at least true in the Standard Model. For example, large b → sg or b → $c\tau\nu$ rates which are absolutely out of scale in the Standard Model but can arise in models with charged Higgs[50], could fix the problem. More conventionally a large W-exchange contribution (with gluon emission) could conceivably be evoked[49]. But this explanation would require $\tau^-/\tau^0 \simeq 1.2$, which at present is rather marginal, but not excluded. Other people may simply find that a 20% violation of the parton model picture with inert spectators for the b-quark system is not surprising[51].

Going back to $B_{SL}\Gamma_b/\Gamma_h$ the other possible path is to take B_{SL} from PEP/PETRA plus the measurement by L3 of B_{SL} from $N_{\ell\ell}/N_\ell$ and extract Γ_b/Γ_h from the results of Table 4. With $B_{SL} = 11.8 \pm 0.6\%$ one obtains[48]:

$$\frac{\Gamma_b}{\Gamma_h} = 0.240 \pm 0.012 \text{ (LEP AVERAGE)} \qquad (54)$$

This ratio is in principle interesting because it is independent of $\alpha_s(m_Z)$ and sensitive to the large Z → $b\bar{b}$ vertex corrections of order $G_F m_t^2$ [42]. (The b quark is special because $V_{tb} \simeq 1$). With a precise enough measurement one could determine m_t independent of $\Delta\rho$. In fact $\Delta\rho$ depends on m_t through vacuum polarization diagrams where other new physics effects could mask the m_t dependence. In practice there is not enough precision (and there will not be in the near future) for such a test.

I conclude by recalling that there is a persistent small discrepancy between the Standard Model and the measured values of the τ-lifetime and the τ leptonic branching ratio. One can compute the τ-lifetime by:

$$(\tau_\tau)_{th} = \tau_\mu \left(\frac{m_\mu}{m_\tau}\right)^5 B(\tau \to \nu_\tau \ell \nu) \qquad (55)$$

where $B(\tau \to \nu_\tau \ell \nu)$ is taken from experiment. The effects of differences in phase space and in radiative corrections are insignificant, and are easily taken into account. The theoretical number $(\tau_\tau)_{th}$ can be compared with the experimental value. The result is a deficit in the τ width that can be described by an equivalent ratio of Fermi couplings given by[47]:

$$\frac{G_\tau}{G_\mu} = 0.975 \pm 0.010 \qquad (56)$$

The fact that perhaps $\Gamma_{exp} < \Gamma_{TH}$ suggests that possibly $\Gamma_{exp} = \cos^2\theta \Gamma_{TH}$. In other words there could be a mixing of ν_τ with some heavy neutrino. One possibility is a fourth lepton generation with ν_4 so heavy that it does not affect the Z invisible width, or $\nu_\tau = \cos\theta\nu_1 + \sin\theta\nu_2$ with the orthogonal combination $-\sin\theta\nu_1 + \cos\theta\nu_2$ being a singlet under SU(2). ν_1 and ν_2 are mass eigenstates. If ν_2 is very heavy, then the Z invisible width is suppressed by $\cos^4\theta$ (because there are two τ-neutrinos in the final state of $Z \to \nu_\tau\nu_\tau$). Thus a 5% (3%) suppression of Γ_τ would imply $N_\nu = 2.90$ (2.94), i.e. a 2σ stretching of the present results would be needed. If $m_{\nu_2} > m_\tau$ but $m_{\nu_2} \ll m_Z$ it is difficult to keep ν_2 invisible and to escape present bounds on a heavy neutral lepton. A more trivial way out is the possibility that the precise measurement of m_τ (which was done only by DELCO) is slightly incorrect.

REFERENCES

[1] S.L Glashow, Nucl. Phys. **22** (1961) 579.
S. Weinberg, Phys. Rev. Lett. **19** (1967) 1264;
A. Salam, Proceedings of the 8th Nobel Symposium, Aspenäsgarden, ed. N. Svartholm (Almqvist and Wiksell, Stockholm, 1968), p. 367.

[2] Z. Physics at LEP 1, eds. G. Altarelli, R. Kleiss and C. Verzegnassi, CERN Yellow Report 89-08 (1989).

[3] G. Altarelli, Proceedings of the 1989 International Symposium on Lepton and Photon Interactions at High Energies, Stanford, 1989, ed. M. Riordan (World Scientific), p. 286;
G. Altarelli, Proceedings of the Neutrino 90 Conference, Geneva, Nucl. Phys. **B19** (Proc. Suppl.) (1991) 354.

[4] M. Consoli, W. Hollik and F. Jegerlehner, Electroweak radiative corrections for Z Physics, Ref. 2, Vol. 1, p. 7.

[5] J. Carter, Proceedings of the LP-HEP 91 Conference, Geneva, July 1991.

[6] M. Davier, Proceedings of the LP-HEP 91 Conference, Geneva, July 1991.

[7] See for example: H.P. Nilles, Physics Report **C110** (1984) 1;
H.E. Haber, and G.L. Kane, Phys. Reports **C117** (1985) 75;
R. Barbieri, Riv. Nuovo Cimento **11** (1988) 1.

[8] P.J. Franzini, P. Taxil et al., Higgs Search, in Ref. [2] Vol. 2, p 59.

[9] See, for example, M. Sher, Phys. Reports **179** (1989) 273, and L. Maiani, Rome preprint 775 (1991), lectures given at the Cargèse Summer School, 1990.

[10] T. Hebbeker, Proceedings of the LP-HEP 91 Conference, Geneva, July 1991.

[11] See, for example, V.A. Novikov, L.B. Okun, and M.I. Vysotsky, CERN-TH 6053 (1991).

[12] A. Sirlin, Phys. Rev. **D22** (1980) 971;
W.J. Marciano and A. Sirlin, Phys. Rev. **D22** (1980), 2659, Phys. Rev. **D29** (1984) 75, 945.

[13] G. Burgers, F. Jegerlehner et al., Δr or the relation between the electroweak couplings and the weak vector boson masses, Ref. [1], Vol. 1, p. 55.

[14] H. Plothow-Besch, Proceedings of the LP-HEP 91 Conference, Geneva, July 1991.

[15] R.G. Stuart, Z Phys. **C34** (1987) 445;
A. Blondel, CERN preprint EP/89-84 (1989).

[16] J.V. Allaby et al., Phys. Lett. **B177** (1986) 446, Z. Phys. **C36** (1987) 611.

[17] H. Abramowicz et al., Phys. Rev. Lett. **57** (1986) 298;
A. Blondel et al., Z. Phys. **C45** (1990) 361.

[18] G.L. Fogli, Private communication. I am grateful to G.L. Fogli for providing me with these curves.

[19] U. Amaldi, A. Böhm et al., Phys. Rev. **D36** (1987) 1385;
G. Costa, J. Ellis et al., Nucl. Phys. **B297** (1988) 244.

[20] P. Langacker, in Review of Particle Properties, Phys. Lett. **239** (1990) 1;
W.J. Marciano, BNL-45999 (1991).

[21] F.A. Berends et al., Ref. [1], Vol. 1, p. 89;
the authors of ZSHAPE are W. Beenakker, F.A. Berends and S. Van der Marck.

[22] D.C. Kennedy, B.W. Lynn, C.J.-C In and R.G. Stuart, Nucl. Phys. **B321** (1989), 83 (EXPOSTAR).

[23] A. Borrelli, M. Consoli, L. Maiani and R. Sisto, Nucl. Phys. **B333** (1989) 357.

[24] D. Bardin et al., Comp. Phys. Commun. 59 (1990) 303.
A. Akhundov et al., Nucl. Phys. **B276** (1986) 1;
D. Bardin et al., Z. Phys. **38** (1990) 493.

[25] W. Hollik, Fortsch. f. Physik **38** (1990) 65.

[26] S. Jadach, B.F.L. Ward and Z. Was, CERN preprint TH.5994/91 (1991) to be published in Comp. Phys. Commun.

[27] V. Nikolaenko, Proceedings of the LP-HEP Conference, Geneva, July 1991.

[28] M. Böhm, and W. Hollik, in Ref. [2], p. 203;
J.H. Kühn, and P.M. Zerwas, in Ref. [2], p. 267.

[29] P. Langacker, and M. Wu, VPR-0466T;
S. Banerjee, S.N. Ganguli, and A. Gurter, TLFR/EHEP 91-1;
J. Ellis, Proceedings of the LP-HEP 91 Conference, Geneva, July 1991;
J. Carter, Ref. 5.

[30] G. Altarelli, R. Barbieri, and S. Jadach, CERN-TH.6124/91.

[31] M.E. Peskin and T. Takeuchi, Phys. Rev. Lett. **65** (1990) 964.

[32] B. Holdom and J. Terning, Phys. Lett. **B247** (1990) 88;
M. Golden and L. Randall, Fermilab-Pub. 90-83-T (1990);

A. Dobado et al., Phys. Lett. **B255** (1991) 405;

R.D. Peccei and S. Peris, UCLA preprint UCLA-TEP/91/13 (1991).

[33] G. Altarelli and R. Barbieri, Phys. Lett. **B253** (1190) 161.

[34] W.J. Marciano and J.L. Rosner, Phys. Rev. Lett. **65** (1990) 2963.

[35] D.C. Kennedy and P. Langacker, Phys. Rev. Lett. **65** (1990) 2967 and preprint UPR-0467T;

A. Ali and G. Degrassi, DESY preprint, DESY 91-035 (1991).

[36] B. Grinstein and M. Wise, HUTP-91/A015 (1991).

[37] M. Consoli, S. Lopresti and L. Maiani, Nucl. Phys. **B223** (1983) 472.

[38] D.C. Kennedy and B.W. Lynn, Nucl. Phys. **B322** (1989) 1;

D.C. Kennedy et al., Nucl. Phys. **B231** (1989) 83;

B.W. Lynn, SU-ITP-867 (1989).

[39] A. Sirlin, CERN preprint TH.5506/89 (1989).

[40] M. Veltman, Nucl. Phys. **B123** (1977) 89;

M.S. Chanowitz et al., Phys. Lett. **78B** (1978) 285.

[41] G. Passarino and M. Veltman, Nucl. Phys. **B160** (1979) 151.

[42] A.A. Akundov et al., Nucl. Phys. **B276** (1988) 1;

F. Diakonov and W. Wetzel, HD-THEP-88-4 (1988);

W. Beenakker and W. Hollik, Z. Phys. **C40** (1988) 569;

J. Bernabeu, A. Pich and A. Santamaria, Phys. Lett. **B200** (1988) 569; CERN preprint, CERN-TH.5931/90 (1990);

B.W. Lynn and R.G. Stuart, CERN preprint TH.5786/90 (1990).

[43] D. Geigerat et al., CHARM II Collaboration, Phys. Lett. **B232** (1989) 539.

[44] M.C. Noecker et al., Phys. Rev. Lett. **61** (1988) 310;

M. Bouchiat, Proceedings of the 12th International Atomic Physics Conference (1990).

[45] S.A. Blundell et al., Phys. Rev. Lett. **65** (1990) 1411;

V. Dzuba et al., Phys. Lett. **A141** (1989) 147.

[46] R. Casalbuoni et al., Phys. Lett. **B258** (1991) 161.

R.N. Cahn and M. Suzuki, LBL-30351 (1991);

C. Roisnel, and Tran N. Truong, Phys. Lett. **B253** (1991) 439.

[47] M. Danilov, Proceedings of the LP-HEP 91 Conference, Geneva, July 1991.

[48] P. Wells, Proceedings of the 4th International Symposium on Heavy Flavour Physics, Orsay, France, 1991.

[49] G. Altarelli, and S. Petrarca, Phys. Lett. **B261** (1991) 303.

[50] B. Grzadkowski, and W.S. Hou, Paul Sherrer Inst. preprint PSI-PR-91-20 (1991).

[51] A. Dobrovolskaya, Proceedings of the 4th International Symposium on Heavy Flavour Physics, Orsay, France, 1991.

CHAIRMAN: G. Altarelli

Scientific Secretaries: P. Hernandez, M. Neubert,
Z. Pluciennik, H.J. Schulze, M. Wadhwa

DISCUSSION I

– *Malik:*

1) You have given the range of mass of the top from 100 GeV to 180 GeV or so, while the L3-lineshape group has given the m_t range from 179 GeV to 209 GeV (1991) by fixing the value of $\alpha_s = 0.115 \pm 0.009$. Can you please comment on that?

2) How do you combine the result of different experiments?

– *Altarelli:*

1) The number I have presented here is an overall summary of the 1990 data from all the LEP experiments, and other information is also included (e.g. the CDF limit). There is the possibility that some results from the 1991 data will be presented next week at the Geneva conference, but I think that the 1990 numbers will not change much until substantially more statistics are collected. There are also little things that can slightly modify the limits. For example, differences in the set of data which are considered, the way the errors are combined or the way the CDF limit is included.

2) In general I use my common sense in trying to understand the strength and weakness of each experiment. In this analysis I start with the input parameters α, G_F and M_z that are in general measured with almost ∞ precision and I fix α_s in the range given by LEP. Then I consider M_w, the low energy results (e.g. νN scattering data) and the LEP averages for Z widths and asymmetries and just get the best fit separately for two extreme Higgs masses.

– *Hasan:*

When you talked about τ-polarization, you mentioned that we cannot measure e and μ polarization. Could you explain why?

– *Altarelli:*

The e and μ do not decay in the detector while the τ does. By studying in a given τ-decay mode (e.g. $\tau \to \pi\nu$) the energy spectrum or the angular distribution of the τ decay products, you can measure the average τ polarization, while for the other fermions this measurement is impossible or (e.g. for the b-quark) much more difficult.

– *Hasan:*

What would be the effect of the beam polarization on precision measurements at LEP?

– *Altarelli:*

With beam polarization, it is possible to make a much simpler precision test of the electroweak theory by measuring left-right asymmetry:

$$A_{LR} = \frac{\sigma_L - \sigma_R}{\sigma_L + \sigma_R}$$

where

$$\sigma_L : e_L^- e_{unpolarized}^+ \to \text{all}$$
$$\sigma_R : e_R^- e_{unpolarized}^+ \to \text{all}$$

The τ-polarization asymmetry is conceptually very similar, but experimentally it is much more difficult to measure. In fact for A_{LR} we can use all channels, while for A_{POL}^τ one is restricted to the τ channel and besides one has to identify some particular decay modes of the τ, which introduces additional inefficiencies and systematic errors and further reduces statistics.

– *Hernandez:*

Which radiative corrections due to the top quark are included in LEP analysis? How can a very heavy top quark affect LEP measurements and lead to bounds on the top mass?

– *Altarelli:*

We have seen that m_t appears in radiative corrections ($\sim m_t^2$) and thus we can get some information about it without actually producing it (high precision can substitute high energy). This is a result of the non-decoupling property of the theories with spontaneous symmetry breaking. In unbroken gauge theories like QED or QCD the decoupling theorem ensures that in the limit of $m_t \to \infty$, the resulting theory can be obtained by just crossing out all diagrams involving heavy particles. This is true if the couplings do not grow with masses and the resulting theory is still renormalizable. Then there is no trace of the heavy particles at low energies. The reason why this does not apply to theories with spontaneous symmetry breaking is because in this case there are couplings that increase with masses (Higgs, longitudinal W and Z modes).

In particular in the Electroweak Theory there are two sources of large m_t-effects:

1) Vacuum polarization diagrams that are universal (i.e. independent of the final state in $e^+e^- \to f\bar{f}$)

2) Vertex corrections to Z − b$\bar{\text{b}}$, e.g.:

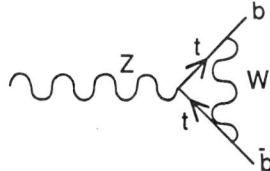

Technically, the full 1-loop corrections have been computed. For the main terms in the vacuum polarization diagrams the Dyson resummation can be implemented. Also the α_s corrections to the large terms of order $G_F m_t^2$ have been computed.

I think that for practical purposes we know all we need about large m_t terms.

– *Hernandez:*

We have seen that the bounds on m_t depend on the value of m_H. Could a non-standard symmetry breaking sector (e.g. strongly interacting Higgs) affect the present bounds on m_t that we get from radiative corrections?

– *Altarelli:*

It is certainly true that these limits change if you go out from the minimal Standard Model. However, I do not think that the bounds I have presented here can change by very much, unless one accepts a lot of ad-hoc fine-tuning. In fact the top effects are included in vacuum polarization diagrams leading to corrections to the so-called ρ parameter. Other terms can arise from new physics:

$$\delta\rho = \delta\rho_{\text{top}} + \delta\rho_{\text{Higgs}} + \delta\rho_{Z'} + \delta\rho_{\text{SUSY part.}} + \cdots$$

If the new terms are of positive sign, the limit on m_t is even more stringent. However, the sign can in some cases be negative (e.g. from non-doublet Higgs, etc.). Even in this case, a big deviation from the limit valid in the Standard Model would need an accidental large cancellation between $\delta\rho_{\text{top}}$ and $\delta\rho_{\text{New Physics}}$. In conclusion I think the limit could be stretched up to 250-300 GeV but no more.

– *Goobar:*

You have shown how you can constrain m_t from measurements of M_W/M_Z partial widths of the Z-decays and the asymmetry measurements at LEP. If m_t is measured next year at the Tevatron, do you see any chance that the accuracy of the ongoing LEP measurements can allow you to set an experimental upper limit on the SM Higgs mass?

– Altarelli:

It is possible to constrain m_H when m_t is known provided the experiments are precise enough. By using the asymmetry measurements one plans to reach a precision that is comparable to the variation produced by the uncertainty in m_H.

In the minimal Standard Model, there is a theoretical limit to the maximum acceptable m_H value (~ 1 TeV) due to the occurrence of a Landau pole. This result is common to all theories that are not asymptotically free, like the scalar sector in the Standard Model or QED. At 1-loop the renormalization group equation for the quartic coupling $\lambda(t)$ as function of $t = \ell n \frac{\Lambda^2}{v^2}$ where Λ is some large scale and v is the Higgs vacuum expectation value gives:

$$\lambda((t) = \frac{\lambda_0}{1 - \beta_0 \lambda_0 t}$$

With β_0 being the β-function coefficient in lowest order.

As m_H grows, λ_0 grows and eventually there is a point in which $t = 1/\beta_0 \lambda_0$ and $\lambda(t) \to \infty$. Although one cannot trust perturbation theory for $\beta_0 \lambda_0 t \sim 1$, lattice non-perturbative studies also support this result. This phenomenon has first been considered by Landau within QED. In that case, the energy where the Landau pole occurs is greater than the Planck scale and, for example, by extending the gauge group to SU(5), the problem is solved. Assuming that the Landau pole does not come as low as m_H (together with vacuum stability) gives the following theoretically allowed region:

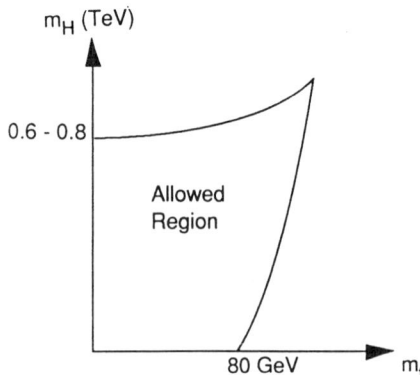

– Gourdin:

The various processes observed have obviously a different sensitivity to m_t and, of course, different accuracy. Improved LEP data will come from the 1991 run. Which are the observables that will be more relevant for the reduction of the uncertainty on m_t?

– *Altarelli:*

LEP 1991 data will not bring drastic changes because the error is not going down as $1/\sqrt{N_{events}}$ due to the systematic errors. The real narrowing down of m_t will come, we hope, with the discovery of the top by CDF and D0. If this really happens we will be able to do very stringent tests of the Electroweak Theory at LEP.

– *Sivaram:*

Is there any information or evidence as yet about triple boson couplings, i.e. WWZ type of couplings? Presumably it would be easier to study this if the energies are raised to $\approx 2M_W$.

– *Altarelli:*

This question is mainly for LEP 200. Of course there are already indirect limits on WWZ couplings. If this vertex is changed a divergence appears and a cut-off must be introduced. Calculations become somewhat model dependent but one could have large effects on radiative corrections.

On the other hand, experimental limits can also be obtained at hadron colliders. In particular before LEP 200, CDF can put a limit on the departure from the standard coupling.

– *Shabelski:*

You discussed α_s in your lecture, but the scale of α_s should not be M_Z but the transverse momentum of the jet. Can you comment on this?

– *Altarelli:*

The ambiguity in the choice of the scale of α_s, due to the truncation of the perturbation series, is already included in the error presented here.

– *Khoze (comment):*

It seems to me that the range of m_t you presented here is rather close to that obtained sometime ago from $B^0 - \overline{B}^0$ mixing data.

– *Altarelli:*

That depends on what you assume for f_B.

– *Khoze:*

Despite all the success of QCD, there were some statements about CDF data on $b\bar{b}$ production differing by a factor of two from the standard QCD prediction. Could you comment on this?

– Altarelli:

I do not believe that there is a real disagreement. B production at Tevatron can only be predicted with a large error (a factor of 3-5). In order to be able to apply perturbation theory to the process pp $\to$ b$\bar{\text{b}}$X, not only m_b must be large enough, but also $m_b/\sqrt{s}$ must not be too small. This ratio is, in fact, too small at Tevatron energies and a reliable prediction would need the solution of the diffraction problem. At present B production can be better predicted at the energy of fixed target experiments.

– Khoze:

Are you refering to the uncertainty in the structure functions at small x?

– Altarelli:

Both the knowledge of the structure function at small x and the solution of the diffraction problem (multigluon exchange contribution) are needed in the prediction of this cross-section.

– Lisi:

Concerning the vector boson mass relation, did you consider in your analysis the so-called modified resummation of leading corrections proposed by Consoli et al. and the corrections of order $\alpha\alpha_s$ computed more recently by Kniehl et al.?

– Altarelli:

Yes.

– Wang:

How far can we reduce the errors on the measurements of Γ_h and Γ_ℓ at LEP?

– Altarelli:

The main systematics on the widths is due to point to point calibration of energy and to the luminosity measurements. By increasing statistics at LEP1 I imagine that the total error will not be reduced much more than by a factor of 1.5 - 2 in the partial widths. We can still expect big improvements in the measurement of the asymmetries.

– Passalacqua:

1) The LEP experiments measured α_s in two ways: using EEC and jets, which means using hadronization models, and using Γ_h/Γ_ℓ. The error quoted from the first kind of measurement is smaller than the error from the second one, but still the difference is significant.

It seems to me that the second one is more reliable because it is free from the dependence of the hadronization models. Can you give some insight of this question?

2) What changes will there be if one uses the second one?

– *Altarelli:*

I think that the two methods lead to consistent results. More or less one has $\alpha_s(m_Z) \simeq 0.115 \pm 0.01$ from jets and 0.14 ± 0.02 from Γ_h/Γ_ℓ. The experiments tend to underestimate the theoretical errors on the first method, in my opinion. The second method is less affected by systematics and will become very good with more statistics.

CHAIRMAN: G. Altarelli

Scientific Secretaries: H. Hernandez, M. Neubert,
Z. Pluciennik, H.J. Schulze, M. Wadhwa

DISCUSSION II

– *Khoze:*

I would like to make two comments:

1) An interesting possibility is the direct study of an exotic non-doublet Higgs sector. It could be done by searching for the $W^\pm Z H^\mp$ vertex, which is absent in the minimal SM. Phenomenological consequences were discussed already 10 years ago by A.A. Iogansen, N.G. Vraltgev and myself.

– *Altarelli:*

If this vertex exists, it could well have escaped detection up to now.

– *Khoze:*

2) I would like to make the "Russian-style comment" that the associated production of Higgs and gauge bosons above threshold (Bjorken-mechanism) has first been discussed by B.L. Ioffe and myself in 1976.

– *Pluciennik:*

1) In the plots which you presented, the values of the parameters $\epsilon_1, \epsilon_2, \epsilon_3$ are consistent with zero. Can one say then that the radiative corrections in the SM have not been verified yet?

2) Have models with a non-linear σ-model Lagrangian for the Higgs been considered? (They give no observable Higgs). What is wrong with those models?

– *Altarelli:*

1) Non pure-QED radiative corrections have at least been grossly verified. A better verification clearly requires more precise measurements, but already the data constrain m_t. One of the virtues of the proposed type of analysis is to focus on the essence of the problem, which is the structure of the radiative corrections.

2) They are not renormalizable. Such models require a cutoff, which might indicate new physics. However, the non-linear σ-model by itself is not consistent and hence cannot be a solution to the problem of electroweak symmetry breaking.

– *Rothstein:*

Presently, as far as I know, there are no consistent technicolour models. As such, what assumption go into your calculation in "technicolour" models?

– *Altarelli:*

I agree, one cannot kill a theory which does not exist. Most technicolour models are more or less copies of QCD, i.e., they are based on a SU(N) gauge group and assume prominent vector resonances. These models lead to predictions strongly disfavoured by the data. However, these are examples of peculiar technicolour models, assuming prominent axial-vector resonances, which even give $\Delta\varepsilon_3 < 0$.

– *Neubert:*

Physics beyond the SM could include additional sources of CP-violation not related to the Kobayaski-Maskawa matrix. Examples are extended Higgs models with a neutral Higgs particle of indefinite parity. Would the associated effects show up in the parameters ε_i?

– *Altarelli:*

The observables I have introduced are CP conserving and specific for LEP1, where there is no chance to see CP-violation. Eventually one may be able to study CP-violation at LEP2 by considering the WWZ vertex, but this requires the measurement of new, non CP-conserving, observables.

– *Sivaram:*

1) The new physics, if any, at $\Lambda \gg m_z$, should give rise to contact-interaction effective terms at low energies. Is there any evidence from the existing data about the presence of such terms?

2) You said that for parity violation in atomic transitions, $\Delta\varepsilon_3$ small and negative is favoured. Could you explain this?

– *Altarelli:*

1) What you say is not necessarily true, but possible. Large contact-

interactions would influence particular channels. For example a $e\bar{e}b\bar{b}$ 4-fermion interaction would affect the $e^+e^- \to b\bar{b}$ cross-section, and you would find that the relevant data would not fit the SM predictions.

2) If one measures parity-violation in Cs, accidentally the result is almost only sensitive to ε_3 and not to ε_1. The result is then $\varepsilon_3 = -(1.8 \pm 1.8)\%$. This is one reason why the ellipse labelled "all data" is slightly below the SM level.

– Chizhov:

What are the restrictions for the compositeness of elementary particles that can be obtained from the analysis of the latest data from LEP?

– Altarelli:

If there is something that really does not exist, it is a compositeness model. What is called a limit on compositeness is usually a bound on the size of some sort of ad-hoc interaction. A composite-model of quarks and leptons must explain the family replication, and no such model has been built. There is no basis for compositeness at the Fermi scale.

– Hernandez:

In renormalizable theories the only source of non-decoupling we know is due to the generation of masses by spontaneous symmetry breaking, but in this case, taking $m_t \to \infty$ means taking a dimensionless Yukawa coupling to infinity, so eventually a perturbative approach is not reliable.

Thus, if new physics appear at a scale $\Lambda \gg v$ would it not be more natural to consider only extensions of SM that decouple at low energies?

– Altarelli:

First of all, my treatment is completely open minded and does not restrict to any particular type of non-standard physics. If I have considered non-decoupling examples it is because they are those that are treated in the literature. Moreover since the Standard Model has no decoupling I do not see why this could not apply to new physics as well.

– McNulty:

The size of the statistical error is generally well known; the size of the systematic error is much more subjective, worse than that, the value of the quantity measured is only good somewhere within the systematic and not necessarily at the mid point. It is my opinion that systematic errors cannot simply be averaged. How do you account for this?

– *Altarelli:*

That is true, but nobody averages systematic errors. In the quoted averages a common systematic error is added in quadrature to the combined statistical error.

NOVEL NEUTRINO PHYSICS

D.H. Perkins

University of Oxford, United Kingdom

1. Introduction: the Kinematics of Beta Decay

There are two areas in which the field of neutrino physics has led to significant new activity during the past year or so: solar neutrino experiments, and the "17 keV neutrino" phenomenon in nuclear beta decay. There is not time to discuss both topics in the depth required and I will therefore concentrate on the 17 keV effect.

The existence of a neutrino of mass about 17 keV was first claimed by Simpson (1985) during studies of tritium beta decay. His interpretation was criticized and independent experiments showed no effect. The matter seemed to be dead. However, during the last year, experiments at Berkeley, Oxford and Zagreb have again claimed a significant effect, while one at Cal Tech claims no effect. The matter is clearly open. The consequences of a 17 keV neutrino for particle physics, and for astrophysics and cosmology, would be tremendous and can hardly be over-emphasized: so, the matter is worth discussing.

The 17 keV neutrino experiments hardly fall into the theme of this school — physics at the highest energy and luminosity — although they presumably do have a connection with the origin of mass. Their scale is different from what is usual in particle physics. The total number of people currently in this field is less than 1% of those on LEP. These are "table-top" type experiments, with an average author list, per experiment, of 4 rather than 400.

Let me start by recalling the form of the electron spectrum in beta decay. For an allowed transition, where a neutron changes into a proton in exactly the same quantum state, the rate is simply the product of the phase space factors of electron and neutrino. For a massless neutrino, the electron spectrum has the form

$$\frac{dN}{dT} = \text{const.} pEF(Z, E)(Q - T)^2 \qquad (1)$$

where T =electron kinetic energy, $Q = T(\text{max})$ is the end-point energy, p and E are momentum and total energy of the electron. $F(Z, E)$ is a factor describing the effect of the nuclear Coulomb field on the $e^-(e^+)$, with Ze the charge on

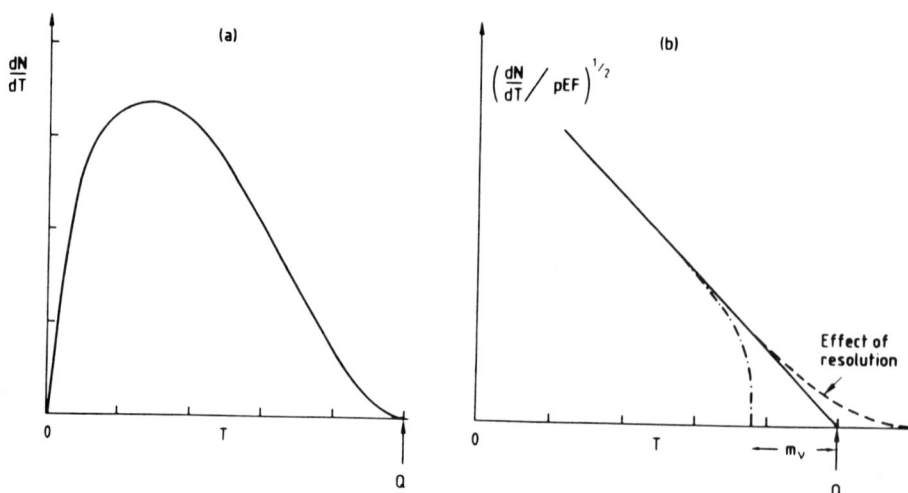

Fig.1 (a) Electron spectrum in allowed transition
(b) Kurie plot, for zero and non-zero neutrino mass.

the daughter nucleus. In the non-relativistic limit — all that concerns us here — F is given by the barrier penetration factor

$$F(Z,E) = \frac{2\pi\eta}{(1-e^{-2\pi\eta})} : \qquad (2)$$

where
$$\eta = +\alpha Z E/p \quad \text{for } e^-$$
$$= -\alpha Z E/p \quad \text{for } e^+$$

From (1) we see that

$$K(T) = \left(\frac{dN/dT}{pEF}\right)^{\frac{1}{2}} \propto (Q-T) \qquad (3)$$

giving a straight line when plotted against T as in Fig.1. This is called a **Kurie plot**. If there are two neutrino components, one of mass $m_1 = 0$ and the other of $m_2 > 0$, with θ the mixing angle, (1) generalises to

$$\frac{dN}{dT} = pEF(Q-T)^2 \cos^2\theta + pEF(Q-T)^2 \left[1 - \frac{m_2^2}{(Q-T)^2}\right]^{\frac{1}{2}} \sin^2\theta \qquad (4)$$

where the first term applies for $0 < T < Q$ and the second, for $0 < T < Q - m_2$. Note that the second term gives a Kurie plot which is straight for $Q - T \gg m_2$ (or $T \ll Q - m_2$) but bends over to cut the T-axis vertically when $T = Q - m_2$. Fig.2(a) shows the Kurie plot of Eqn.(4), for a largish value of θ ($\sim 30°$) for clarity. Since we shall be discussing values of $\sin^2\theta \simeq 0.01$, the actual change in slope and the "kink" in the Kurie plot cannot be detected by eye in the raw data: one has to display the deviation, or the ratio S, of the 2-component distribution to that expected for a single, massless neutrino. Two methods are shown in Fig.2. In Fig.2(b), the quantity $(S-1)$ is plotted, based on a fit to the dominant, single-component distribution for $Q > T > Q - m_2$. In Fig.2(c), S is the ratio of the 2-component distribution to the best fit single straight-line

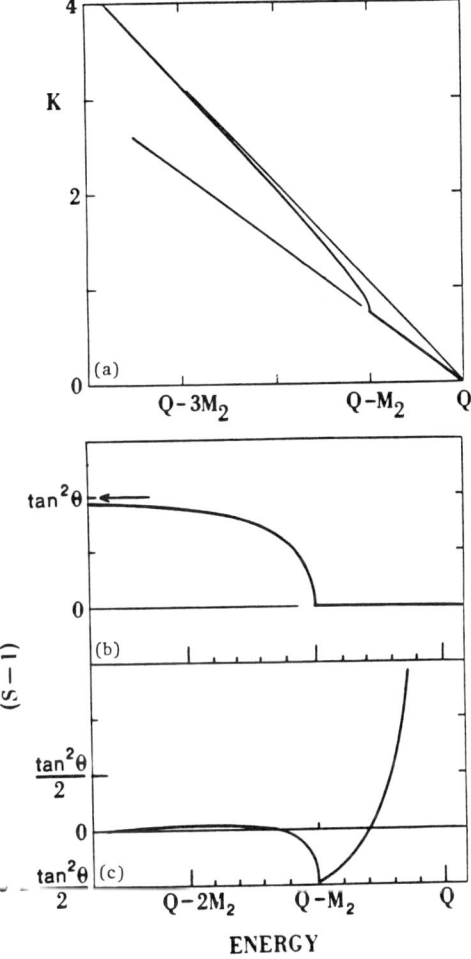

Fig.2 (a) Kurie plot for one massless neutrino mixed with one of mass m_2 (Eq.(4)).
(b) Ratio, S, of 2 component distribution to that expected for a massless neutrino, fitted to the distribution for $Q - m_2 < T < Q$.
(c) Ratio S of the 2 component distribution to the best-fit single straight-line Kurie plot.

Kurie plot through all the data. The best straight line will clearly overshoot the 2-component distribution in the "kink" region and start to undershoot almost half-way between the kink and the end-point. A disadvantage of this plot is that $S \to \infty$ and then changes sign as one approaches the end-point.

To end this section, I should emphasize that the beautiful straight-line Kurie plot is a purely text-book phenomenon. No one has ever seen a straight (within 1%) Kurie plot in practice. First, there are deviations at the end-point due to the apparatus resolution, but that can be measured and allowed for (see Fig.1). Then there are other deviations from strict linearity — euphemistically called "shape factors" — in all experiments to date. In the magnetic spectrometer experiments some of these deviations are quite large (several per cent) and fitted by empirical terms linear or quadratic in energy. Their origin is presumably the scattering of electrons from collimators, walls, thick source effects etc. Finally, there could be deviations due to finite neutrino mass of the type described above. The main task in assessing experiments is to try to decide what are real physical effects intrinsic to nuclear beta decay, and what are artefacts of the experimental method and apparatus.

2.   The Early Experiments on the 17 keV Neutrino

Figs.3 and 4 show the first results by Simpson (1985) on tritium decay, using tritium ions accelerated to 10–15 MeV with a Van der Graaff machine and implanted into a Si(Li) crystal. The radiation damage to the crystal was cured by annealing. Simpson claimed a deviation from the distribution expected for a single massless neutrino, and ascribed it to the effect of a second neutrino of mass $m_2 = 17$ keV and $\sin^2 \theta = 0.03$ (see Eqn.(4)). Since for tritium $Q = 18.6$ keV, the effect appears at low electron energy, $T \simeq 1.5$ keV. These results were criticised because of an incorrect calculation of the atomic screening of the nuclear Coulomb field, as well as neglect of crystal lattice effects. When these were taken into account, the value of $\sin^2 \theta$ was reduced from 0.03 to 0.01. These atomic effects are important only for tritium decay, because of its low Q value. In the other radioisotopes with higher Q values discussed below, they are expected to be negligible.

Following the Simpson paper, a Bombay group (Datar et al 1985) used a Si(Li) detector to measure the spectrum from decay of ^{35}S ($Q = 167$ keV). Fig.5 shows the data. The authors claim $\sin^2 \theta < .006$ at 90% CL, although from the statistical errors indicated it seems hard to draw such a conclusion. Indeed the curve shown for $\sin^2 \theta = 0.008$ appears to be almost as good a fit as the straight line ($\sin^2 \theta = 0$). Ohi et al (1985) also presented negative results from ^{35}S decay with Si detectors. They sandwiched the source between 2 detectors. By requiring a signal in only one crystal they hoped to alleviate the effects of backscattering of electrons which enter one crystal, are scattered through a large angle and re-emerge having lost only a part of their energy, the rest being deposited in the other crystal. Since in fact Si(Li) detectors have energy thresholds due to surface dead-layers and noise, the advantage of this method is questionable. Their results are given in Fig.6. They quote $\sin^2 \theta < .003$ at 90% CL for a 17 keV neutrino. Simpson has re-analysed their data, and claims to be able to fit it with a 17 keV neutrino by restricting himself to that part of the data bridging the "kink" region. I have to say that I do not support this re-analysis by Simpson.

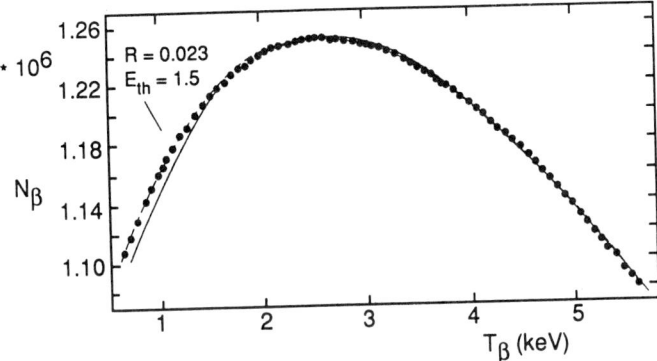

Fig.3 Beta decay spectrum of tritium measured by Simpson (1985). The full curve is the expectation for a single massless neutrino, the dotted curve that with a massive neutrino in addition.

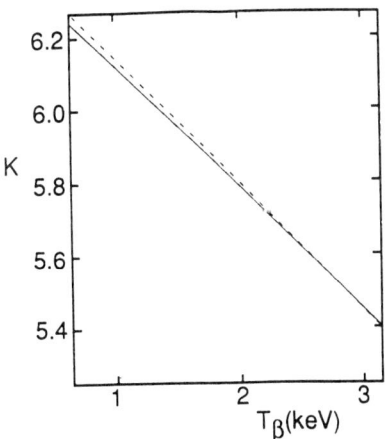

Fig.4 Kurie plot of the data of Fig.3.

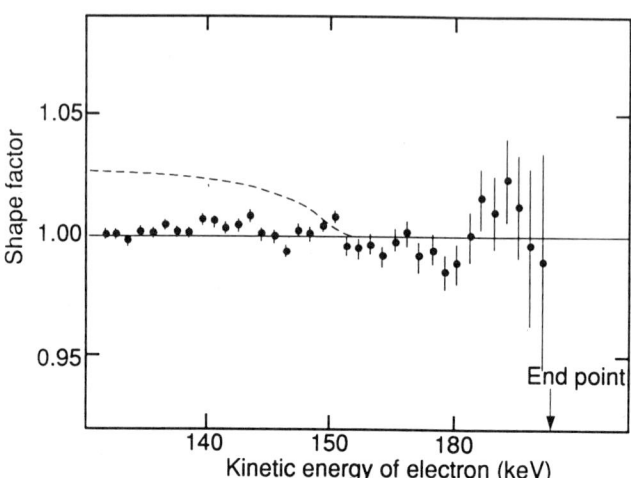

Fig.5 Shape factor of the ^{35}S spectrum for Datar et al (1985). The --- curve is the variation expected for $m_2 = 17$ KeV, $\sin^2\theta = 0.03$

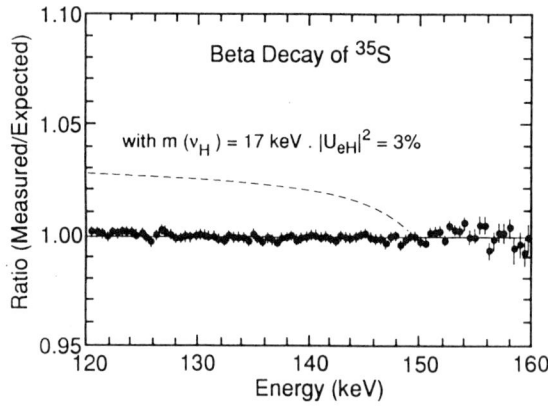

Fig.6 Data from Ohi et al (1985) on ^{35}S decay, using a source sandwiched between two Si(Li) counters.

In 1989, Simpson and Hime presented new data on both tritium ($Q = 18.6$ keV) and ^{35}S ($Q = 167$ keV) using Si(Li) detectors, and also criticized some of the previous experiments finding negative results. Fig.7 shows the Hime-Simpson data on ^{35}S using a source placed outside the detector. The effects of backscattering and energy loss in the source were assessed by measuring the apparent line-shape of internal conversion sources, as described below. Again, a "kink" was observed and they concluded that $m_2 = 16.9 \pm 0.4$ keV and $\sin^2 \theta = .0073 \pm .0009 \pm .0006$.

Another obvious technique is the classical one of measuring the momentum spectrum with a magnetic spectrometer. The difference is that, for a crystal detector, all electrons of any energy are measured contemporaneously, while with a magnetic spectrometer one can only measure a small momentum slice at a time. The solid angle acceptance is also very restricted. Hence, despite the use of strong sources, these spectrometer experiments can be very time consuming and stability over long periods as well as source decay can be a problem. The other feature of magnetic spectrometers is that, unless very carefully designed there are significant distortions of the spectrum from the effects of scattering off collimators etc used to define the momentum bite, as well as energy loss and scattering in the source or detector. Use of an extended strip source compensated by a suitably graded accelerating potential increases the luminosity but introduces an energy-dependent acceptance. All these effects are described by a "shape factor" in the form of a polynominal in electron energy or momentum, with the coefficients determined empirically to give the best fit to experiment.

Fig.8 shows the toroidal spectrometer of Altzitzoglou *et al* (1985) and Fig.9 their ^{35}S spectrum. They fit the observations by a Kurie plot multiplied by a shape factor of the form

$$S = a + bT + cT^2 + d\left(1 - \frac{m_2^2}{(Q-T)^2}\right)^{\frac{1}{2}} \tag{5}$$

Here, 'a' is a normalization constant of order unity, the last term is the effect of a neutrino mass m_2, and $(d/a) = \tan^2 \theta$ in the terminology of (4). The best fit coefficients were (with $m_2 = 17$ keV)

$$b = 3.10^{-3} \text{ keV}^{-1}$$
$$c = -9.10^{-6} \text{ keV}^{-2}$$
$$d = -(3 \pm 2).10^{-3}$$

Note that, over the fitting interval $T = 110 - 160$ keV, the second term in (5) change S by $\Delta S = 15\%$ and the third term, by $\Delta S = -14\%$. So, both empirical shape factor terms are an order of magnitude larger than the effect of a 17 keV neutrino with $\sin^2 \theta \sim 1\%$. Hence their conclusion that $\sin^2 \theta < .004$ needs to be treated with reserve.

Fig.10 shows a diagram of the double-focussing $\pi\sqrt{2}$ iron-free spectrometer of Hetherington *et al* (1987) together with the multi-strip source employed, and Fig.11 the end portion of their spectrum from decay of ^{63}Ni ($Q = 67$ keV). They

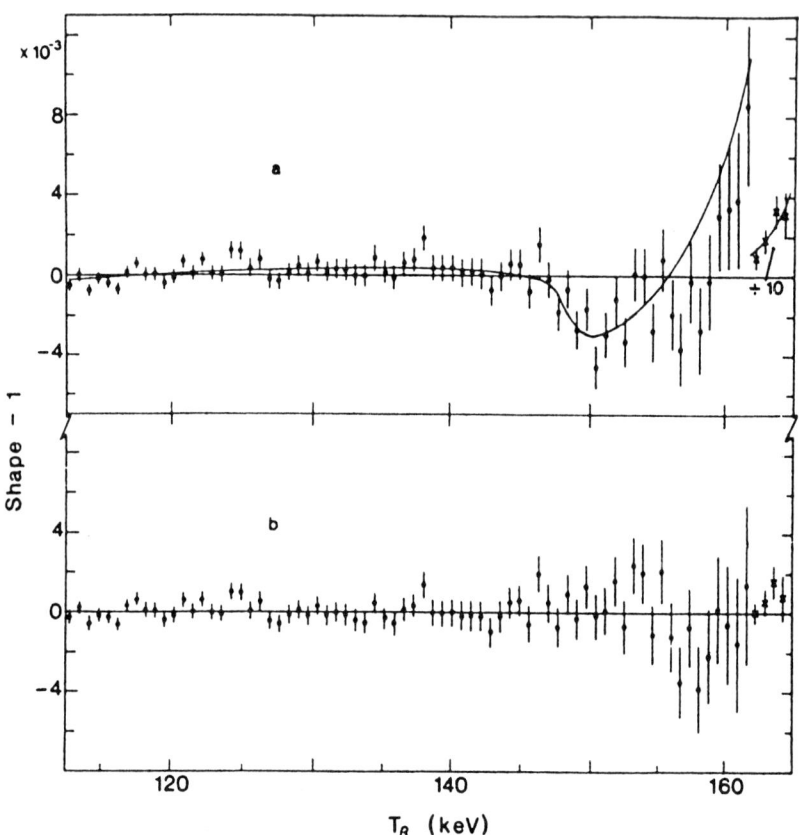

Fig.7 Data for Simpson and Hime (1989) on the ^{35}S spectrum measured with a Si(Li) detector. The lower graph shows the deviations from the expected 2 component distribution including a massive neutrino of $m_2 = 17$ keV and $\sin^2\theta = 0.0073$.

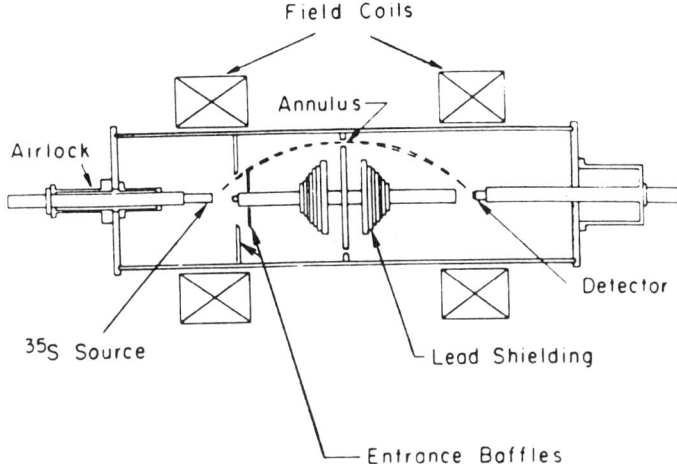

Fig.8 Beta spectrometer employed by Altzitzoglou *et al* (1985) in their investigation of the ^{35}S beta spectrum.

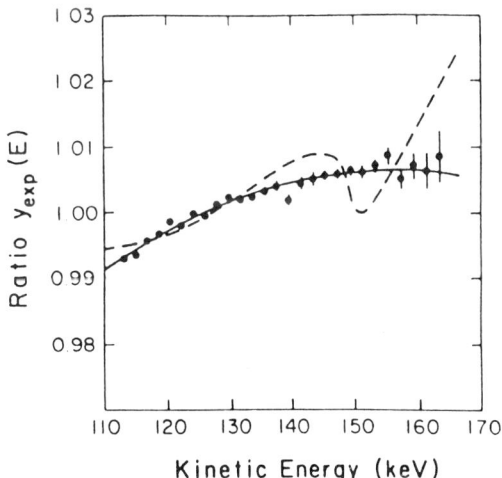

Fig.9 Ratio of Altzitzoglou *et al* data to the fit of Equation (5). The full line is for $\sin^2 \theta = 0$, the dashed line that for $\sin^2 \theta = 0.03$.

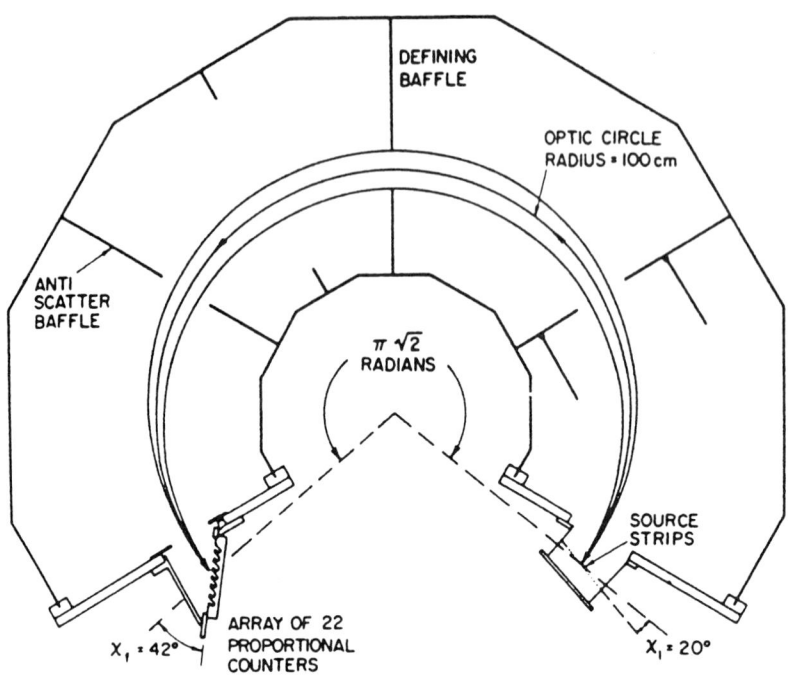
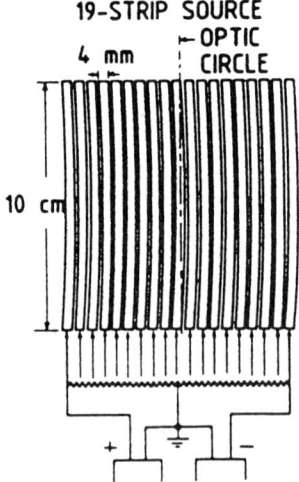

Fig.10 (top) $\pi\sqrt{2}$ beta spectrometer of the Hetherington *et al* (1987), with (bottom) detail of the strip source.

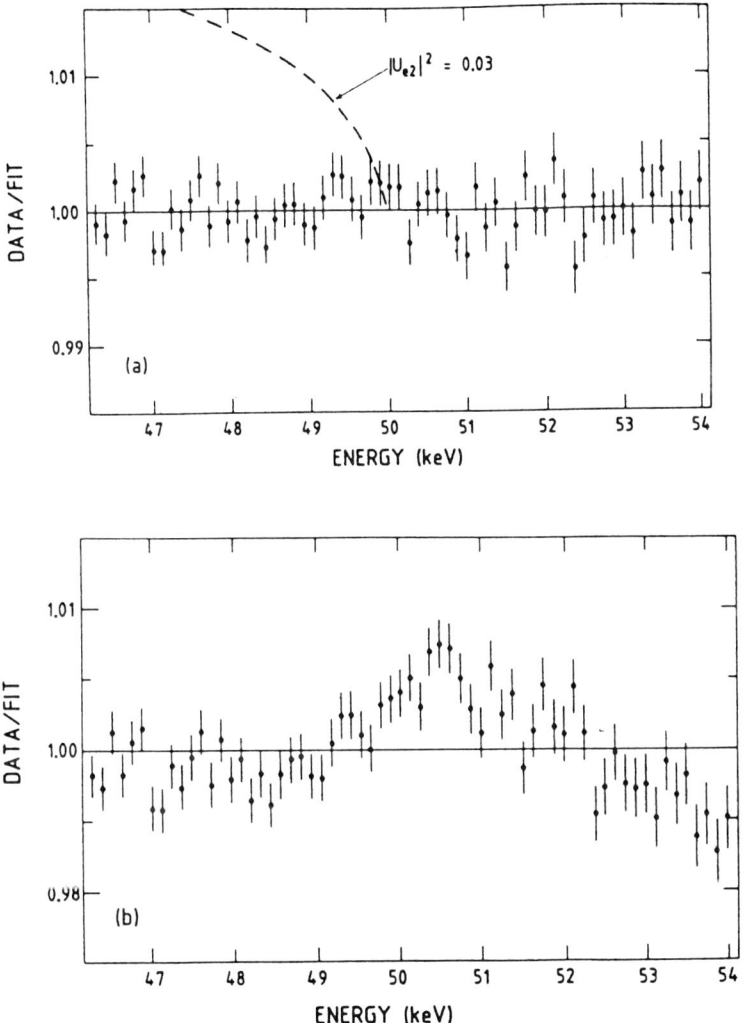

Fig.11 Ratio of data of Hetherington *et al* to fit of equation (6), for (top) $\sin^2 \theta = 0$, and (bottom) $\sin^2 \theta = 0.03$.

employed a shape factor of the form

$$S = 1 + bT + d\left(1 - \frac{m_2^2}{(Q-T)^2}\right)^{\frac{1}{2}} \tag{6}$$

with no second order term (in T^2) being required to get an acceptable fit. Clearly b and d will be correlated: they find for $m_2 = 17$ keV:

$$\text{for} \quad d = 0 \quad , \quad b = 7 \cdot 5.10^{-3} \text{ keV}^{-1} \quad \chi^2 = 1.15$$
$$d = 0.03, \quad b = 1 \cdot 5.10^{-3} \text{ keV}^{-1} \quad \chi^2 = 3.45$$

Thus the χ^2 greatly prefers no 17 keV neutrino, and they set $d = \sin^2\theta < 0.003$ at 90% CL. Again, it should be pointed out that the b term changes by 6% over the fitted interval ($T = 46 - 54$ KeV). Although a quadratic term does not improve the fit, one could also ask what would be the effect of cubic, quartic, etc terms. Nevertheless if the 17 keV neutrino exists, this experiment had enough statistics and resolution to detect a "kink" at the 0.5% level. Its non-observation in this experiment is for me the best evidence **against** a 17 keV neutrino.

3. Recent Experiments in the 17 KeV Neutrino

Fig.12 shows the apparatus employed in a high statistics experiment on ^{35}S by Hime and Jelley (1991) at Oxford. The beta source is in the form of a monolayer of barium sulphate deposited on a 110Å gold layer mounted on a 2.8μm mylar substrate. The geometry is determined by circular apertures of copper and aluminium, so that electrons are accepted from a cone of half angle 10°. Two runs were made with slightly different geometry. Internal conversion sources, of ^{57}Co (115 and 129 keV electrons) and ^{109}Cd (63 keV) were mounted in identical fashion, so that the "response function" of the detector to electrons at fixed energies can be measured. Fig.13 shows the result for one line: a Gaussian peak of 0.6 keV resolution (FWHM) plus a low-energy tail from backscattering, energy losses in source and detector etc.. The shape and magnitude of the backscattering distribution can only be measured to about half the peak electron energy, but is consistent with extensive measurements in the literature. With the geometry employed, the total backscattering is 12%, almost independent of energy at these energies. Furthermore, the differential distribution scales with energy, that is, is a function of the ratio of the backscattered energy T, to the peak energy T(peak). Therefore, the measurement of the response function at 3 electron energies allows one to interpolate fairly reliably for energies over the region of interest ($T = 140 - 165$ keV). The procedure was in fact to parametrize the backscattering and fold this response function into the theoretical beta spectrum. Of course, the relevant backscatter fraction falling within the range $T = Q$ to $T = Q - 17$ keV is only about 1% and the uncertainty is estimated to be $< 0.1\%$.

The ratio S of the data to the predicted spectrum, or rather $(S - 1)$, is shown in Fig.14. (a) and (b) show the data compared with expectations for no 17 keV neutrino (straight horizontal line) while the curves show the effect of a 17 KeV neutrino with $\sin^2\theta = 0.008$ (as in Fig.2(c)) while plot (c) shows the

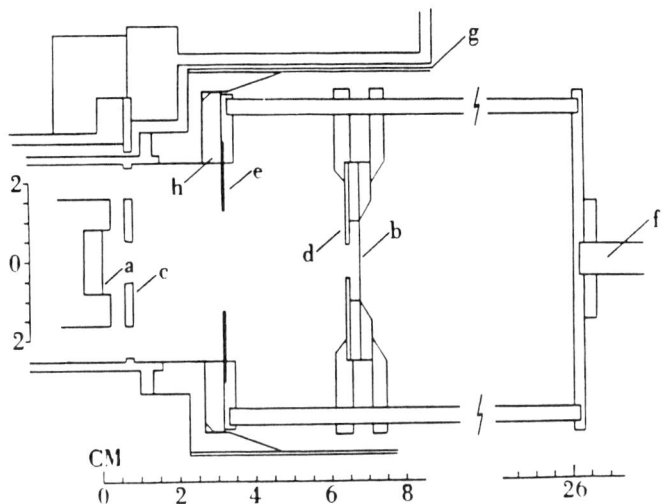

Fig.12 Apparatus of Hime and Jelley (1991) for investigating ^{35}S spectrum with Si(Li) detector, (a).
(b) source substrate (c) Al detector aperture (d) Cu source aperture (e) Al anti-scatter baffle (f) linear motion feed-through (g) liquid nitrogen cryo-panel

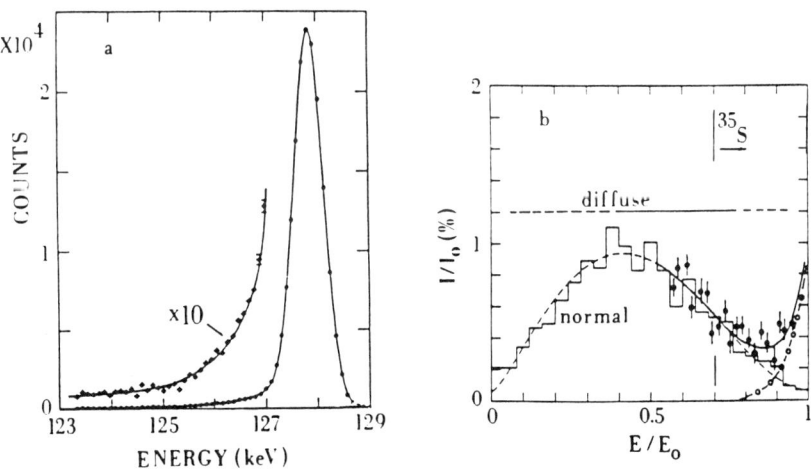

Fig.13 Backscattering/target energy loss distributions for 129 keV internal conversion electrons from γ-decay of ^{57}Co.

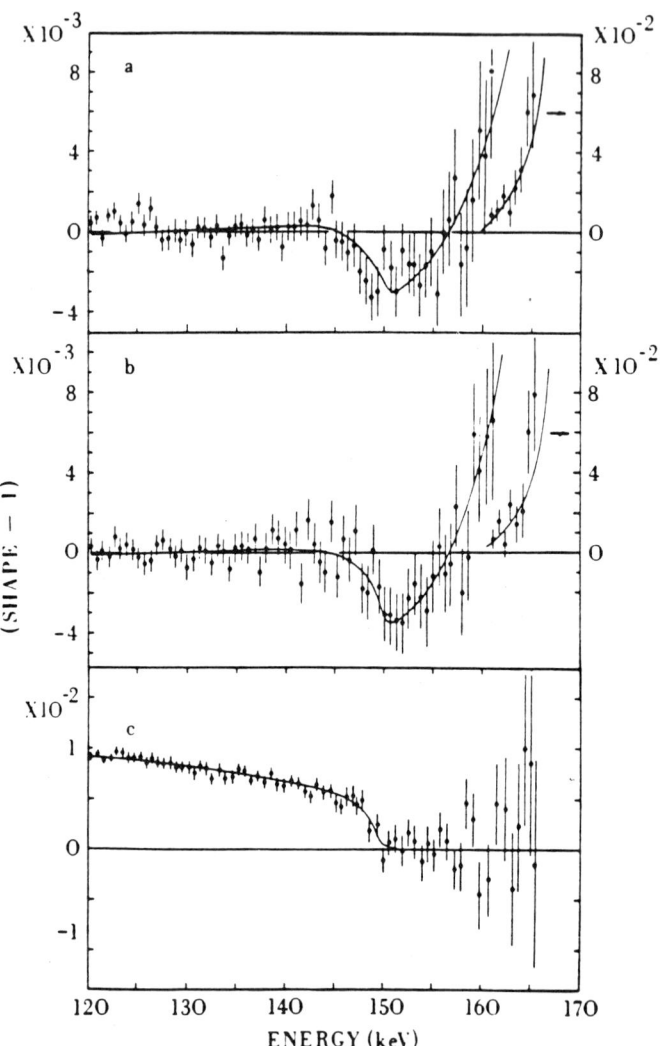

Fig.14 $(S-1)$, as defined in Fig.2(b) and 2(c), from the Hime and Jelley experiment of Fig.12. (a) and (b) are two runs with different geometry showing the ratio of data to the best-fit single straight-line Kurie plot. (c) combines both runs and normalizes to the data above the end-point for a heavy neutrino, with $m_2 = 17$ keV.

result of fitting to the last 17 keV of the spectrum, as in Fig.2(b). Hime and Jelley quote the values

$$m_2 = 17.0 \pm 0.04 \text{ keV}$$
$$\sin^2 \theta = 0.0084 \pm .001$$

It should be emphasized that the ratio of data to predictions (for $m_\nu = 0$) can also be fitted by a polynomial with terms in T, T^2, T^3 and T^4 (no T^5 term is needed). However, one then has to give an explanation of the physical origin of these empirical coefficients and their magnitude.

Contemporaneously, a Berkeley group (Sur *et al* 1991) have measured the spectrum of ^{14}C grown in a Ge crystal detector. The ^{14}C is a part of the crystal structure and the whole is surrounded by a "guard ring" crystal, so that backscattering effects can be eliminated. To fit the data ($Q = 156$ keV) requires a shape factor relative to that for a single massless neutrino

$$S = 1 + b(Q - T) + d\left(1 - \frac{m_2^2}{(Q-T)^2}\right)^{\frac{1}{2}} \tag{7}$$

where
$$b = 0.00117 \pm .00003 \text{ keV}^{-1}$$
$$d = \sin^2 \theta = 0.013 \pm .003$$
$$m_2 = 17.0 \pm 0.5 \text{ keV}$$

It turns that the b term has a physical origin. ^{14}C is an allowed transition, but because the nuclear matrix element is small ($\tau = 5000$yr) there is a contribution also from a first-forbidden transition, producing a distortion of the Kurie plot. According to Commins this has magnitude $b = .0012$ keV^{-1}, in good accord with the empirical value above. The statistical weight of the data is hardly overwhelming and the "kink" is barely visible. Nevertheless, it is an experiment with a different technique and with a different isotope (see Fig.15).

Two other experiments have recently reported data. One measures the γ-ray spectrum arising from inner bremsstrahlung in electron capture (IBEC) by a nucleus A:-
$$e^- + A(Z) \rightarrow \nu + \gamma + A(Z-1) + Q$$

If k denotes the photon momentum, the spectral form for $1s$-state (K) capture and a massless neutrino is:

$$\frac{dF}{dk}(K - \text{capture}) \approx \frac{\alpha}{\pi} \frac{k}{Q} \left(1 - \frac{k}{Q}\right)^2 \tag{8}$$

Hence, if one plots $(k^{-1}kF/dk)^{\frac{1}{2}}$ against k, a straight-line plot results — called a Jauch plot and analogous to a Kurie plot. All the effects of neutrino mass in the electron spectrum in beta decay are mirrored in the IBEC spectrum. Unfortunately there are complications — for example L-capture as well as K-capture, as is capture of $2s$ and $3s$, in addition to $1s$, electrons. All the atomic physics can be folded in, but the experimental situation is, even so, not so straight forward as in beta decay. The big reduction factor of α/π in (8) can be compensated by using thicker sources, but the efficiency of γ detection is energy-dependent. Fig.16 shows the results on the ^{71}Ge IBEC spectrum by a Zagreb group (Zlimen et al 1990) which suggest a 2σ effect of $\sim 1\%$ mixing of a heavy neutrino of mass $m_2 \sim 17.5$ keV. The ^{55}Fe IBEC spectrum has been measured by an LBL group (Norman et al 1990) and they claim a possible effect, with $m_2 = 21 \pm 2$ KeV, $\sin^2\theta = 0.009 \pm .005$. Both of these IBEC experiments are statistically very weak.

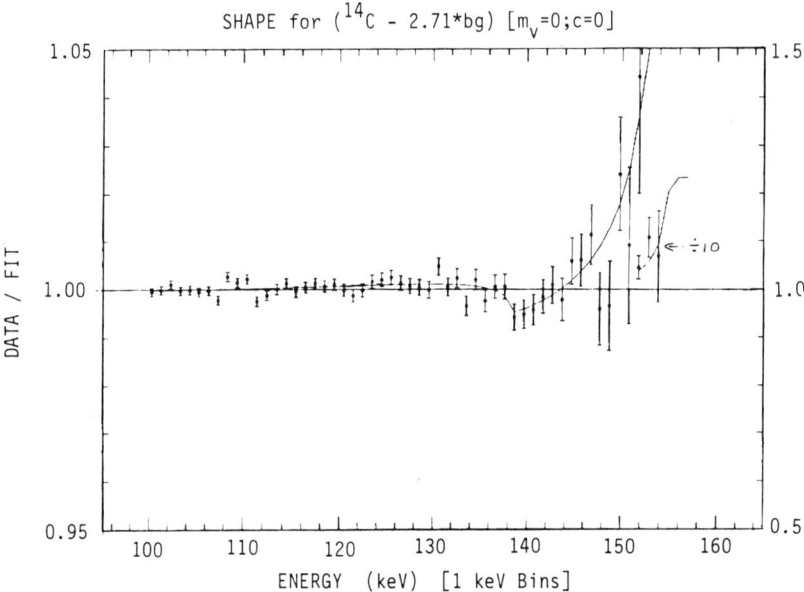

Fig.15 Data of Sur et al (1991) on the ^{14}C decay spectrum. The straight horizontal line, as in Fig.14(a) and (b) is the best fit line for a massless neutrino; the curve shows the deviation for an additional neutrino of $m_2 = 17$ keV, $\sin^2\theta = 0.013$.

Finally, a Cal Tech experiment (Becker et al 1991) using a magnetic spectrometer of the same type as that of Hetherington et al, and a ^{35}S source, has presented data. Fig.17 shows the plot of the residuals to the best fit to the spectrum. This best fit requires linear and quadratic correction terms. It is actually made to electron momentum p rather than energy. The shape factor is

$$S = 1 + b(p_{\max} - p) + c(p_{\max} - p)^2 + d\left[1 + \frac{m_2^2}{(Q-T)^2}\right]^{\frac{1}{2}}$$

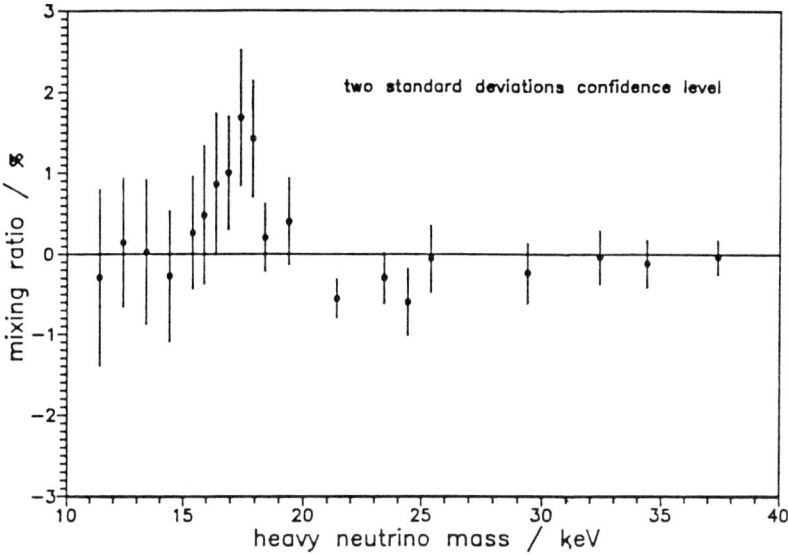

Fig.16 Scan of IBEC spectrum of ^{71}Ge for a heavy neutrino signal, showing the 2σ value of the mixing probability as a function of neutrino mass.

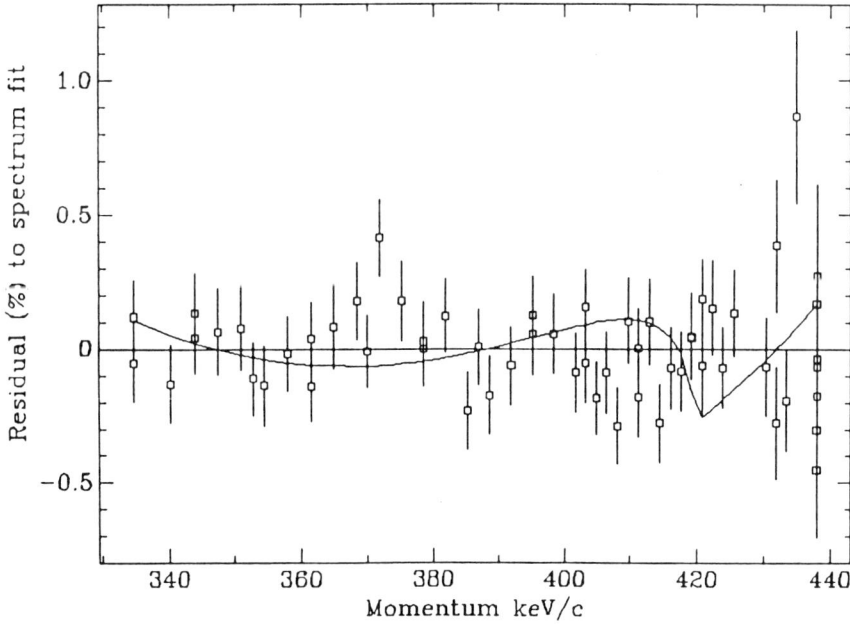

Fig.17 Residuals in the momentum spectrum of ^{35}S measured by Becker *et al* (1991) with the Cal Tech spectrometer. The straight line is the fit for $\sin^2\theta = 0$, the curve for $\sin^2\theta = .0085$, $m_2 = 17$ keV.

where

$$b = 1 \times 10^{-3} (\text{keV}/c)^{-1}$$
$$c = -3.8 \times 10^{-6} (\text{keV}/c)^{-2}$$
$$d = 0$$
$$m_2 = 17 \text{ keV}$$

Again, the variation of the second and third terms over the fit region is much larger than the last term, if $d \simeq .01$. Worse still, all the fits are unsatisfactory ($\chi^2 \sim 4$ per degree of freedom). This is apparently due to the fact that the ^{35}S was deteriorating in an irregular way with time ("bits" coming off it) and as a result all the statistical errors had to be increased by a factor of 1.9 in order to achieve an acceptable χ^2. Although the authors quote $d = \sin^2\theta < 0.006$ at 90% CL, I believe the result should be treated with caution because of the uncertainties in the errors.

4. Discussion

As we have seen, some experiments claim to see a 17 keV effect at a high level of significance, while others see nothing. The positive results are all from solid state detectors, the negative one mainly (but not only) from magnetic spectrometers. Clearly any real effect must be visible in all types of detector and with different isotopes.

It has been suggested that the resolution in the crystal detectors can be at fault. It is obvious that, as shown in Fig.18, an underestimate of the resolution will result in a characteristic "tail" to the distribution observed and an upturn towards the end-point when comparing the data with the prediction (best fit straight line). However, such a deviation occurs over an energy interval $\Delta T \sim 2\sigma$ where σ characterises the resolution (the exact value depends clearly on whether the resolution function is symmetric or asymmetric). The main point however is that $\Delta T \simeq m_2 = 17$ keV implies that one needs $\sigma \sim 8$ keV, whereas it is measured (see Fig.13) to be only 0.6 keV.

It is clear that many of the experiments are unsatisfactory in the sense that the fits require postulating empirical "shape factors" (with or without a 17 keV neutrino), and until one understands the physical origin of such factors, any conclusions have to be premature. Of course it can be argued that no smoothly-varying shape factor can remove a "kink", but this is only true in the case of adequate resolution and statistics in the "kink" region. Clearly in future experiments, high statistics are essential and this, in turn, implies long runs and good instrumental stability.

If the discrepancy between the solid-state and magnetic spectrometers persists, then of course one will have discovered, not a new neutrino, but some novel solid-state effects of almost unbelievable subtlety. However, let us suppose for a moment that the 17 keV neutrino really does exist and discuss some of the consequences.

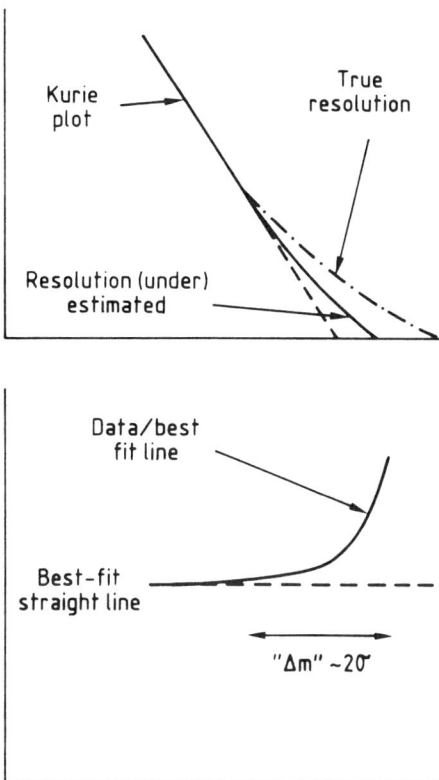

Fig.18 Kurie plot for zero neutrino mass, with effect (full curve) of resolution underestimated and (dot-dash curve) true resolution included. The ratio of these two curves gives a characteristic deviation near the end-point, corresponds to a "mass" $m \sim 2\sigma$ where σ is the true width of the resolution function.

5. Consequences of a 17 keV Neutrino in Astrophysics and Cosmology

(a) Neutrino oscillations: Dirac and Majorana Neutrinos

Suppose that the 17 keV neutrino is one of the known types ν_e, ν_μ, ν_τ. Since ν_e must be the dominant neutrino in beta decay ($m_{\nu_e} < 9\text{eV}$) then $\nu_{17} = \nu_\mu$ or ν_τ. The limits from oscillation experiments are as follows:

Table 1.   Oscillation limits ($X = \mu, \tau$)

Mixing	90% CL upper limit	Method
$\bar{\nu}_e \to \bar{\nu}_X$	$\sin^2 2\theta < 0.16$	reactor
$\nu_e \to \nu_X$	< 0.14	accelerator
$\nu_e \to \nu_\tau$	< 0.12	accelerator
$\nu_e \to \nu_\mu$	< 0.0034	accelerator
$\nu_\mu \to \nu_\tau$	< 0.0040	accelerator

Since the ^{35}S result gives $\sin^2 \theta \simeq 0.008$, or $\sin^2 2\theta \approx 0.03$, it is clear that the present limits would only allow $\nu_e \to \nu_\tau$ so that the 17 keV neutrino would be dominantly ν_τ. Neutrino oscillation experiments in the future to detect $\nu_e \to \nu_\tau$ at the level $\sin^2 2\theta = 0.03$ would be quite difficult, but this sensitivity is expected to be attained within the next 5 years or so. (The difficulty of course is that the signature for ν_τ: $\nu_\tau + N \to \tau^\pm + \cdots$ requires $E_\nu > 5$ GeV for efficient production; and that ν_e constitute only about 1% of the total neutrino flux generated at an accelerator).

If $\nu_{17} \equiv \nu_\tau$ then one would require an extension of the Standard Model to accommodate massive neutrinos — so, an extension of the Higgs sector — but one could retain all the existing couplings (as well as having some new ones). Then clearly ν_{17} would be unstable since it mixes with ν_e:

$$\nu_{17} \to \nu_e + \nu_x + \bar{\nu}_x \tag{10}$$

where ν_x is any light neutrino (see for example the diagram of Fig.19(b)). Comparing with

$$\mu^+ \to e^+ + \nu_e + \bar{\nu}_\mu$$

we get for the lifetime using the Sargent rule

$$\tau_{17} = \frac{\tau_\mu}{\sin^2 \theta} \left(\frac{m_\mu}{m_{17}}\right)^5 = 2.10^{15} \text{ secs} \tag{11}$$

This lifetime turns out to be embarrassingly long from the viewpoint of cosmologists, as indicated below. Radiative decay, $\nu_{17} \to \nu + \gamma$, as in Fig.19(a), is also much too slow and will not be discussed further. Of course, we have assumed here that ν_{17} is a Dirac-type neutrino, that is with distinct particle and antiparticle and both left-handed and right handed states. An alternative possibility is that it is a Majorana neutrino, with no distinction between particle and antiparticle, and simply two helicity states. In this case, neutrinoless double beta decay is possible, via the reactions

(a) $n \to p + e^- + \bar{\nu}$ (b) $\nu(\equiv \bar{\nu}) + n \to p + e^-$

that is,

$$2n \to 2p + 2e^-$$

The lower limit on the lifetime gives an upper limit on the Majorana mass. Since $\nu(\equiv \bar{\nu})$ from (a) is RH there is a helicity suppression factor in (b) of $(1 - v/c) = m_M^2/2E_\nu^2$ in the rate because the neutrino has the "wrong" helicity for $V - A$ coupling. The lifetime limit yields for the Majorana mass $m_M \leq 3$eV. Contrasting this with $m = 17$ keV one concludes that the heavy neutrino must be overwhelmingly a Dirac-type neutrino. Nevertheless, it could be distinct from the ν_e, ν_μ, ν_τ family; for example it could be a weak isospin singlet particle, so not directly coupled to the Z° ($Z^\circ \not\to \nu_s \bar{\nu}_s$), which has $I = 1$, $I_3 = 0$.

(b) Big Bang Nucleosynthesis

We now discuss the consequences of ν_{17} in astrophysics and cosmology. First, ν_{17} will be generated, together with all other fundamental fermions, in the Big Bang. Quarks formed hadrons, hadrons decayed to photons and leptons, the heavy leptons (μ, τ) decayed to electrons and neutrinos, nucleons annihilated with antinucleons and after $t = 1$ second, or $kT \sim 1/(t \text{ secs})^{\frac{1}{2}} \sim 1$ MeV, there would be left a sea of neutrinos (ν_e, ν_μ, ν_τ ...) electrons and photons (plus their antiparticles) and a tiny (10^{-9}) proportion of nucleons. The cross-section for $e^+e^- \to \nu\bar{\nu}$ varies as T^2 and for $kT < 1$ MeV, the collision mean free path exceeds the radius parameter R of the expanding universe. Neutrinos would therefore decouple, and if very light (that is, relativistic) and stable, the density would follow Stefan's Law $\rho \propto T^4$ just as for photons. The density of relic neutrinos would therefore today be similar to that of the microwave (2.7°K) background photons (400/cc). So the ratio of neutrinos to baryons would be of order $N_\nu/N_b \simeq N_\gamma/N_b \approx 10^9 - 10^{10}$. Since the baryon density today is of order 10^{-2} of the critical density for closure of the Universe, it follows that a neutrino mass of order $m_\nu c^2 \approx 100 \times 10^{-10} \times M_p c^2 \sim 30$eV would provide enough "dark mass" to close the Universe. Clearly then, a stable neutrino of mass 17 keV would be disastrous: the Big Crunch would have followed the Big Bang billions of years ago.

The above difficulty can be avoided if ν_{17} undergoes fairly rapid decay. How rapid? For $kT < 17$ keV ($t > 1$ hour) the heavy neutrinos become non-relativistic and the density (if stable) would vary as R^{-3} or T^3 (since $R \propto T^{-1}$) instead of T^4. So, the massive neutrino should decay to relativistic particles before the temperature falls to $T \simeq (17 \text{ keV}/30 \text{ eV})T_0 \approx 600T_0$, where $T_0 = 2.7°$K is the present radiation temperature. Since $t \propto T^{-2}$, this implies decay before a time $\tau < t_0/(600)^2 \sim 10^{12}$ secs, where $t_0 = 3.10^{17}$ secs is the present age of the Universe. This estimate clearly contradicts (11), so that one has to postulate some extra interactions in addition to (10) to make ν_{17} decay faster. There are other, even more restrictive, limits on τ. The development of galaxies and galactic clusters depends on density fluctuations in the early universe, which would be ironed out if a flood of hot radiation (relativistic neutrinos) were to be released from ν_{17} decay. But since there are problems in understanding large scale structure in the Universe in any case, this last argument is hardly compelling.

The identification of ν_{17} with a new type of neutrino, presumably with both LH and RH helicity states, is also claimed to pose difficulty for primordial nucleosynthesis, in the Big Bang. Nucleosynthesis bounds have been quoted as setting a limit on the number of distinct neutrino helicity states, $N_\nu < 3.4$ (Olive et al 1990), with no room for any extra states above the number determined at LEP, of $N_\nu = 2.9 \pm .10$. This limit arises from the dependence of the helium/hydrogen ratio on the cooling rate (and hence N_ν) for a given value of N_b/N_γ. If one takes the measured ratio $N_b/N_\gamma = 6.10^{-11}$ the corresponding limit ($N_\nu \leq 10$) is hardly useful. Only less stringent arguments, based on local Li/H and D/H ratios, suggest that the true N_b/N_γ ratio could be much larger, $\sim 5.10^{-10}$, then corresponding to $N_\nu \leq 3.4$.

(c) **Supernova Neutrinos**

There are astrophysical limits regarding massive neutrinos from SN1987A. If a neutrino is massive, it can undergo spin-flip neutral current scattering off a nucleon as in Fig.19(d)

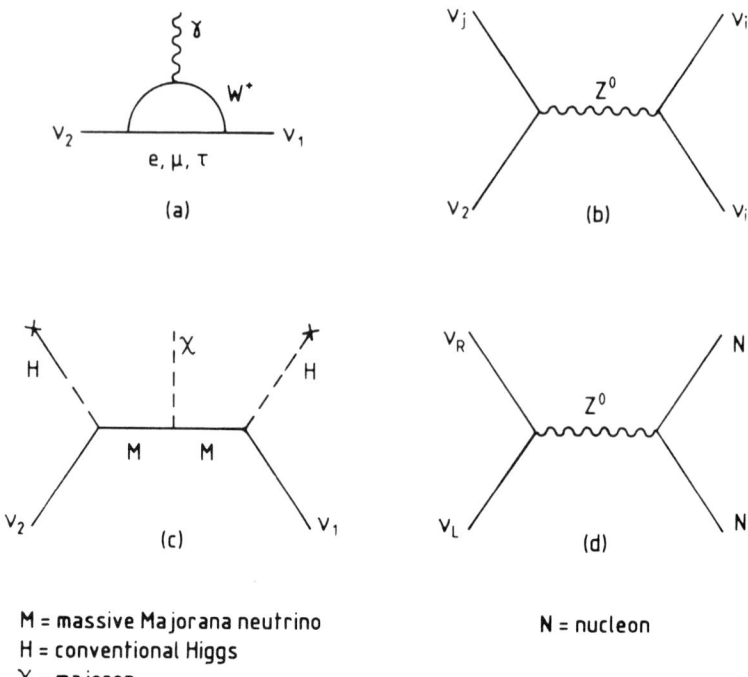

M = massive Majorana neutrino
H = conventional Higgs
X = majoron

N = nucleon

Fig.19 Various mechanisms for ν_{17} decay: (a) radiative decay, $\nu_{17} \to \nu\gamma$, (b) decay to 3 light neutrinos, (c) decay to a majoron of $I = 0$, (d) spin-flip scattering of massive Dirac neutrino.

The cross-section depends on the vector and axial vector couplings C_V and C_A, of the nucleon to the Z° and on the helicity factor $1 - v/c = m_\nu^2/2E^2$. Ghandi and Burrows (1990) give for the differential cross-section

$$(d\sigma/d\Omega)_{\text{spin-flip}} = \frac{G^2}{4\pi}\left(\frac{m_\nu^2}{4s}\right)[M^2(C_V^2 + 3C_A^2) - M^2(C_V^2 - C_A^2)\cos\theta$$
$$+ p^2(C_V - C_A)^2(1 - \cos^2\theta)]$$

where p =CMS momentum, s =(CMS energy)^2, M =nucleon mass, θ =CMS angle of scattering. Since $s \sim M^2 \gg p^2$ for MeV energy neutrinos, the total cross-section

$$\sigma(\text{flip}) \approx G^2 m_\nu^2 = 2.10^{-47}(m_\nu/50 \text{ keV})^2 \text{cm}^2$$

The significance of the spin-flip reaction is that it converts an "active" (LH) neutrino to an "inert" (RH) particle. Such neutrinos can escape the supernova with high probability. Ghandi and Burrows show that the main effect is to shorten the length of the burst of active (normal LH) neutrinos: $m_\nu = 14$ keV reduces the burst length by a factor of 2 (see Fig.20). The only obvious way to avoid this result is to make the "wrong helicity" (RH) neutrino states more active by postulating new interactions.

For further discussion of the 17 keV neutrino in cosmology, see Kolb and Turner (1991).

6. Conclusions

The above examples show that, if the 17 keV neutrino really does exist, there will need to be a rather thorough re-assessment of our whole picture of particle physics and of cosmology, at a basic level. There is little doubt that models must be found to make ν_{17} decay much faster, to avoid problems in cosmology. The very existence of a massive neutrino means an extension of

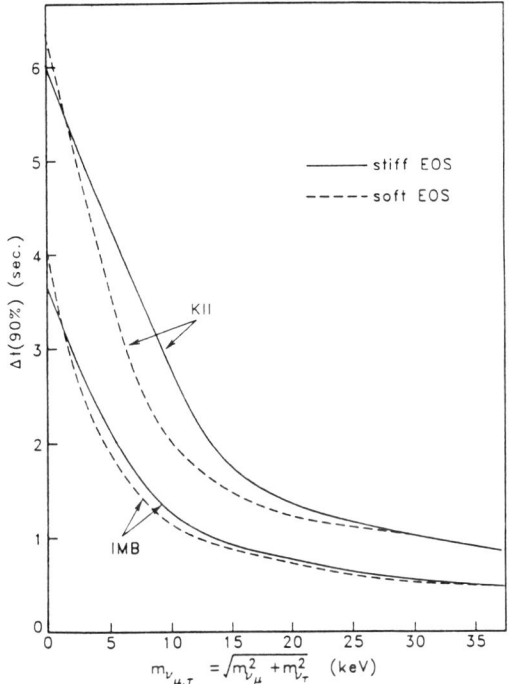

The time taken for the accumulated number of events to reach 90% of the final total number of events as a function of $m_{\nu_{\mu,\tau}}$, for both detectors and both models.

Fig.20 Length of supernova neutrino burst versus mass of heavy neutrino, as a result of spin-flip scattering (Ghandi and Burrows 1990).

the minimal Higgs sector, and possible decay to a massless scalar Goldstone boson J — usually called a "majoron" — $\nu_{17} \to \nu + J$, as in Fig.19(c). The coupling is arbitrary and can be anything to suit experiment. The prospect of more particles and more couplings in physics is either exciting or depressing, depending on one's viewpoint.

Equally, one may need a re-assessment of the way in which we do research in particle physics. It is a rather humbling thought that the world of particle physics and cosmology can be turned upside down by a few table top experiments.

References

Altzitzoglou, T. et al, *Phys. Rev. Lett.* **55** 799 (1985)
Becker, H.W. et al, Cal Tech preprint (1991): Proc. XIth Moriond Workshop, Les Arcs (Jan 1991)
Datar, V.M. et al, *Nature* **318** 547 (1985)
Ghandi, R. and A. Burrows, *Phys. Lett.* **246B** 149 (1990)
Hetherington, D.W. et al, *Phys. Rev.* **C36** 1504 (1987)
Hime, A. and N. Jelley, *Phys. Lett.* **B257** 441 (1991)
Kolb, E.W. and M.S. Turner, *Phys. Rev. Lett.* **67** 5 (1991)
Norman, E.B. et al, Proc. of 14th Conf. on Nuclear Physics, Bratislava, Czechoslovakia (October 1990)
Ohi, T. et al, *Phys. Lett.* **B160** 3221 (1985)
Olive, K.A. et al, *Phys. Lett.* **236B** 454 (1990)
Simpson, J.J., *Phys. Rev. Lett.* **54** 1891 (1985)
Simpson, J.J. and A. Hime, *Phys. Rev.* **D39** 1825 (1989)
Sur, B. et al, Berkeley preprint (December 1990); *Phys. Rev. Lett.* **66** 2444 (1991)
Zlimen, I. et al, Zagreb preprint 11-9/90. LEI (199)

CHAIRMAN: D.H. Perkins

*Scientific Secretaries: R. Malik, N.E. Moulai, I. Rothstein,
H.J. Schulze, M. Wadhwa*

DISCUSSION

– Sivaram:

1) Assuming the 17 KeV neutrino exists, what can you say about its magnetic moment? Standard Model would give $\sim 3 \times 10^{-19} \left(\frac{m_\nu}{1\,\text{eV}}\right) \mu_B$, which for 17 KeV gives $\sim 10^{-14} \mu_B$. What about its implications for supernovae or red giants etc.?

2) Can we say whether the 17 KeV neutrino is a Dirac or Majorana neutrino?

– Perkins:

1) The constraints on the magnetic moment depend on the neutrino type. The constraint on the Dirac type moment is $\mu_\nu^D < 10^{-12}$ from supernovae observation, whereas the cooling of red giants gives the constraint $\mu_\nu^M < 4 \times 10^{-12}$.

2) (Rothstein) From double β decay constraints we know that $\sum_i \xi_i U_{e_i}^2 m_i < 6\,\text{eV}$, where ξ_i are the CP eigenvalues. This implies that 17 KeV neutrino is most probably pseudo-Dirac.

– Wadhwa:

Is it possible to do measurement of the neutrino mass with B-factories?

– Perkins:

In principle this is possible, but why do you want to make your life more difficult?

– Goobar:

You mentioned that there could be other "explanations" to the Simpson-Kurie plot. You said that one could fit a polynominal to the data and achieve the same level of agreement to the data as the 17 KeV hypothesis. What would then be the physical meaning of such a fit?

– Perkins:

We spent a year trying to devise other means of explaining these results and failed. The most convincing explanation to me at the time is that the 17 KeV neutrino exists. The polynominal fit is fine, but where do the 4 empirical coefficients come from? It has no physical meaning.

– *Malik:*

Why do the magnetic spectrometers not give the 17 KeV neutrino mass result, why only semiconductor spectrometers?

– *Perkins:*

Only about 1/2 of the semiconductor experiments see the effect. I can only say that magnetic spectrometer experiments are much more difficult because one has to use a "shape factor" which expresses one's ignorance of the detailed experimental conditions. These correction factors are much larger than the 17 KeV effect on the one-component plot.

– *Moulai:*

Why are the limits on the mixing angle for $\nu_e \leftrightarrow \nu_\tau$ transitions so much worse than for $\nu_\mu \leftrightarrow \nu_e, \nu_\mu \leftrightarrow \nu_\tau$?

– *Perkins:*

Simply because it is much more difficult to measure, and to make $\tau^\pm$ from ν_τ requires $E_\nu > 5$ GeV. No one has ever seen the tau-neutrino.

– *Gourdin:*

1) You were presenting this morning an impressive set of experiments. Which one is the simplest one from the point of view of Nuclear Physics and Atomic Physics corrections?

2) In the data analysis two parameters are inspected: the mass m_ν and the mixing angle θ. Apparently m_ν is stable and θ very uncertain, even if small. I got the impression that $m_\nu = 17$ KeV was more or less constraint and θ a free parameter.

– *Perkins:*

1) The problem with the atomic and nuclear physics is only significant for Tritium. I do not believe that nuclear and atomic effects are crucial for the experiments we have done.

2) I think you are right, most experiments looked at the hypothesis, is the 17 KeV neutrino there? And determined the mixing angle.

– *Schulze:*

Are there possible explanations of the "17 KeV neutrino" by a solid state effect in the detectors?

– *Perkins:*

No, I am not aware of any such explanation either regarding the material or the electronics. The typical energy scale in solids is eV compared to KeV.

– *Passalacqua:*

1) Why have the data been fitted with two neutrinos, not with three? Furthermore, why is the mass of one neutrino zero?

2) Could you please describe the calibration of the SSD?

– *Perkins:*

1) It is just very convenient to use a zero mass neutrino and the end-point is consistent with one component having zero mass. You can use 10 mass values if you like, but it should satisfy data.

2) The counters were calibrated with the pulsers and with gamma ray sources of known energy.

– *Hasan:*

Somehow the observation of 17 KeV neutrinos does not fit well into the unification theories. Do you know any theory which can accomodate 17 KeV neutrinos?

– *Perkins:*

I do not know. I am not concerned with grand unification.

– *Rothstein:*

Many models have been built which incorporate the 17 KeV neutrino, although not necessarily in a GUT scenario.

– *Pluciennik:*

In the Hetherington-experiment, how can you explain that you can get rid of the 17 KeV with a shape factor?

– *Perkins:*

You cannot get rid of a "kink"; but a small change in slope of the Kurie plot can be easily simulated by a shape factor.

– *Junk:*

Is the recoil energy of the radiating nucleus significant at this level? In particular, if a Mössbauer-like effect is going on in the source a small fraction of time, can it fake a 17 KeV neutrino signal?

– *Perkins:*

The emitter is a monolayer of salt, so a Mössbauer effect will not take place. The recoil energy is on the order of eV, so it would not make any difference anyway.

– *Giannotti:*

Could you explain why you need an isotropic source to calibrate the solid state detector?

– *Perkins:*

It is not necessary to have an isotropic source but it has to cover the same angle as the beta source, because otherwise we cannot look at the backscattering with the same geometry. A gun with a parallel electron beam, would be unsuitable.

– *Barletta:*

We find it difficult to make good parallel beams. It is much easier to make a wide angle divergent source.

A SOLUTION TO THE TIME VARYING SOLAR NEUTRINO PROBLEM

I.Z. Rothstein[1]

Department of Physics and Astronomy
University of Maryland, College Park, MD 20742

1. Introduction

Perhaps I should start by explaining what I mean by this title. I don't mean to say that sometimes there is no solar neutrino problem and sometimes there isn't! By the solar neutrino problem (SNP) I am referring to the fact that the present solar neutrino experiments see only a fraction of the expected number[1] of neutrinos as predicted by the standard solar model (SSM). When I use the term " time varying solar neutrino problem" (TVSNP) I am referring to the fact that there seems to be an anti-correlation of the observed solar neutrino flux (as measured in the Chlorine experiment, more on this later) with the solar sun spot cycle.

The existence of the the SNP is on firm footing, as the flux of solar neutrinos reaching the earth has been measured by Davis and collaborators [3] for more than twenty years using a Chlorine detector, and the theoretical expectations have recently been corroborated to be below the measured flux[2]. More recently, the Kamiokande II collaboration has measured this quantity using a water Cherenkov detector[4]. On the average, both of these experiments report a measured neutrino flux which is about 30 to 50 percent of the expected value. The theoretical predic-

---

[1]Work supported by a grant from the National Science Foundation

tions are based on the standard solar model [1,2] and the neutrino properties based on the standard electro-weak model. Therefore, if our understanding of the sun is not faulty, this discrepancy between theory and experiment has profound significance for particle physics beyond the standard model. Presently the most popular solution to the SNP is the MSW solution[5]. In this scenario the electron neutrino (which must be massive if MSW is to work) [2] goes through a matter enhanced resonant oscillation into another species of neutrino which either does not interact at all in either detector or at least interacts more weakly than the electron neutrino does. This solution is quite elegant because there is little fine tuning involved and requires minimal extension of the Weinberg-Salam model. However, the MSW mechanism is unable to address the TVSNP.

The TVSNP on the other hand is not on as firm a ground. Table 1. shows the capture rate in Chlorine as a function of the solar activity. Presently, the statistics are not good enough to make definitive statements about the anti-correlation[13]. However, as is, the data seems to be quite compelling and has attracted a lot of attention in the theoretical community. This is mainly because if such an anti-correlation exists it would seem to indicate that the electron neutrino has a large (where by "large" I mean large compared to the value which would be generated in the standard model if a right handed neutrino is added to the spectrum) , i.e. $> 10^{-11}\mu_B$, magnetic moment[3].

The idea that the paucity of solar neutrinos could be do to a large neutrino magnetic moment was first introduced in ref [6]. In this scenario the conversion to the sterile neutrino takes place in the magnetic field present in the convective layer of the sun, due to ordinary spin precession. Depending on the nature of the magnetic moment interaction, the sterile neutrino can be a right-handed ν_{eR} or the anti-particle of the observed left-handed muon-neutrino $\bar{\nu}_\mu$ (which is of course right-handed too). The first situation occurs if the magnetic moment is of the Dirac type i.e., $\bar{\nu}_{eL}\sigma_{\alpha\beta}\nu_{eR}F^{\alpha\beta}$, whereas the second situation arises if we have instead an interaction of the Majorana type: $\nu_{eL}^T C^{-1}\sigma_{\alpha\beta}\nu_{\mu L}F^{\alpha\beta}$ - the so-called transition magnetic moment. We will specialize here to the discussion of the transition magnetic

---

[2]Recently there have been several proposals for an MSW-like solution with massless neutrino, but new flavour changing neutral currents are needed.

[3]Here I will not address the issue of generating such large magnetic moments, but instead refer the reader to refs [7,8,9].

Table 1. ν_e Capture rate in Chlorine as a function of the solar activity.

Solar Activity	Date	Chlorine Capture Rate in SNU's
Minimum	1977	4.1±0.9
Maximum	1979.5-1980.7	0.7±0.6
Minimum	1986.8-1988.3	4.2 ±0.7
Maximum	1988.4-1989.5	0.8±0.6

moment and its effect on the solar neutrino flux, as it seems to be the most plausible choice from the theoretical point of view as well as from cosmological and astrophysical constraints.

There are cosmological and astrophysical arguments which disfavor a large Dirac type magnetic moment. Specifically, SN1987A observations require $\mu_\nu < (10^{-12} - 10^{-13})\mu_B$ for a Dirac neutrino [10], which make it much too small to be relevant for solar neutrinos. There are ways to evade the supernova bound by postulating exotic Higgs interactions of ν_R [11], which are in any case required to generate a large μ_ν. However, these new interactions, which trap the ν_R's inside the supernova core also keep them in equilibrium with the primordial plasma at the epoch of nucleosynthesis, so that it counts as one extra species. The most recent analysis of primordial $^4He, ^2D$ and ^7Li abundance allows at most 0.3 extra species of neutrinos [12]. Therefore, a Dirac type magnetic moment is unlikely to be relevant for the solar neutrino puzzle. These constraints do not apply to the transition magnetic moment.

At this point we could say to ourselves, fine, there is a TVSNP, but we can understand it quite simply by attributing a large magnetic moment to the neutrino. However, this is not the end of the story. Though the chlorine experiment seems to see an anti-correlation of the solar neutrino flux with the solar sun spot cycle, the Kamiokande experiment shows little or no time variation. Therefore, simply positing the existence of a large neutrino magnetic moment won't do if we want to solve the TVSNP.

In this talk I would like to discuss some recent work that my collaborators (R.N. Mohapatra and K.S. Babu) and I have done on solving the TVSNP. We take the time variation in the Chlorine data seriously and search for parameter ranges of neutrino masses, mixings and magnetic moment for which there would not be significant time variation in the Kamiokande (KII) neutrino signal, while there would be large variations in the Chlorine signal. Our study was motivated by two obvious differences between the Chlorine and Kamiokande experiments. (i) Due to the large threshold energy (7.5 MeV) in the water detector compared to the Chlorine detector, approximately 25 percent fewer neutrinos is expected at KII as compared to Chlorine. Therefore, if one could find a range of parameters for which the variation in neutrino flux with the magnetic field remains confined to those low energy neutrinos which are not expected to be seen at KII, one could perhaps explain the discrepancies between the two experiments [14]. (ii) For the case of the transition magnetic moment, the converted ν_e's become $\bar{\nu}_\mu$'s which do not give any signal in Chlorine, but they do contribute a small amount to the signal at KII via $\bar{\nu}_\mu + e \to \bar{\nu}_\mu + e$ neutral current scattering ($\sigma_{\bar{\nu}_\mu e \to \bar{\nu}_\mu e} \approx \frac{1}{7}\sigma_{\nu_e e \to \nu_e e}$). In order to see how significant these two effects are, a detailed analysis is needed.

The main result of our work is that we've shown that by including the matter and magnetic field effects in the the solar interior on the neutrino propagation, values of the parameters can be found which explain the time variation in Chlorine data without causing any significant variation in the KII data.

This talk is organized as follows. In section 2, I analyze the canonical case of oscillations in a constant magnetic field between two neutrinos species including matter effects. In Section 3, the analysis is extended to the more realistic case of a spatially varying magnetic field. Section 4 introduces the wave equation for the complete case of a 4 by 4 system so that we may include mixing as well as spin precession. In Section 5, I discuss fitting the data from Chlorine and Kamiokande. In Section 6, I give the prediction for the expected results for the case of the Gallium detector. Section 7 contains some concluding remarks.

2. Spin precession in a Constant Magnetic Field

We begin by considering the simplified case of zero mixing angle and a constant magnetic field for Majorana neutrinos with off diagonal magnetic moments. The wave equation for this system is [15]

$$i\frac{d}{dt}\begin{pmatrix} \nu_e \\ \bar{\nu}_\mu \end{pmatrix} = \begin{pmatrix} N_1 - \Delta m^2/4E & \mu B \\ \mu B & -N_2 + \Delta m^2/4E \end{pmatrix} \begin{pmatrix} \nu_e \\ \bar{\nu}_\mu \end{pmatrix} \quad (1)$$

where

$$N_1 = \sqrt{2}G_F(N_e - N_n/2), \quad N_2 = -G_F N_n/\sqrt{2}. \quad (2)$$

N_e and N_n are the electron and neutron number density respectively. $\Delta m^2 = m_2^2 - m_1^2$, where m_1, m_2 are the in vacuum masses. In terms of an instantaneous mass eigenstate basis, eq. (1) is written as

$$i\frac{d}{dx}\left(U\begin{pmatrix} \nu_1 \\ \nu_2 \end{pmatrix}\right) = U\begin{pmatrix} M_1(x) & 0 \\ 0 & M_2(x) \end{pmatrix}\begin{pmatrix} \nu_1 \\ \nu_2 \end{pmatrix} \quad (3)$$

where we have used the fact that the neutrinos are relativistic. $M_1(x)$ and $M_2(x)$ are the instantaneous mass eigenvalues given by

$$M_{1,2} = \frac{1}{2}\left[N_1 - N_2 \pm \left[(N_1 + N_2 - \frac{\Delta m^2}{2E})^2 + (2\mu B)^2\right]^{1/2}\right]. \quad (4)$$

U is the instantaneous mixing matrix

$$U = \begin{pmatrix} cos\theta_m(x) & sin\theta_m(x) \\ -sin\theta_m(x) & cos\theta_m(x) \end{pmatrix} \quad (5)$$

with

$$tan 2\theta_m = \frac{2\mu B}{[(\Delta m^2/2E) - (N_1 + N_2)]}. \quad (6)$$

In terms of this mixing angle we may write the wave equation as

$$i\frac{d}{dx}\begin{pmatrix} \nu_1 \\ \nu_2 \end{pmatrix} = \begin{pmatrix} M_1(x) & -id\theta_m/dx \\ id\theta_m/dx & M_2(x) \end{pmatrix}\begin{pmatrix} \nu_1 \\ \nu_2 \end{pmatrix}. \quad (7)$$

The neutrino oscillations only depend on the two parameters, $d\theta_m/dx$ and $M_1 - M_2$. This can be seen from the fact that we are free to define

$$\begin{pmatrix} \nu_1' \\ \nu_2' \end{pmatrix} = exp(-i(M_1 + M_2)x/2)\begin{pmatrix} \nu_1 \\ \nu_2 \end{pmatrix} \quad (8)$$

in terms of which the wave equation is

$$i\frac{d}{dx}\begin{pmatrix}\nu'_1\\\nu'_2\end{pmatrix}=\begin{pmatrix}M_1(x)-M_2(x) & -id\theta_m/dx\\ id\theta_m/dx & M_2(x)-M_1(x)\end{pmatrix}\begin{pmatrix}\nu'_1\\\nu'_2\end{pmatrix}. \quad (9)$$

Here

$$d\theta_m/dx = \frac{\mu B}{[(2\mu B)^2+(\Delta m^2/2E-N_1-N_2)^2]}\frac{d}{dx}(N_1+N_2). \quad (10)$$

From eq. (10) we see that $d\theta_m/dx$ is maximized when $\Delta m^2/2E = N_1+N_2$ (resonance condition). Therefore, if $|\,d\theta_m/dx\,|_{resonance} \ll |\,M_1-M_2\,|_{resonance}$, then the neutrino propagation through the sun will be adiabatic. That is, if the neutrino starts in an eigenstate of the Hamiltonian it will track that eigenstate while the mixing angle changes due to changes in the matter density. If the mixing angle passes through a resonance then there may be complete conversion from one flavor neutrino to another. This is the usual MSW mechanism. For the case we are discussing here, the adiabaticity condition is

$$\frac{d}{dx}(N_1+N_2)\,|_{resonance} \ll 4(\mu B)^2. \quad (11)$$

Notice that although explicitly the adiabaticity condition is independent of the energy of the neutrino, there is implicit dependence on $E/\Delta m^2$ because the derivative is evaluated at the resonance point inside the sun. As $E/\Delta m^2$ gets larger, the resonance condition (if it is met at all) will be satisfied farther away from the core. Assuming that the adiabaticity condition is satisfied, we can write down an analytic form for the survival probability at the edge of the sun (which for the case of small mixing will be the same as the probability at earth) [15]:

$$P(\nu_e \to \nu_e) = 1/2(1+cos2\theta_{edge}cos2\theta_{core}). \quad (12)$$

Here θ_{edge} and θ_{core} are the matter mixing angles at the edge and core of the sun. This is the so-called classical probability in that it assumes that all phase information is lost, and therefore all quantum mechanical cross terms are washed out. The conditions for loss of phase information may be due to decoherence of the wave function or other affects such as the finite size of the production or detection region (we will discuss the details shortly). One must be careful that these conditions are met if one is to use the analytic expression given above.

Even if the adiabaticity condition is not satisfied, there will be a finite probability for a jump from one mass eigenstate to the other. This can be taken into account in the above equation for the survival probability by including the jump probability [16]

$$P(\nu_e \to \nu_e) = 1/2 + (1/2 - P_c)cos2\theta_{edge} cos2\theta_{core}. \tag{13}$$

The jump probability is given by

$$P_c = exp\left[-\gamma\pi/2\left\{1 - tan^2(1/2(tan^{-1}[\frac{\mu B}{\Delta m^2/4E}]))\right\}\right] \tag{14}$$

where

$$\gamma = \frac{4(\mu B)^2}{d/dx(N_1 + N_2)|_{resosnance}}. \tag{15}$$

We integrated eq. (1) numerically to generate survival probability graphs for electron neutrinos. We used the neutron matter densities given by Bachall and Ulrich [1] which can be parametrized as [17]

$$N_n = N_n(0) \; exp(-\frac{1}{x_0}\frac{x^2}{x+b}) \tag{16}$$

$$x = R/R_\odot \; ; \; x_0 = 0.09 \; ; \; b = 0.02; \; N_n(0) = 48.4 N_A \tag{17}$$

where N_A is Avogadro number. The electron density used is [1]

$$log(N_e/N_A) = 2.32 - 4.17x - 0.000125/[x^2 + .5^2]cm^{-3} \quad x = R/R_\odot < 0.25$$

$$N_e/N_A = 245 exp(-10.54x)cm^{-3} \quad x > 0.25 \tag{18}$$

In order to get a feeling for the qualitative features of the survival probability, we integrate eq. (1) numerically with a constant magnetic field $B = 10^6$ Gauss throughout the sun and $\mu = 2 \times 10^{-11}\mu_B$. Since only the product μB is relevant, the results for other values of μ (or B) can be obtained by a rescaling. The survival probability $P_{\nu_e \to \nu_e}$ as a function of $E/\Delta m^2$ is plotted in Figure 1a. The region of complete depletion is the region where the resonance condition is satisfied.

In comparing Figure 1a with the MSW solution, we notice that for constant magnetic field, the probability does not return to 1 for large $E/\Delta m^2$, as it does in the case of standard MSW. To understand this, we must first consider why the MSW graph returns to 1 for large values of $E/\Delta m^2$. In the case of MSW the adiabaticity condition is

$$1 \ll \gamma \equiv \frac{N_e \Delta m^2 sin^2 2\theta}{2E cos2\theta \; | \; dN_e/dx \; |_{resonance}}. \tag{19}$$

We see that for large values of $E/\Delta m^2$ this condition will no longer be met. We must therefore consider the jump probability for this case, which is

$$P_c = exp\left[\frac{-\pi}{2}\gamma\left(1 - tan^2\theta\right)\right] . \qquad (20)$$

Therefore for large values of $E/\Delta m^2$, the jump probability is 1, as such, $P(\nu_e \to \nu_e)$ is 1 as can be seen from eqs. (13), (19) and (20). (It is interesting to note that even if we ignore the fact that adiabaticity is lost for large values of $E/\Delta m^2$ the probability will still return to 1, except at larger values of $E/\Delta m^2$.)

In the case under study, we can see that the adiabaticity condition, if it is met at all, will be met for all values of $E/\Delta m^2$. This is due to the following reasons. Firstly, as mentioned before, the adiabaticity condition does not explicitly depend on $E/\Delta m^2$, as it does in the MSW case, so it is not clear that the jump probability will go to one for large values of that parameter. However, there is implicit dependence on this parameter because the adiabaticity condition is evaluated at resonance. As $E/\Delta m^2$ gets larger the resonance is met farther away from the core. However, as can be seen from eq. (18), the derivative of the matter density decreases as we go away from the core. Therefore, the adiabaticity condition is effectively independent of $E/\Delta m^2$.

It remains to understand why the conversion probability in Figure 1a goes to a constant value for large values of $E/\Delta m^2$. Since the adiabaticity condition is effectively independent of $E/\Delta m^2$, this limit does not affect adiabaticity. Therefore, it is clear in light of equations (6) and (12) that in this limit the mixing becomes maximal and the probability will become 1/2. Figure 1b shows a plot of eq. (12) for the case of adiabatic conversion. Note its agreement with Figure 1a.

To see the dependence of the survival probability on the magnitude of B, we plot in Figure 2a an analogous plot for a larger value of $B = 10^7$ Gauss. In this plot we see that there are spikes in the region of parameter space for values of $E/\Delta m^2$ greater than 10^{11} eV^{-1}. These spikes will not be born out in the analytical calculations because they are a consequence of oscillations which have been averaged out in the analytic calculation. This can be seen from Figure 2b which shows the survival probability through the sun for a fixed value of $E/\Delta m^2 = 10^{12}$ eV^{-1}. The wavelength of oscillation is given by

$$\lambda_m = \frac{2\pi}{\sqrt{(N_1 + N_2 - \frac{\Delta m^2}{2E})^2 + (2\mu B)^2]}} . \qquad (21)$$

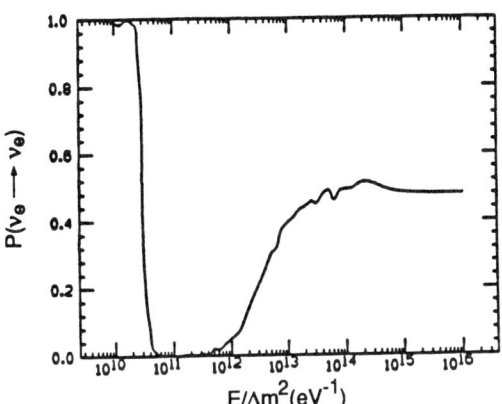

Fig. 1a. Numerically generated probability of a ν_e created at the center of the sun emerging at the surface of the sun as a ν_e ($P(\nu_e \to \nu_e)$) as a function of $E/\Delta m^2$ for a constant magnetic field throughout the sun of 10^6 Gauss and $\mu = 2 \times 10^{-11} \mu_B$.

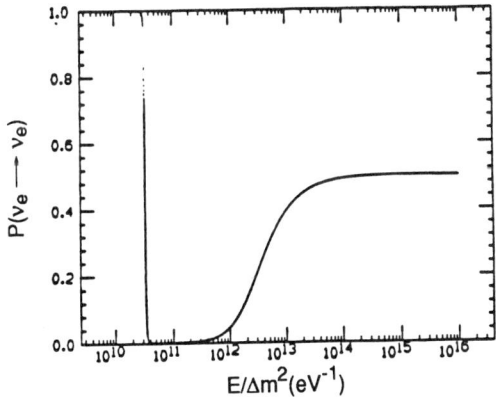

Fig. 1b. Plot of eq. (12) for B=10^6 Gauss throughout the sun and $\mu = 2 \times 10^{-11} \mu_B$.

We see that when $\frac{\Delta m^2}{4E}$ is comparable to μB there will be energy dependence of the oscillations which causes the spikes in Figure 2a. As $\frac{\Delta m^2}{4E}$ decreases further, the frequency of oscillation becomes independent of the energy and hence the leveling off in Figure 2a. It is also important to point out that the amplitude of the oscillations is not completely determined by the effective mixing angle. If it were, we would expect these spikes to occur for any value of μB which is much larger than $N_1 + N_2$ at the outer surface of the sun. For instance, we would expect these spikes in Figure 1a also. However, the amplitude of oscillation also depends on the wave function when the neutrino reaches the region of large mixing in the sun. For instance, if the neutrino is 50 percent ν_e and 50 percent $\bar{\nu}_\mu$, then the amplitude of oscillation will be zero, as any amount being converted to ν_e will be compensated by conversion to ν_e from $\bar{\nu}_\mu$.

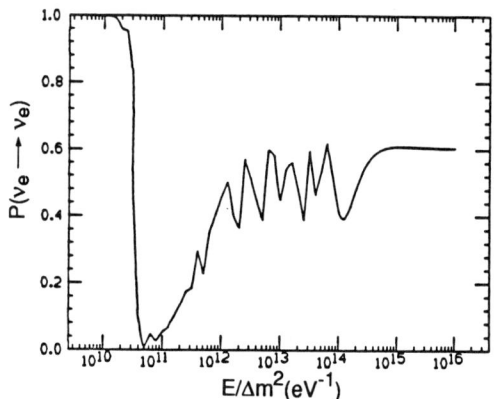

Fig. 2a. $P(\nu_e \to \nu_e)$ for B=10^7 Gauss throughout the sun and $\mu = 2 \times 10^{-11} \mu_B$.

The next obvious question is: which should we take seriously, the analytical result or the numerical calculations? In order to determine which is correct, one must check each case individually. The numerical calculations should only be taken seriously if the following conditions are met: (i) the survival probability at the edge of the sun should not depend sensitively on the radius of the sun. This is not quite equivalent to the usual statement that the number of oscillations should not be too large within the sun, because matter effect may cause the oscillation amplitude to vary within the sun. (ii) There should not be large oscillations in the region of production. This is the so-called "depolarization effect". Since the production region is finite, large oscillations will lead to a strong dependence of the probability at the edge of the sun, on the exact position of production. Therefore, one must average over the oscillations. (iii) The probability at the edge should not change appreciably when the spread in the energy of the wave packet is taken into account.

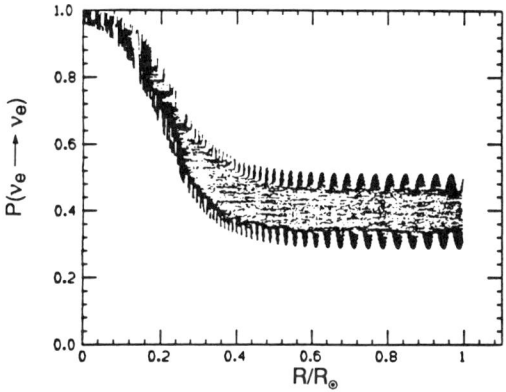

Fig. 2b. $P(\nu_e \to \nu_e)$ as a function of $R/R_\odot$ for $E/\Delta m^2 = 10^{12}~eV^{-1}$, B=$10^7$ Gauss

In the case here the spread in the energy is approximately O(Kev) [18]. (iv) If the neutrino is a linear superposition of the two mass eigenstates then as the neutrino propagates the components of the wave function in the two mass eigenstate will begin to separate spatially due to the difference in velocities. If they separate a distance which is greater then the width of the wave packet, then they will no longer interfere with each other, thus wiping out the oscillations [18,19]. In the adiabatic case this is not a consideration since the wave function only contains one mass eigenstate component during its propagation. For each case one must make sure that the computer simulations pass all the above mentioned tests before one can take the results seriously. For example, when there are large oscillations as in Figure 2b, the survival probability at the edge of the sun must be averaged, and the true probability will be the same as that determined in Figure 1b, where there are no large oscillations.

3. Case of Varying Magnetic Field

We now take into account the variation of the magnetic field in the sun. There is no direct observational data on the magnetic field inside the sun. However, there seems to be some general consensus regarding the gross pattern in the solar core ($R/R_\odot > 0.65$). The core, due to a higher magnetic permeablity (by some 5 orders of magnitude compared to outside the core), is not affected by changes in the magnetic field in the convective zone ($R \geq 0.65 R_\odot$). Furthermore, the core magnetic field is believed to be much larger in magnitude compared to the convective zone. Both fields are spatially decreasing. In this article we use the parameterization given in [14] which incorporates all these features[4]:

$$B = B_1 \left(\frac{0.1}{x+0.1}\right)^2 \qquad x = R/R_\odot \leq 0.65$$
$$B = B_0 \left[1 - \left(\frac{x-0.7}{0.3}\right)^2\right] \qquad 0.65 \leq x \leq 1 \; . \qquad (22)$$

The core magnetic field amplitude B_1 is expected to be less than 10^8 Gauss, (but is probably $> 10^5$ Gauss) and the convective zone field amplitude B_0 is expected be of order $10^3 - 10^4$ Gauss. Also, the core magnetic field is not time dependent whereas, the convective zone field will vary at least by an order of magnitude during a sun spot cycle. Here we will study the survival probability using the parameterization given above, for several different values of the convective zone magnetic field amplitude B_0, keeping $\mu = 2 \times 10^{-11} \mu_B$. Figure 3a shows the survival probability curve with $B_1 = 10^7$ Gauss and $B_0 = 2 \cdot 10^4$ Gauss. The first region of complete depletion between 10^{11} and $10^{12.5}$ eV^{-1} is the usual MSW resonance whose width is determined by the size of the effective mixing angle of eq. (6). The second region of complete depletion, which is the region we shall use in our solution, is more like an VVO[6] solution. That is, it is due to a node in a long wavelength oscillation. Figure 3b shows the survival probability through the sun for $E/\Delta m^2 = 10^{14.2}$ eV^{-1} so that it is in second depletion region. Notice that the probability is essentially constant throughout the core where the matter effects dominate. Then once the neutrino reaches the convective zone, the magnetic off-diagonal mixing terms become important and large oscillation sets in. The amplitude of the oscillation as well as the wavelength is set by the size of the magnetic field. We therefore expect that as we vary the magnetic field, the probability at the edge of the sun (at least for the value of

---

[4]For an explanation of the dependence of the survival probability on the magnetic field profile see ref [23]

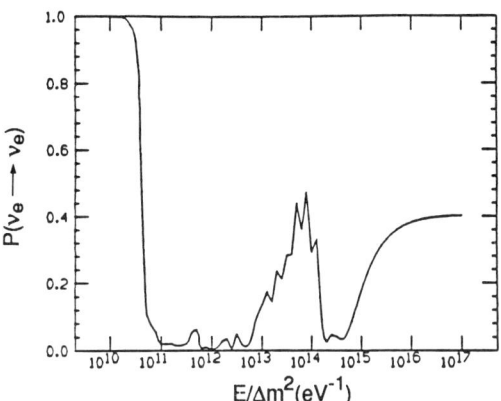

Fig. 3a. ν_e survival probability for a core field amplitude $B_1 = 10^7$ Gauss and a convective zone amplitude $B_0 = 2 \times 10^4$ Gauss, and $\mu = 2 \times 10^{-11} \mu_B$.

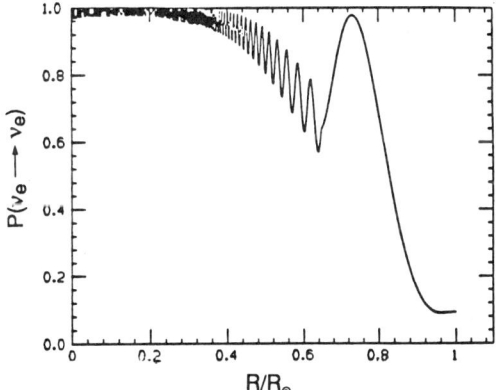

Fig. 3b. $P(\nu_e \to \nu_e)$ as a function of $R/R_\odot$, with $E/\Delta m^2 = 10^{14.2}\ eV^{-1}$, and the same magnetic field and μ as in Fig. 3a.

$E/\Delta m^2$ chosen here) will also vary. From eq. (21) we see that as $E/\Delta m^2$ gets large, the frequency depends only on μB, hence the leveling off in Figure 3a.

4. The four by four Case

In this section we include effects of in vacuum mixing. That is, the gauge eigenstates are no longer mass eigenstates. This is the more natural possibility as there is no reason why the gauge eigenstates should simultaneously be eigenstates of the mass matrix. When we include vacuum mixing we must also include two additional two-component spinors. The wave equation is now given by

$$i\frac{d}{dt}\begin{pmatrix} \nu_e \\ \bar{\nu}_e \\ \nu_\mu \\ \bar{\nu}_\mu \end{pmatrix} = \begin{pmatrix} N_1 - \delta c2 & 0 & \delta s2 & \mu B \\ 0 & -N_1 - \delta c2 & -\mu B & \delta s2 \\ \delta s2 & -\mu B & N_2 + \delta c2 & 0 \\ \mu B & \delta s2 & 0 & -N_2 + \delta c2 \end{pmatrix} \begin{pmatrix} \nu_e \\ \bar{\nu}_e \\ \nu_\mu \\ \bar{\nu}_\mu \end{pmatrix} \quad (23)$$

Here $c2 = cos2\theta$, $s2 = sin2\theta$, and $\delta = \Delta m^2/4E$. There are now four resonance conditions:

$$(N_1 - N_2) = \sqrt{2}G_F N_e = 2\delta c2, \quad (\nu_e \leftrightarrow \nu_\mu) \quad (24)$$

$$(N_1 + N_2) = \sqrt{2}G_F(N_e - N_n) = 2\delta c2, \quad (\nu_e \leftrightarrow \bar{\nu}_\mu) \quad (25)$$

$$(N_1 + N_2) = \sqrt{2}G_F(N_e - N_n) = -2\delta c2, \quad (\bar{\nu}_e \leftrightarrow \nu_\mu) \quad (26)$$

$$(N_1 - N_2) = \sqrt{2}G_F N_e = -2\delta c2, \quad (\bar{\nu}_e \leftrightarrow \bar{\nu}_\mu) \quad (27)$$

Of these four conditions, only two can be satisfied for the same set of parameters. Assuming Δm^2 to be greater than 0 and noting that $N_e > N_n$, we see that only the first two resonance conditions are possible. The first condition corresponds to the usual MSW resonance while the second condition is the spin-flavor resonance. Previous attempts at trying to understand the anti-correlation [20,21] with the sun spots have concentrated efforts in trying to have the resonance condition met in the convective zone, where there is a time variation in the magnetic field.

As has been pointed out in references [20,21], it is possible for both resonance conditions to be satisfied. Depending on their relative proximity they may be treated individually. The case of overlapping divergences is discussed in reference [13]. The degree to which spin precession dominates over flavour precession (or visa versa) depends upon the adiabaticity of the respective resonance. This is demonstrated in Figure 4a which gives the survival probability for the case of $sin2\theta=0.1$, $B_1 = 10^7$ Gauss and $B_0 = 2 \times 10^4$ Gauss. In this figure the depletion region is

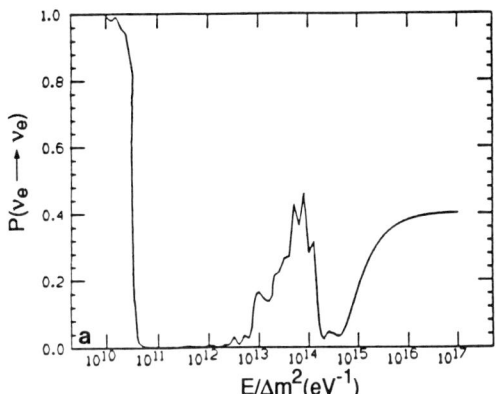

Fig. 4a. $P(\nu_e \to \nu_e)$ for the 4 by 4 case with $\sin 2\theta = 0.1$, a core magnetic field amplitude of 10^7 Gauss and a convective zone amplitude of 2×10^4 Gauss and $\mu = 2 \times 10^{-11} \mu_B$.

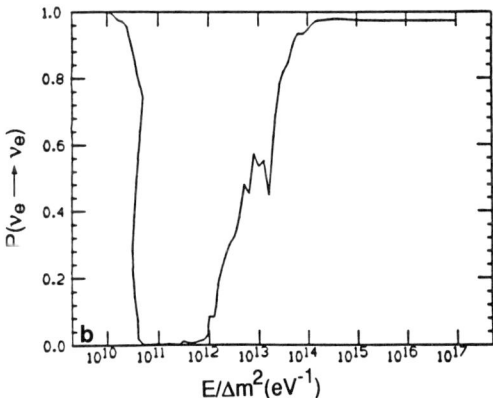

Fig. 4b. $P(\nu_e \to \nu_e)$ for the 4 by 4 case with $\sin 2\theta = 0.1$, a core magnetic field amplitude of 10^6 Gauss, convective zone amplitude of 10^3 Gauss and $\mu = 2 \times 10^{-11} \mu_B$.

dominantly due to spin precession. The second depletion region is again due to a node of the convective zone oscillation as discussed in the previous section. If we decrease the size of the magnetic field in the core then the conversion will no longer be dominated by spin flavor resonance, but by the MSW resonance, as can be seen from Figure 4b, which corresponds to $B_1 = 10^6$ Gauss, $B_0 = 10^3$ Gauss and $sin2\theta = 0.1$.

At first glance there seems to be a simple minded solution using the magnetic resonance. That is, from eq. (15) we see that as the magnetic field is decreased adiabaticity is lost. Therefore, one might hope that by choosing Δm^2 appropriately one could have a large suppression due to the resonance when the field is large, and as the field decreases adiabaticity will be lost and the suppression will decrease. Unfortunately life isn't so simple, and this idea is doomed. The reason being that as adiabaticity is lost, the resonance region narrows, from left to right on the curves. That is resonance is lost for the lower energy neutrinos. However, we want to suppress the low energy neutrinos, not the high energy neutrinos as it is the Chlorine detector with the lower threshold energy which sees the large suppression

5. Fitting the Data

We would now like to find a region in parameter space which explains simultaneously the time dependence of the data from the Chlorine and Kamiokande experiments. Results from the Chlorine experiment (see Table 1) show suppression factors of approximately 1/2 during solar minima and 1/8 during solar maxima (the SSM predicted rate is 7.9 SNU), whereas the Kamiokande data show a maximum variation of about 50 percent. The KII experiment has a threshold of 7.5 MeV, whereas the Chlorine experiment threshold is 0.8 MeV. Therefore, we would like to find a region in parameter space where there is a large variation in the low energy neutrinos (E<7.5 MeV) and a smaller variation for the high energy neutrinos. Also, we must have the proper amount of overall suppression at the times of solar minima (i.e., small convective zone magnetic fields) for both experiments.

Figures 5 (a-d) show the survival probability at the edge of the sun for an increasing convective magnetic field and fixed values of $B_1 = 10^7$ Gauss, $\mu = 2 \times 10^{-11} \mu_B$, $sin2\theta = 0.1$. Note that if we choose Δm^2 to be around 8×10^{-9} eV^2, the lower energy neutrinos will be caught in the region of large suppression when the magnetic field in the convective zone is large (B=2×10^4 Gauss). (Recall this region of suppression is not an MSW resonance, but a long wavelength node.) Also, the higher energy neutrinos will see a smaller time variation, yet there will still be an overall suppression, due to the core magnetic field. If the core field is made less then 10^7 Gauss then Figure 6 shows that there will not be enough suppression when the convective zone magnetic field is at its nadir (10^3 Gauss).

We calculated the number of expected events at KII and Chlorine, utilizing the program written for the survival probability. The Kamiokande event rate was calculated by integrating over the Boron neutrino spectrum and folding in the survival probability as well as the detector efficiency. For the Chlorine experiment we fitted the absorption cross section and the energy spectrum for the Boron, Nitrogen and Beryllium neutrinos given in Bahcall and Ulrich [1]. (The Oxygen and Nitrogen spectra were given to us by Bachall.) For the KII case an extra integration over the electron energy must be performed, and the detector efficiency and energy resolution (given to us by the KII group) must be folded in. In Figures 7a and 7b we show the event rates at KII and Chlorine respectively for the case of zero flavor-mixing. Here we chose Δm^2 to be $7.8 \times 10^{-9}\ eV^2$ and $\mu_\nu = 2 \times 10^{-11}\mu_B$. It is clear from these figures that while the event rate in Chlorine can vary from 5 SNU to about 1 SNU as the convective magnetic field changes by an order of magnitude, the corresponding variation in the KII signal is not significant. The results are little changed if the mixing angle is made non zero and small.

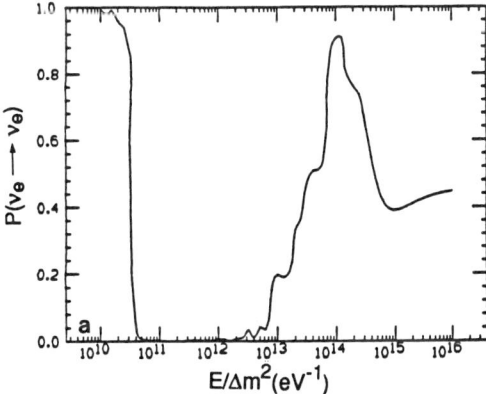

Fig. 5a. Expected event rate in Chlorine as a function of the convective magnetic field for the case of zero flavor-mixing. Here $\Delta m^2 = 7.8 \times 10^{-9}\ eV^2$, the core amplitude $B_1 = 10^7$ Gauss and $\mu = 2 \times 10^{-11}\mu_B$.

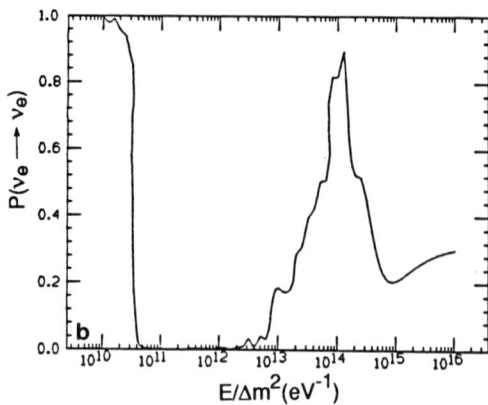

Fig. 5b. Expected event rate in Kamiokande versus the convective magnetic field for the same set of parameters as in Fig. 5a.

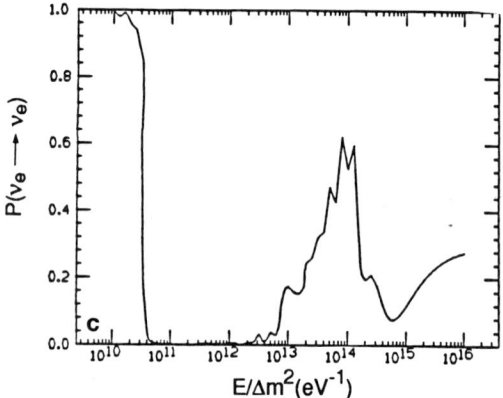

Fig. 5c. Expected event rate in Chlorine as a function of the convective zone magnetic field for the 4 × 4 case with sin2θ=0.1, $\Delta m^2 = 7.8 \times 10^{-9}$ eV^2, a core field amplitude of 10^7 Gauss and $\mu = 2 \times 10^{-11} \mu_B$.

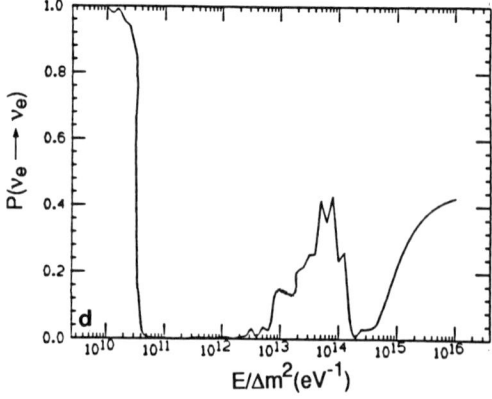

Fig. 5d. Expected event rate for Kamiokande versus the convective zone magnetic field for the same parameters as in Fig. 5c.

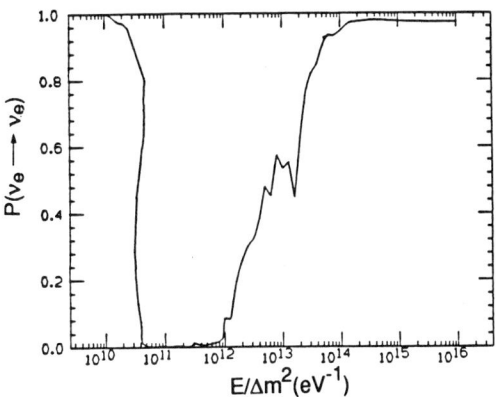

Fig. 6. $P(\nu_e \to \nu_e)$ with a core field amplitude of 10^6 Gauss and convective zone amplitude of 10^3 Gauss, $\mu = 2 \times 10^{-11} \mu_B$ and $sin 2\theta = 0.1$.

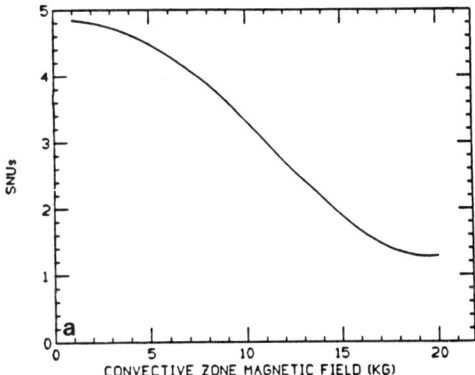

Fig. 7a. Expected event rate in Chlorine as a function of the convective magnetic field for the case of zero flavor-mixing. Here $\Delta m^2 = 7.8 \times 10^{-9}~eV^2$, the core amplitude $B_1 = 10^7$ Gauss and $\mu = 2 \times 10^{-11} \mu_B$.

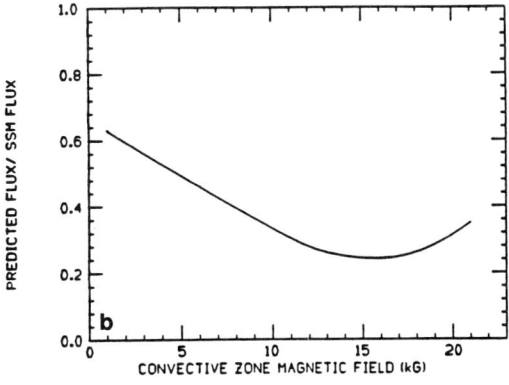

Fig. 7b. Expected event rate in Kamiokande versus the convective magnetic field for the same set of parameters

6. Expected Gallium Result

Using the parameters we found to fit the data from Davis and KII, we calculated the expected event rate in a Gallium detector (see Figure 8). The differential cross section was fitted using the the results given in Bahcall and Ulrich [1]. The neutrino spectrum was integrated over using the previously calculated survival probability. The p-p neutrino spectrum and spatial production distribution were also fitted from Ref. [1] and folded into the integration [22]. We see that we expect a large time variation. Given that we are passing through a solar maximum, the preliminary results from SAGE of a solar neutrino deficit is consistent with our expected results. We also expect to see a rise in the event rate during the quiet sun, with a maximum value of 80 SNU during the time of minimal sun spot number.

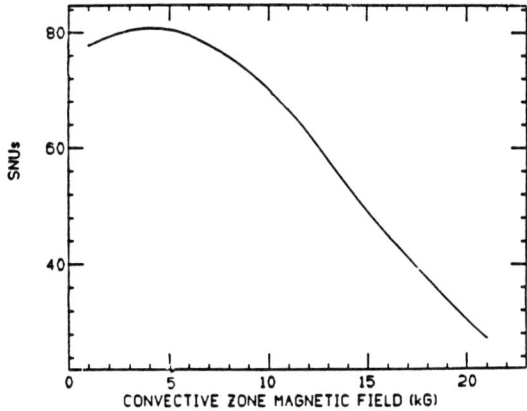

Fig. 8. Expected event rate for Gallium as a function of the convective zone magnetic field for the same set of parameters

7. Conclusion

In this talk I have presented an explanation for the possible time variation on the solar neutrino data in the chlorine experiment, moreover, this scenario can simultaneously explain the lack of time variation in the Kamiokande data. The solution dictates that the parameters be $\Delta m^2 \approx 10^{-8}$ eV^2 and $\mu_\nu = 2 \times 10^{-11} \mu_B$. The flavor mixing does not play an important role in the combined fit, and can be in the 10-20% range. For this set of parameters, we can predict the Gallium experiment to show anti-correlation with sun spots with a maximum flux of 80 SNU's during the sun spot minimum. This solution also implies that the Chlorine experiment should see a semi-annual variation in the neutrino flux, whereas Kamiokande should not. Present observational limits are consistent with this prediction.

Acknowledgements

I would like to thank A. Zichichi and the entire staff at the Ettore Majorana for their gracious hospitality.

References

[1] J. Bahcall and R. Ulrich, Rev. Mod. Phys. 60, 297 (1988)
J. Bahcall "Neutrino Astrophysics" (Cambridge Univ. Press, 1989)

[2] For a more recent analysis see S. Turek-Chieze, S. Cohen, M. Casse and C. Doom, Ap. J. 335, 415 (1988)

[3] R. Davis et. al., Proceedings of the XIX International Conference on Neutrino Physics, "Neutrino '88", ed. by H Schneps, T. Kafka, W. Mann and P. Nath, p. 518 (World Scientific, 1989)
R. Davis, K. Lande, C.K. Lee, B.T. Clevlend, J. Ullman, "Inside the Sun", ed. G. Bortholomiu, (Kluwer Publishing, Boston)

[4] K. Hirata et. al., Phys. Rev. Lett. 63, 16 (1989); 65, 1297 (1990); 65, 1301 (1990)

[5] S.P. Mikheyev and A.Y. Smirnov, Yad. Fiz. 42, 1441 (1985) [Sov. J. Nucl. Phys. 42, 913 (1985)]
L. Wolfenstein, Phys. Rev. D17, 2369 (1978)

[6] M.B. Voloshin and M.I. Vysotsky, Yad. Fiz. 44, 845 (1986) [Sov. J. Nucl. Phys. 44, 544 (1986)]
M.B. Voloshin, M.I. Vysotsky and L.B. Okun, Yad. Fiz. 44, 677 (1986) [Sov. J. Nucl. Phys. 44, 440 (1986)]

L.B. Okun, M.B. Voloshin and M.I. Vysotsky, Zh. Exp. Theor. Fiz. 91, 754 (1986) [Sov. Phys. JETP 64, 446 (1986)]

R. Cisneros, Astrophys. Space. Sc. 10, 87 (1971)

[7] M.B. Voloshin, Yad. Fiz. 48, 804 (1988) [Sov. J. Nucl. Phys. 48, 512 (1988)]

[8] K.S. Babu and R.N. Mohapatra, Phys. Rev. Lett. 63, 228 (1989)

[9] K.S. Babu and R.N. Mohapatra, Phys. Rev. Lett. 64, 1705 (1990) and Phys. Rev. D42, 3778 (1990)

M. Leurer and N. Marcus, Phys. Lett. B237, 81 (1990) and Phys. Rev. D43,

[17] V. Barger, N. Deshpande, P.B. Pal, R.J.N. Phillips and K. Whisnant, Phys. Rev. D43, 1759 (1991)

[18] S. Nussinov, Phys. Lett. B63, 201 (1976)

[19] B. Kayser, Phys. Rev. D24, 110 (1981)

[20] C.S. Lim and W. Marciano, Phys. Rev. D37, 1368 (1988)

E. Akhmedov, Phys. Lett. B213, 64 (1988)

[21] E. Akhmedov and O.V. Bychuk, Sov. Phys. JETP 68, 250 (1989)

E. Akhmedov, Sov. Phys. JETP 68, 690 (1989)

H. Minakata and H. Nunokawa, Phys. Rev. Lett. 63, 121 (1990)

J. Pulido, Phys. Lett. B251, 305 (1990)

[22] For a discussion of the effects of a finite size production region see J.M. Gelb and S.P. Rosen, Phys. Rev. D34, 969 (1986)

[23] R.N. Mohapatra, K.S. Babu and I.Z. Rothstein Phys. Rev. D, (1991)

CHAIRMAN: I. Rothstein

Scientific Secretaries: N.E. Moulai, M.A. Niaz

DISCUSSION

– *Altarelli:*

Your curves refer to fixed values of μ and Δm^2. What happens when these parameters are varied?

– *Rothstein:*

A change of μ corresponds to a rescaling of the magnetic field, because the Hamiltonian only depends upon μB. Δm^2, on the other hand, is a critical parameter and must be chosen in a specific range. However, I would say that if you are concerned with a fine tuning problem, this case is no worse than in the MSW scenario.

– *Sivaram:*

I have two questions:

a) Was the variation in the magnetic field with the solar radius taken into account?

b) Have you taken into account the HEP and Be neutrinos, which have large relative energy?

– *Rothstein:*

a) Yes, we have included the effect of the varying magnetic field, however, I should say that the functional dependence of the magnetic field on distance is not well known.

b) Yes, we included the entire neutrino spectrum as given by Bahcall.

– *Lindholm:*

How many events have been seen at Davis and Kamiokande?

– *Rothstein:*

Davis has been running for approximately 20 years. The average number of events per month is of the order of 10. Kamiokande sees less because the cross section is smaller for νe scattering than for inverse beta decay.

– *U. Vikas:*

Do you solve the equations exactly or are these numerical solutions?

– *Rothstein:*

You can not solve the propagation equation exactly because it is highly nonlinear. However, if you are willing to set the magnetic field constant and approximate the matter density as being a pure exponential, then it is possible to write down closed form solutions for the probability .

– *Junk:*

Why are neutrinos which flip in the supernovae considered sterile, can't they flip back?

– *Rothstein:*

First off, they are only considered sterile if they flip to a sterile species. For instance, in the case of a transition moment it is possible for the neutrino to flip from one active species to another. In the case that the neutrino makes a transition from an active species to a sterile one, we don't worry about the spin reflipping because the cross-section for spin flip is usually very small.

– *Cifarelli:*

Do you know if there are any proposed experiments that would help clarify the situation?

– *Rothstein:*

Yes, there have been several proposed experiments which would give us more information than the present experiments. For instance, the proposed SNO experiment will be able to separate neutral current from charged current events. I believe that there is also a proposed experiment to study specifically Be neutrinos which may have novel effects because they are relatively monochromatic.

– *Lindholm:*

Have there been other calculations of the predicted solar neutrino flux?

– *Rothstein:*

Yes, there is a calculation that has recently been done by Turck-Chieze et al. which quotes a smaller expected result than given by Bahcall, however their results are still well above the measured flux.

SEARCHING FOR THE HIGGS BOSON AT A PHOTON-PHOTON COLLIDER

Douglas L. Borden

Department of Physics
University of California, Santa Barbara, 93106

One of the prime pursuits of the particle physics community over the next decade will be the exploration of the spontaneous symmetry breaking sector the Standard Model of elementary particle physics. The initial task in such an exploration will of course be the search for, and hopefully study of, the Higgs boson or bosons.

We concentrate on a Higgs boson in the intermediate mass region of 80-180 GeV. A Standard Model Higgs boson below ~48 GeV has already been ruled out by LEP,[1] and a Higgs with a mass between 48 and 80 GeV will either be ruled out or found at LEP-II.[2] For Higgs masses above twice the Z mass, almost certain detection is assured at the SSC or LHC in the 'goldplated' decay mode $H \rightarrow ZZ \rightarrow 4$ leptons.[2] It is the intermediate mass region — from 80 to 180 GeV — which is the most difficult region to access experimentally.

Discovery of an intermediate mass Higgs boson at a hadron collider will most probably occur through the decay of the Higgs to two photons.[2] This $\gamma\gamma$ discovery channel seems tenable for Higgs masses above ~120 GeV, but studies indicate that only a detector with superb electromagnetic calorimetry and photon angle resolution will be capable of extending the range any further down in mass. Associated W production of the Higgs (resulting in the decay mode $HW \rightarrow \gamma\gamma l\nu$) allows for significant background suppression, extending the accessible range all the way down to 80 GeV, but suffers from a very low rate.[3]

Production of an intermediate mass Higgs boson at an e^+e^- collider proceeds through either the Bjorken Process ($e^+e^- \rightarrow ZH$), or through WW fusion ($e^+e^- \rightarrow \nu\nu H$). At center of mass energies below about 500 GeV the Bjorken process dominates, while above that the WW fusion process is most important.[2] The absence of an energy constraint in WW fusion limits the Higgs mass discovery region accessible at high energy (> 500 GeV) linear colliders to greater than about 130 GeV. At lower $\sqrt{s}$ machines the discovery potential is better, and the mass region accessible extends all the way down to LEP-II limits, but a Higgs boson in the W or Z mass region poses special problems.[2,4]

Physics at the Highest Energy and Luminosity, Edited by A. Zichichi
Plenum Press, New York, 1992

In addition to being the most difficult to access experimentally, the intermediate mass region is also a particularly intriguing region from a theoretical perspective. Weak scale supersymmetry, one of the most compelling solutions to the hierarchy problem and one of the most attractive extensions to the Standard Model, predicts the presence of a Higgs boson near this region. At tree level the theory predicts a Higgs boson with a mass below that of the Z, but radiative corrections push this upper limit well into the intermediate mass region.[5] If a Higgs is not found at LEP-II, the presence of a Higgs in the intermediate mass region would be a crucial test of Weak Scale supersymmetry.

In the future, in addition to e^+e^- and hadron colliders, there may be another facility available for the exploration of the spontaneous symmetry breaking sector of the Standard Model. A Photon-Photon Collider, capable of high energy $\gamma\gamma$ collisions will allow a possible search for, and certainly study of, an intermediate mass Higgs boson through the production mechanism $\gamma\gamma \rightarrow$ Higgs.

A PHOTON-PHOTON COLLIDER

In the past, photon-photon collisions have been obtained via virtual bremsstrahlung emission at e^+e^- storage rings. Virtual bremsstrahlung is the emission of a virtual photon by an electron in a 'grazing' (low momentum transfer) collision with a positron or another electron.[6] Although an extremely useful tool in the study of the strong interactions of the Standard Model,[7] the invariant mass reach of virtual bremsstrahlung photons is extremely limited, and so is an ineffective tool for probing physics in the 10's or 100's of GeV. Instead, two-photon physics of the future will rely on a new mechanism for creating and colliding photon beams, namely the backscattering of laser light off of the electron beams in a linear e^+e^- collider.

The idea is now rather old and has many variations but all revolve around the collision of a high-power, approximately optical wavelength laser beam with an intense, high-energy electron beam (Fig. 1).[8-13] Through the process of Compton backscattering, shown in Fig. 2, the result is a high-energy photon beam which closely follows the original trajectory of the electron beam. If a laser is scattered off each beam in a linear collider, the resulting photon beams collide at the interaction point, and a Photon-Photon Collider results.

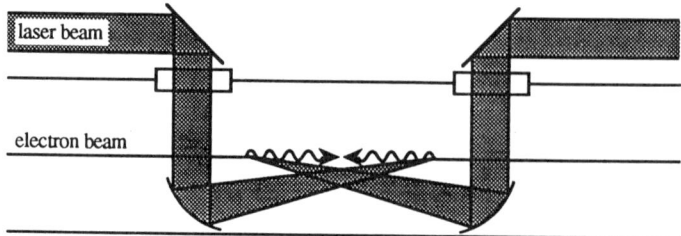

FIG. 1 A schematic design for a Photon-Photon Collider. Laser beams are brought into the beampipe and are focused on the electron beams from a linear collider; the laser photons are Compton backscattered, resulting in intense beams of high energy photons.

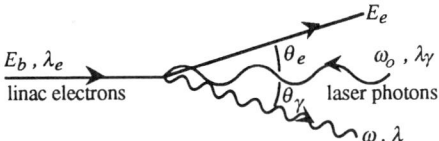

FIG. 2 Compton scattering of an electron of energy E_b and longitudinal polarization λ_e ($=\pm 1$) with a laser photon of frequency ω_o ($\ll E_b$) and circular polarization λ_γ ($=\pm 1$).

Compton Cross Section and Kinematics

An optical laser beam of frequency ω_o and circular polarization λ_γ is focused on a beam of linac electrons of energy E_b and longitudinal polarization λ_e a few centimeters upstream of the interaction point (IP). The laser light Compton backscatters off the high energy electrons, resulting in a collimated beam of very high energy photons (of order the original e^- beam energy). Both the scattered electrons and the backscattered photons follow approximately the original e^- beam direction within a few μrad and are therefore incident on the interaction point in a tight final focus. The electrons, however, can be deflected away from the interaction point by a transverse magnetic field in order to reduce backgrounds. It is possible, by varying the polarization of the laser photons and linac electrons, and by exploiting the energy dependence of the photon scattering angle, to obtain a fairly monochromatic luminosity distribution in $\gamma\gamma$ collisions. It is also possible to choose polarization parameters so as to obtain a broad luminosity distribution, allowing for the search for resonances — such as the Higgs — through their $\gamma\gamma$ couplings.

The Compton kinematics are characterized by the dimensionless variable

$$x \equiv (4E_b\omega_o/m_e^2)\cos^2\tfrac{\alpha}{2} \approx 15.32\left(\tfrac{E_b}{\text{TeV}}\right)\left(\tfrac{\omega_o}{\text{eV}}\right), \qquad (1)$$

where α is the angle between the electron beam and laser beam, and where we assume $\cos^2\alpha/2 \approx 1$ (nearly achievable in practice with the use of focusing mirrors to direct the laser beams into the vacuum pipe). The differential Compton cross section (for $\omega \le \omega_{max} = E_b\tfrac{x}{x+1}$) is[11]

$$\frac{1}{\sigma_c}\frac{d\sigma_c}{d\omega} \equiv f(\omega) = \frac{1}{E_b\sigma_c}\frac{2\pi\alpha^2}{xm_e^2}\left[\frac{1}{1-y}+1-y-4r(1-r)-\lambda_e\lambda_\gamma rx(2r-1)(2-y)\right] \qquad (2)$$

where $y = \omega/E_b$, $r = \tfrac{y}{x(1-y)} \le 1$, and where the total Compton cross section σ_c is

$$\sigma_c = \sigma_c^0 + \lambda_e\lambda_\gamma \sigma_c^1$$

$$\sigma_c^0 = \frac{\pi\alpha^2}{xm_e^2}\left[\left(2-\frac{8}{x}-\frac{16}{x^2}\right)\ln(x+1)+1+\frac{16}{x}-\frac{1}{(x+1)^2}\right]$$

$$\sigma_c^1 = \frac{\pi\alpha^2}{xm_e^2}\left[\left(2+\frac{4}{x}\right)\ln(x+1)-5+\frac{2}{x+1}-\frac{1}{(x+1)^2}\right]. \qquad (3)$$

The backscattered photon distribution is plotted in Fig. 3 for unpolarized beams and for various values of the x parameter. The spectrum becomes harder as x increases. One might be tempted to strive for as high an x value as one could by using the highest frequency laser available, and thereby achieve the hardest photon spectrum possible. Unfortunately, as x is increased, processes other than Compton scattering become possible, introducing backgrounds and altering the resulting photon spectrum. The most important of these processes is the pair conversion of a high energy photon in a collision with a laser photon further along in the laser pulse.[13] The threshold ($\omega_{max}\omega_o = m_e^2$) for this reaction occurs at $x = 2(1+\sqrt{2}) \approx 4.83$. The result is a depletion of the highest energy photons from the spectrum, as demonstrated in Fig. 3b. Not only does pair conversion degrade the photon spectrum, but the resulting e^+e^- pairs represent an additional beam background with which to be concerned. In order to avoid such complicating effects, we only consider designs with $x < 4.83$.

Polarizing both the electron beam and the laser results in a substantial change in the photon spectrum: if the electrons and laser are like-polarized — both right-handed or both left-handed — then the resulting photon spectrum is relatively flat; if the electrons and laser are oppositely polarized the resulting spectrum is more monochromatic, peaking at very high energy (Fig. 4(a)). Polarizing the electrons or laser also results in polarized high energy photons. The mean scattered photon helicity depends on the energy of the photon and is given by:

$$\lambda(\omega) = \frac{\lambda_\gamma(1-2r)(1-y+\frac{1}{1-y}) + \lambda_e rx\left[1+(1-y)(1-2r)^2\right]}{1-y+\frac{1}{1-y} - 4r(1-r) - \lambda_e\lambda_\gamma rx(2r-1)(2-y)} \tag{4}$$

where y and r are defined as in Eq. 2.[11] Notice that the backscattered photons are polarized when *either* electrons *or* laser photons are polarized. Fig. 4(a) shows the mean polarization in two extreme cases, overlaid with the resulting photon spectrum. Note that the flat spectrum (obtained by backscattering electrons and laser photons with the same handedness) has nearly maximum

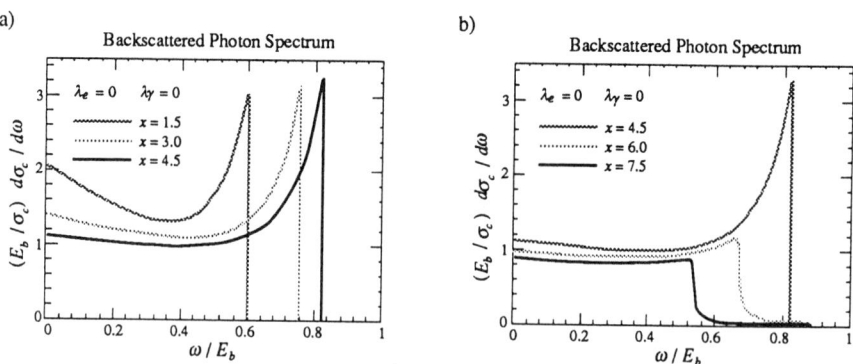

FIG. 3 Energy spectrum of high energy photons from Compton backscattering with unpolarized beams for various x values. At x values above 4.83 the highest energy photons reconvert to an e^+e^- pair in collisions with laser photons, resulting in a depletion of the spectrum at high energies. The reconversion of the high energy photons into e^+e^- pairs is exaggerated by assuming a very dense laser photon pulse, which ensures a high probability that a photon reconverts if kinematically allowed.

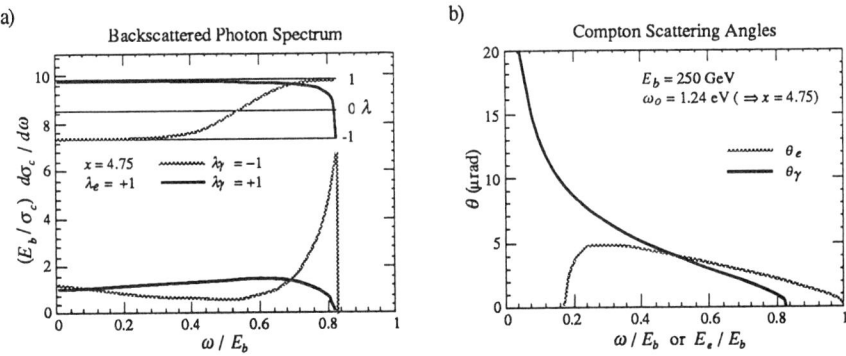

FIG. 4 a) Polarization and energy spectrum of backscattered photons for Compton scattering with polarized beams. b) Electron and photon scattering angles for a typical beam energy and x value.

photon polarization over almost the whole energy range, whereas the peaked spectrum has opposite polarizations for low and high energies.

After Compton scattering, both the electrons and photons travel essentially along the original e^- beam direction. The photon and electron scattering angles are unique functions of the photon or electron energy and (for θ small) are given by

$$\theta_\gamma(\omega) \approx \frac{m_e}{E_b}\sqrt{\frac{E_b x}{\omega} - (x+1)}$$

$$\theta_e(E_e) \approx \frac{m_e}{E_b}\sqrt{\frac{E_b}{E_e}\left(2 + x - \frac{E_b}{E_e}\right) - (x+1)} \tag{5}$$

These functions are displayed in Fig. 4(b) for a typical beam energy of 250 GeV and laser energy of 1.24 eV. The high energy photons scatter at very small angles (~few μrad), while the softest photons scatter the most; the electrons are only slightly deflected from their original direction and scatter into a narrow cone.

γγ Luminosity Distribution

The high energy photon beam produced from one beam of a linear collider can be brought into collision with a similarly produced photon beam from the other electron beam, resulting in γγ collisions. Because of the small, but finite, photon scattering angles, the resulting luminosity distribution depends sensitively on the conversion distance (distance from the conversion point — where the laser pulse intersects the electron beam — to the interaction point) and on the size and shape of the electron beam at the interaction point (in the absence of a backscattering laser).

Define z as the distance from conversion point to interaction point, and consider an electron on a path which intersects the interaction plane at a point (x_o, y_o) (Fig. 5). A photon scattering off this electron at a polar angle θ and azimuthal angle β intersects the interaction plane at a point (x,y) given by

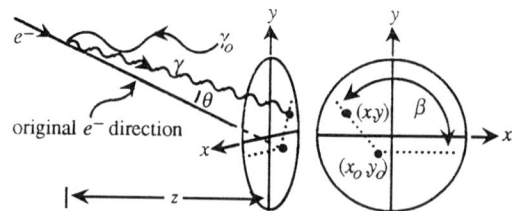

FIG. 5 An electron whose path intersects the interaction plane at (x_0,y_0) encounters a laser photon (γ_0) a distance z prior to the interaction plane. The laser photon backscatters at polar angle θ and azimuthal angle β, intersecting the interaction plane at the point (x,y).

$$\begin{pmatrix} x \\ y \end{pmatrix} = \begin{pmatrix} x_o + z\theta\cos\beta \\ y_o + z\theta\sin\beta \end{pmatrix}. \tag{6}$$

If $\rho_e(x,y)$ is the transverse density of the original electron beam at the interaction point, then the density of photons (which scatter at this polar and azimuthal angle) is

$$\rho_\gamma(x,y;\theta,\beta) = \rho_e(x - z\theta\cos\beta, y - z\theta\sin\beta). \tag{7}$$

If $z\theta$ is much smaller than typical beam spotsize dimensions, the photons arrive at the interaction point in as tight a final focus as the electrons would have had. However, if $z\theta$ is much larger than typical beam spotsize dimensions, then the photon interaction point is much larger than the original electron spotsize, resulting in a loss of luminosity. For beam energies of several hundred GeV and conversion distances of a few cm, $z\theta$ is of order several tens of nm, which is close to the typical spotsize dimensions envisioned for the next generation of linear colliders. Thus, in designing a photon collider neither limit will strictly hold, and the $\gamma\gamma$ luminosity will depend sensitively on the particular machine parameters chosen.

For definiteness consider an electron beam with a cylindrical gaussian profile:

$$\rho_e(x,y) = \frac{1}{2\pi\sigma^2} e^{-\frac{x^2+y^2}{2\sigma^2}}. \tag{8}$$

After Compton backscattering, the probability that a given photon will be found with energy ω intersecting the interaction plane at (r,ϕ) is

$$P_\gamma(r,\phi,\omega) = \frac{f(\omega)}{2\pi\sigma^2} I_o\left(\frac{rz\theta_\gamma(\omega)}{\sigma^2}\right) e^{-\frac{r^2+z^2\theta_\gamma(\omega)^2}{2\sigma^2}} \tag{9}$$

where I_o is the modified Bessel function of order 0.

Consider the collision of two such high energy backscattered photon beams, and define the variables:

$$W = 2\sqrt{\omega_1 \omega_2} = \gamma\gamma \text{ invariant mass}$$

$$y = \tanh^{-1}\left(\frac{\omega_1 - \omega_2}{\omega_1 + \omega_2}\right) = \text{rapidity} \tag{10}$$

The $\gamma\gamma$ differential luminosity is then

$$\frac{1}{L_{ee}} \frac{dL_{\gamma\gamma}}{dW\, dy} = \frac{W}{2} f\left(\frac{We^y}{2}\right) f\left(\frac{We^{-y}}{2}\right) I_0\left(\frac{d_1 d_2}{2\sigma^2}\right) e^{-\frac{d_1^2 + d_2^2}{4\sigma^2}}, \tag{11}$$

where $d_1 \equiv z\theta_\gamma(\frac{We^y}{2})$ and $d_2 \equiv z\theta_\gamma(\frac{We^{-y}}{2})$.

Plotted in Fig. 6 are $\gamma\gamma$ luminosity distributions for various polarization combinations and for various conversion distances at typical values of the machine parameters, integrated over rapidity. For simplicity we have assumed that each electron has scattered once and only once in the laser photon pulse — i.e. we have ignored the effects of both unscattered electrons and multiple scattering of electrons on the luminosity distribution. The primary effect of increasing the conversion distance is to decrease the luminosity at the low-mass end of the spectrum, as the lowest energy photons scatter outside of the interaction point. For most purposes this is a beneficial effect, as a more monochromatic distribution makes for a cleaner initial state. Resonance searches, however, require a broad luminosity distribution; this will necessitate a rather small conversion distance.

$\gamma\gamma \rightarrow$ Higgs

A Photon-Photon Collider offers a unique opportunity to search for and study a neutral Higgs boson in the intermediate mass region. The Standard Model Higgs coupling to two photons is small but results in an observable production cross section; the same is generally true of extended Higgs sectors, and photon-photon reactions give access to all neutral Higgs particles, as well as to charged Higgs via pair production.[2,14] Additionally, and perhaps more importantly, the coupling of the Higgs to two photons, shown in Fig. 7(a), involves loops where any new charged particles

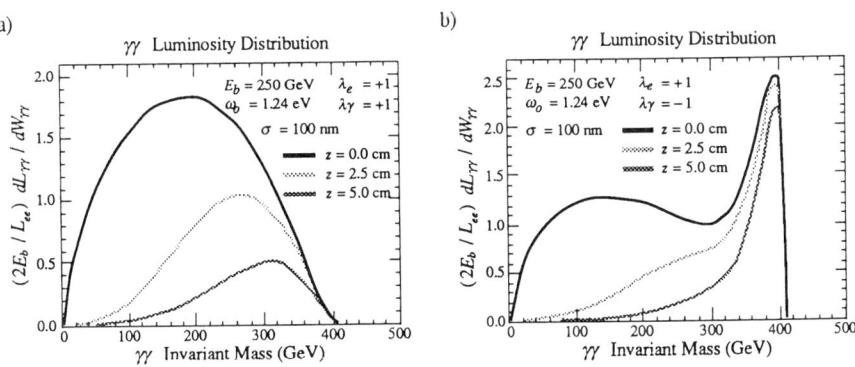

FIG. 6 $\gamma\gamma$ luminosity distributions for various polarization combinations and for various conversion distances. The vertical axis is dL/dW in units of $L_{ee}/2E_b$. The $\gamma\gamma$ rapidity has been integrated out.

with couplings to the Higgs must contribute, giving crucial information on the nature of the Higgs itself; a measurement of this process would be quite sensitive to new physics even at higher mass scales.[14]

$\gamma\gamma \rightarrow$ Higgs: Discovery

A $\gamma\gamma$ collider provides an attractive alternative to e^+e^- and hadron colliders for discovering an intermediate mass Higgs boson. The broad luminosity distribution available allows a search for a Higgs boson as a resonance in $\gamma\gamma$ collisions. We assume a standard model Higgs boson. In the intermediate mass region the Higgs has a very narrow total width (of order a few MeV) and decays predominantly to $b\bar{b}$, so we expect to observe the Higgs as a resonance in $\gamma\gamma \rightarrow b\bar{b}$ production. Because the Higgs is such a narrow resonance, we may safely ignore interference effects between the (u and t channel) continuum diagrams and the (s channel) resonance diagram, so that the

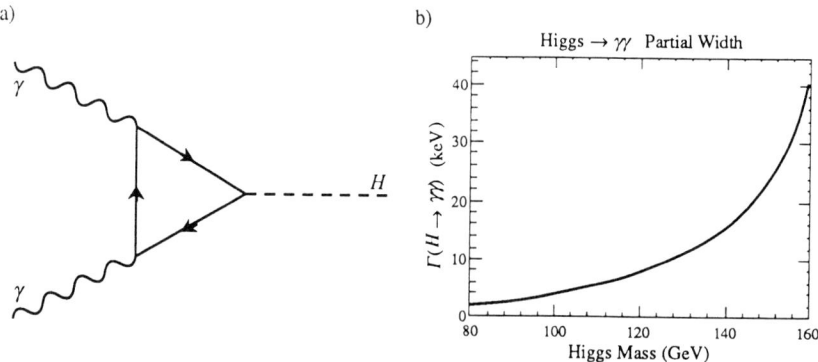

FIG. 7 a) The coupling of a Higgs to two photons. b) Higgs $\rightarrow \gamma\gamma$ partial width as a function of Higgs mass, in the Standard Model.

$\gamma\gamma \rightarrow b\bar{b}$ cross section decomposes into a sum of a $b\bar{b}$ continuum production cross section and a Higgs Breit-Wigner:

$$\sigma(\gamma\gamma \rightarrow b\bar{b}) = \sigma_c(\gamma\gamma \rightarrow b\bar{b}) + \sigma(\gamma\gamma \rightarrow H \rightarrow b\bar{b})$$

$$\frac{d\sigma_c(\gamma\gamma \rightarrow b\bar{b})}{d\cos\theta} = \frac{2\pi\alpha^2 N_c q^4}{W^2} \beta \frac{[2 - 2\beta^4 - (1 - \lambda_1\lambda_2)(1 + \beta^2 \cos^2\theta)(1 - 2\beta^2 + \beta^2 \cos^2\theta)]}{(1 - \beta^2 \cos^2\theta)^2}$$

$$\sigma(\gamma\gamma \rightarrow H \rightarrow b\bar{b}) = \frac{8\pi\Gamma(H \rightarrow \gamma\gamma)\Gamma(H \rightarrow b\bar{b})}{(W^2 - M^2)^2 + \Gamma_T^2 M^2}(1 + \lambda_1\lambda_2)$$

(12)

where $\beta = \sqrt{1 - \frac{4m_b^2}{W^2}}$.[15,16] The Higgs–$\gamma$–$\gamma$ coupling proceeds through loops of charged particles (Fig. 7(a)). In the Standard Model, the $\gamma\gamma$ width is given by:[2]

$$\Gamma(H \rightarrow \gamma\gamma) = \frac{\alpha^2}{256\pi^3} \frac{M^3}{v^2} \left| \sum_i N_{ci} q_i^2 F_i\left(\frac{4m_i^2}{M^2}\right) \right|^2$$

(13)

where v is the vacuum expectation value of the Higgs, the sum is over all charged particles, and N_{ci} and q_i are the color factor and charge of each particle in the sum. The functions F_i depend on the spin of the particle and are given by:

$$F_i(\tau) = \begin{cases} -2\tau[1+(1-\tau)g(\tau)] & \text{fermions} \\ 2+3\tau+3\tau(2-\tau)g(\tau) & W \text{ boson} \end{cases}$$

where $g(\tau) = \begin{cases} [\arcsin(\frac{1}{\sqrt{\tau}})]^2 & \tau \geq 1 \\ -\frac{1}{4}[\ln(\frac{1+\sqrt{1-\tau}}{1-\sqrt{1-\tau}}) - i\pi]^2 & \tau < 1. \end{cases}$

Fig. 7(b) shows $\Gamma(H \to \gamma\gamma)$ as a function of the Higgs mass.

The Higgs to $b\bar{b}$ width, which is very nearly the Higgs total width in the intermediate mass region, is

$$\Gamma(H \to b\bar{b}) = \frac{3m_b^2 M}{8\pi v^2}\left(1 - \frac{4m_b^2}{M^2}\right)^{\frac{3}{2}} \tag{14}$$

and is of order a few MeV.[2] In the narrow width approximation,

$$\frac{1}{\pi}\frac{M\Gamma_T}{(W^2-M^2)^2 + M^2\Gamma_T^2} \approx \delta(W^2 - M^2) \tag{15}$$

so the number of Higgs bosons produced is given by

$$N_H = \left.\frac{dL_{\gamma\gamma}}{dW}\right|_M \frac{4\pi^2 \Gamma(H \to \gamma\gamma)}{M^2}(1+\lambda_1\lambda_2)$$

$$\approx 1.5 \times 10^4 \left(\frac{L_{ee}}{\text{fb}^{-1}}\right)\left(\frac{E_{ee}}{\text{TeV}}\right)^{-1}\left(\frac{\Gamma_{H\to\gamma\gamma}}{\text{keV}}\right)\left(\frac{M}{\text{GeV}}\right)^{-2} F(M)(1+\lambda_1\lambda_2) \tag{16}$$

where $F(M) = \frac{E_{ee}}{L_{ee}}\left.\frac{dL_{\gamma\gamma}}{dW}\right|_M$ is dimensionless and of O(1).

Photon polarization can play a crucial role in background (continuum $b\bar{b}$ production) suppression. Far above $b\bar{b}$ production threshold, the continuum cross section goes as

$$\frac{d\sigma_c(\gamma\gamma \to b\bar{b})}{d\cos\theta} \sim \frac{1+\cos^2\theta}{1-\cos^2\theta}(1-\lambda_1\lambda_2) \quad \text{for } \beta \to 1, \tag{17}$$

while Higgs production goes as $(1+\lambda_1\lambda_2)$. Choosing machine parameters to enforce highly polarized photon beams (in the mass region of interest) and arranging for the colliding beams to have the same polarization significantly reduces the background while enhancing the signal. Also the continuum $b\bar{b}$ are preferentially produced at large dip angle ($|\cos\theta| \approx 1$) while the signal events are distributed uniformly in $\cos\theta$; thus, a cut on $\cos\theta$ also serves as a signal to background enhancer.

In order to study the ability of a Photon-Photon Collider to search for an intermediate mass

Higgs boson, the following method was utilized. A Monte Carlo event generator was used to produce $\gamma\gamma \to H \to b\bar{b}$ and continuum $\gamma\gamma \to b\bar{b}$ events using the $\gamma\gamma$ luminosity distribution (Eq. 11) and the cross sections of Eq. 12. The $b\bar{b}$ partons were then fragmented into jets using JETSET 6.3 (LUND)[17] and the events processed through the Fast Monte Carlo Simulation of the SLD Detector, the detector operating at the SLC at Stanford Linear Accelerator Center.[18] The resulting 'raw data' were then analyzed to try to detect the presence of a Higgs boson. The JADE[19] jet-finding algorithm was applied to the data and all two-jet events were kept as the final data sample. After a cut on the dip angle of the two jets (both jets must have satisfied $|\cos\theta|<0.9$) and a cut on the transverse jet-jet colinearity to reject badly measured events, a jet-jet invariant mass was formed. The presence of a 'bump' in a histogram of invariant masses would then signal the presence of a Higgs boson.

A note on the invariant mass reconstruction is due. A naïve reconstruction — simply adding the measured jet 4-momenta together and squaring — leads to a very poor invariant mass resolution; it is possible, by cleverly rescaling the jet 4-momenta before adding them, to improve the resolution significantly. After passing through the detector, many of the events have a fairly large measured value of p_T. This could be due to two causes: either the jet direction was badly measured (due to particles in the jet escaping down the beampipe or through a crack in the detector, for instance) or neutral particles in the jet were not detected by the calorimeter. In either case the resulting reconstructed invariant mass is too low. In the first case (mismeasured direction), there is not much that can be done except to reject the event. Only those events whose x-y jet projections were less than 10° from being back-to-back were kept in the event sample; fortunately this rejects very few of the events which also pass the dip-angle cut ($|\cos\theta|$ of each jet < 0.9) — mismeasured direction simply is not much of a problem. Losing neutrals, however, is a significant problem, but fortunately for this case the problem can be partially remedied: since the p_T arises because more neutrals escaped in one jet than in the other, before forming the jet-jet invariant mass one of the jet's 4-vectors is rescaled to balance p_T. As losing neutrals can never increase the measured momentum of a jet, the jet with the smaller value of p_T has its 4-momentum scaled upward so that the rescaled p_T matches that of the other jet. This improves the jet-jet invariant mass resolution by a factor of two. This method of improving jet-jet mass resolution is similar in spirit to the one used by Baltay et al. in Ref. 19.

It should be noted that no $b\bar{b}$ flavor tagging was simulated in the analysis; we unrealistically assumed 100% $b\bar{b}$ flavor tagging efficiency and no contamination of the $b\bar{b}$ sample by other flavors of quark (e.g. $c\bar{c}$). Although the assumption of 100% efficiency is optimisic, it is not yet clear how well one can do $b\bar{b}$ flavor tagging. Use of present methods implies that 40%-50% $b\bar{b}$ flavor tagging efficiency is possible without significant contamination by other quark species, but research in this area is ongoing, and it is possible that much higher efficiencies are possible.[21,22] In any case, such an assumption does not significantly affect the conclusions to be drawn from this study. If ultimately it is found that 50% tagging efficiency is the best one can do, our results would remain unchanged except that the integrated luminosites we quoted would need to be multiplied by two.

With regard to the choice of particular machine parameters, we demonstrate a search for an intermediate mass Higgs boson in the following manner: we first consider a collider in the limit of zero conversion distance (the laser beam intersects the linac beam close enough to the interaction point so that we may ignore the backscattered photon's angular divergence) and study the effects of varying electron polarization at $\sqrt{s}$ of 300 and 500 GeV; we then study the effect of moving the conversion point a finite distance from the interaction point. The particular parameters chosen are as follows:

Case 1:
- Electron beams of 150 GeV with 50% and 90% right polarization;
- Laser beams of 2.06 eV and 100% right polarization;
- Higgs masses of 90 and 140 GeV;
- 10 fb^{-1} of integrated *ee* luminosity;

Case 2:
- Electron beams of 250 GeV with 50% and 90% right polarization;
- Laser beams of 1.24 eV and 100% right polarization;
- Higgs masses of 90 and 140 GeV;
- 10 fb^{-1} of integrated *ee* luminosity;

Case 3:
- Electron beams of 250 GeV, 90% right polarization, and a cylindrical gaussian profile with sigma of 100 nm;
- Laser beams of 1.24 eV and 100% right polarization, converting at a distance of 4 mm, 8 mm, and 12 mm;
- Higgs mass of 90 GeV;
- 20 fb^{-1} of integrated *ee* luminosity;

We have chosen to ignore the effect of multiple scattering of the electrons in the laser photon pulse and for simplicity assume each electron scatters exactly once in the laser photon pulse.

The results are displayed in figures 8-10. At $\sqrt{s}$ of 300 GeV (Fig. 8), a clear Higgs signal emerges above the continuum background for both 50% and 90% linac electron polarization, but the advantage of the higher e^- polarization is obvious: the ensuing higher degree of photon polarization greatly reduces the continuum background, enhancing the significance of the Higgs signal. As the energy of the *ee* collider is increased to 500 GeV (Fig. 9), two important effects are evident. The first is that signal/background improves, because the Higgs lies 'further back' on the photon polarization curve, resulting in a higher mean helicity product even at moderate linac electron polarizations. The second effect is a noticeable reduction in rate. This is due to two causes: first, as the total machine energy increases, the luminosity per unit mass interval decreases (the photon collisions are spread out over a greater energy range); second, the final state jets tend to be more boosted along the beam line, resulting in a lower fraction of the events in the detector. At machine energies above 500 GeV there is little further advantage from increased photon polarization, and the event rate begins to drop precipitously. It is for this reason that a moderate energy machine is favored for purposes of searching for the Higgs. If highly polarized electron beams are obtainable and a laser of the appropriate frequency range is available, then an ee machine with center of mass energy near 300 GeV seems most desirable; if only moderate electron polarization is possible then a 500 GeV machine would be indicated.

As the conversion distance is increased (from 4 mm to 1.2 cm for a spotsize of 100 nm) (Fig. 10) the low-end luminosity decreases (recall that the lowest energy photons scatter at the highest angles and so scatter outside of the interaction point). The result is a decreased event rate at low invariant mass, with a corresponding drop in the statistical significance of the Higgs signal. It is obvious that small conversion distances will need to be employed in order to ensure sufficient luminosity at low (~90 GeV) invariant mass. This will pose challenging questions for the design of the interaction region.

It should be noted that the results presented may be considered conservative, as the detectors which will be available for the next generation of linear collider will have considerable advantages over the SLD detector we use in our study. Primarily, better momentum resolution and greater calorimeter depth and segmentation will significantly improve the invariant mass resolution, allowing for a much greater signal to background ratio.

For a Higgs boson nearly degenerate with the Z there are additional backgrounds we have heretofore not considered. These backgrounds represent the production of a $Z + X$ final state with 'X' going undetected and the Z decaying to a $b\bar{b}$ pair. Before discussing these backgrounds, it should be noted that resonant production of a Z (i.e. $\gamma\gamma \to Z$) is *not* a background to $\gamma\gamma \to H$. By Bose symmetry, two massless spin-one objects do not couple to a spin-one resonance, a result known as the Yang-Landau theorem.[23] Consequently, the coupling of the Z to two real photons is identically zero.

Two potential backgrounds which might fake the presence of a 90 GeV Higgs boson are $\gamma\gamma \to \gamma Z$ with the photon disappearing down the beam pipe and the Z decaying to $b\bar{b}$, and $\gamma\gamma \to ZZ \to \nu\bar{\nu}b\bar{b}$. Fortunately, the cross sections for both these processes should be very small. The processes are 4th order in the coupling, so we expect the cross section to go as α^4/W^2 (≈ 0.2 fb at $W = 100$ GeV), more than two orders of magnitude smaller than the Higgs production cross section. We conclude that γZ and ZZ production do not constitute a serious background to a 90 GeV Higgs signal.

The most important background faking the presence of a 90 GeV Higgs is not a two-photon background, but is due rather to the presence of the residual electrons left over from the original Compton backscatter. Recall that if these electrons are not deflected following conversion then they follow very nearly their original path and so intersect the oncoming high energy photon beam at the interaction point. The process $e\gamma \to eZ$ is then possible, and the final state electron is preferentially backscattered down the beam pipe, where it goes undetected. If the Z then decays to $b\bar{b}$, the event mimics a 90 GeV Higgs event, and Monte Carlo studies show that event rates for

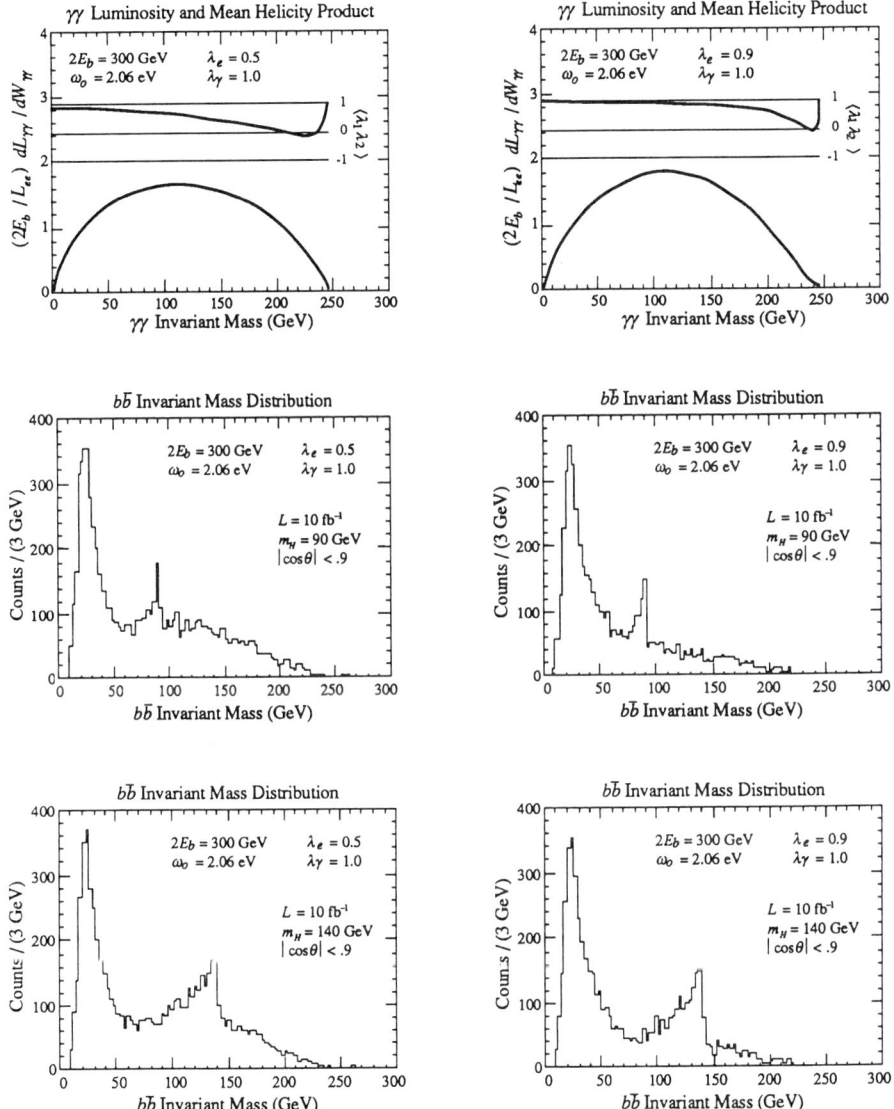

FIG 8 The search for an intermediate mass Higgs boson at a γγ collider — Case 1. At an ee machine energy of 300 GeV, and ignoring the effect of a finite conversion distance, a clear Higgs signal (for Higgs masses of 90 and 140 GeV) appears. The integrated luminosity of 10 fb^{-1} refers to 10 fb^{-1} of effective ee luminosity, and represents 1 'year' (10^7 sec) of running at 10^{33} cm^{-2} s^{-1}. Two different values of the linac electron polarization are demonstrated. The superior background suppression (arising from the higher degree of photon polarization) associated with the higher value of the electron polarization is obvious.

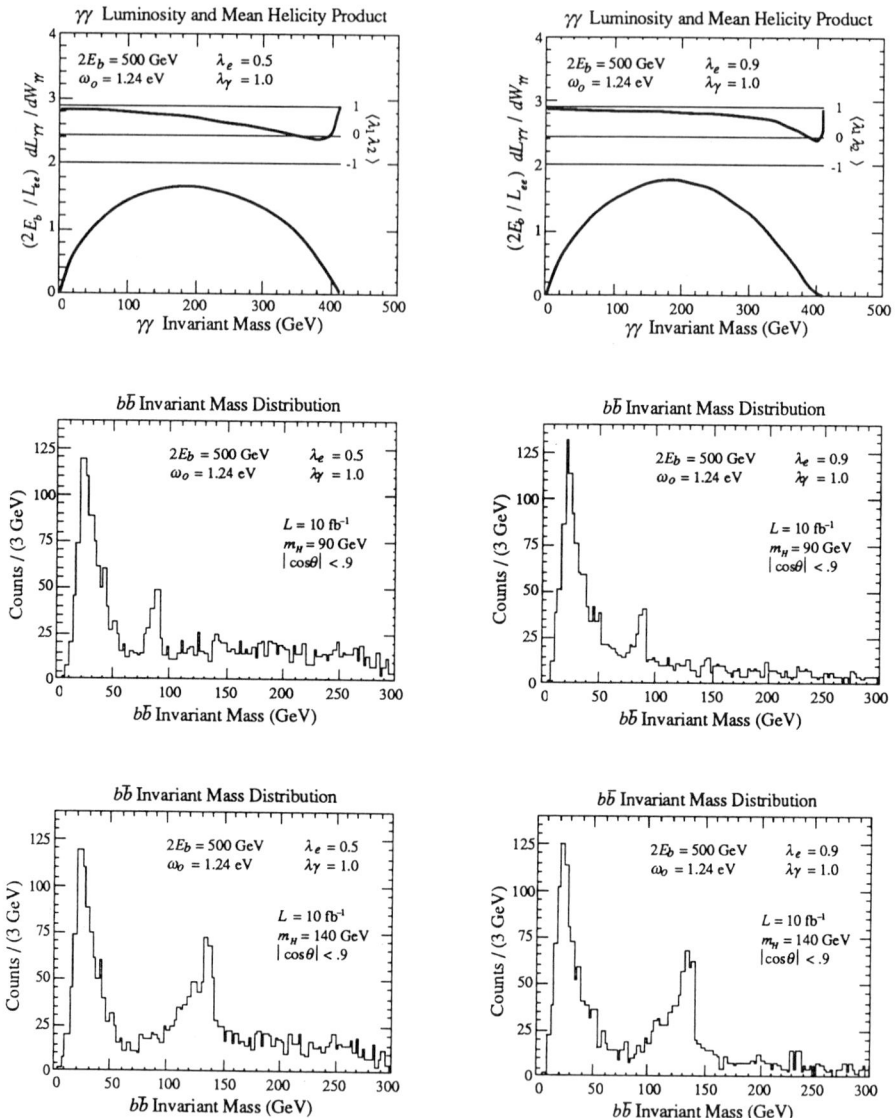

FIG 9 The search for an intermediate mass Higgs boson at a $\gamma\gamma$ collider — Case 2. At an ee machine energy of 500 GeV (and again ignoring the effect of a finite conversion distance) one distinct advantage and one distinct disadvantage of this higher energy machine over the 300 GeV case are apparent. The higher machine energy means that the Higgs boson lies 'further back' on the photon helicity curve, resulting in a high degree of photon polarization for lower linac electron polarization values, and therefore enhanced signal to noise. The event rate, however, is significantly lower at this energy than at 300 GeV, for two reasons: as the machine energy increases, the luminosity per unit mass interval decreases accordingly; also, as the machine energy increases, the amount by which the final state is boosted along the beamline increases, resulting in a lower fraction of the events in the detector.

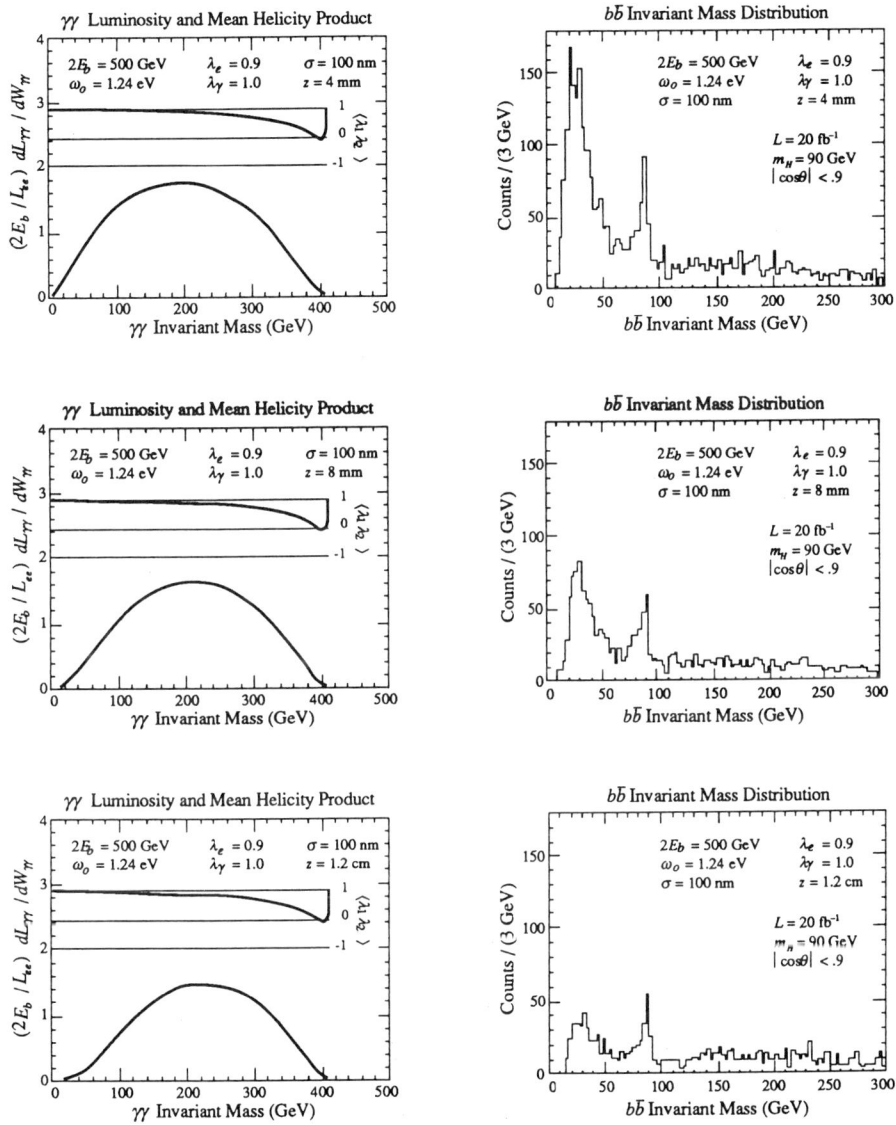

FIG 10 The search for an intermediate mass Higgs boson at a γγ collider — Case 3. At an *ee* machine energy of 500 GeV and linac electron polarization of 90%, the effect of moving the conversion point back from the interaction point is demonstrated. A cylindrical elecron beam with gaussian profile and σ of 100 nm is assumed, and conversion distances of 4, 8, and 12 mm are considered. As the conversion distance is increased, the drop in luminosity (at 90 GeV) is apparent, and the significance of the signal is reduced accordingly. Note that an integrated luminosity of 20 fb^{-1} is assumed, twice that of Figs. 8 and 9.

this process are 2-3 orders of magnitude larger than the Higgs production cross section. In order to minimize this background, it will be essential to remove the residual electrons from the interaction point.

This can be accomplished by the application of a magnetic field transverse to the beam direction at the interaction point. Following Compton scattering, such a field would bend the electrons away from the collision point. The field required to bend electrons of momentum p a distance δ given a conversion distance z is:

$$\frac{B}{\text{Tesla}} = 0.066 \left(\frac{p}{\text{GeV}}\right)\left(\frac{\delta}{\mu\text{m}}\right)\left(\frac{z}{\text{cm}}\right)^{-2} . \qquad (18)$$

At a conversion distance of 1 cm, a 150 GeV electron can be deflected 0.5 μm by a 5 T magnetic field. Such a field might be generated by superconducting Helmholtz coils placed just outside the beampipe (perhaps between the 1st and 2nd layers of the vertex detector).

$\gamma\gamma \rightarrow$ Higgs: Study

While a $\gamma\gamma$ collider serves as an attractive alternative to e^+e^- and hadron colliders in the search for a Higgs boson in the intermediate mass range, perhaps the most compelling argument for a Photon-Photon Collider is its singular ability to study a Higgs boson once it is found.[14] After the discovery of a Higgs boson, it will be paramount to try to learn as much about it as is possible. One of the most important properties to determine is the Higgs partial width to 2 photons: recall that the coupling of the Higgs to $\gamma\gamma$ proceeds through loops of charged particles which couple to the Higgs, so the two-photon width is a sensitive probe of physics beyond the Standard Model. Supersymmetric models, technicolor models, and other extensions of the Standard Model with more complicated Higgs sectors all predict two photon couplings in general different from the Standard Model.[2] As an example in Fig. 11 we plot the Higgs to $\gamma\gamma$ partial width for an extension to the Standard Model with two Higgs doublets.[2] In this model the up-type quarks couple to one doublet while the down-type quarks couple to the other. Tanβ is the ratio of Higgs vacuum expectation values (v_u/v_d), and α is the neutral Higgs mixing angle. Plotted is the ratio of

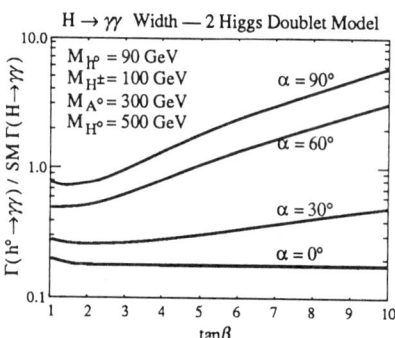

FIG. 11 The H $\rightarrow \gamma\gamma$ partial width in an extension to the Standard Model with two Higgs doublets. Tanβ is the ratio of Higgs vacuum expectation values and α is the neutral Higgs mixing angle.

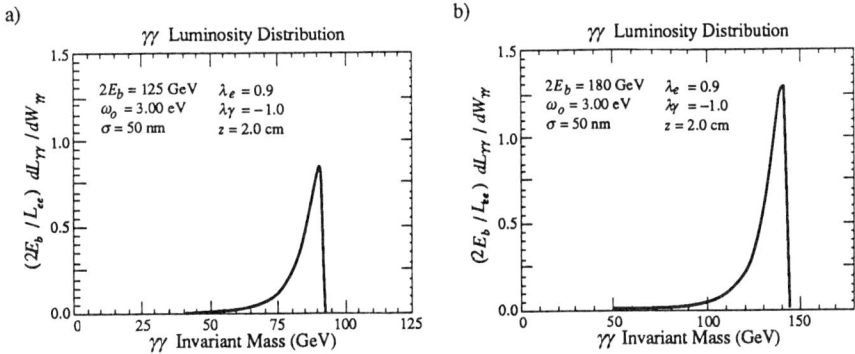

FIG. 12 The luminosity distributions of a $\gamma\gamma$ collider designed to study a 90 and 140 GeV Higgs boson.

$\gamma\gamma$ partial widths in the two-Higgs doublet model to that in the Standard model, as a function of $\tan\beta$, for various values of α, at representative values of the Higgs' masses. The $\gamma\gamma$ width differs significantly from the Standard Model value.

A $\gamma\gamma$ collider provides a unique opportunity to measure the two-photon width of a Higgs boson. We have heretofore stressed the broad luminosity distribution attainable, but it is also possible (and in practice much easier) to attain a more monochromatic 'peaked' luminosity distribution. The machine can be designed so that the peak of the distribution 'sits' on the Higgs resonance. As an example we assume that a Higgs with a mass of 90 GeV has been discovered; machine parameters similar to those shown in Fig. 12a could then be used. Assuming an integrated luminosity of 10 fb^{-1} and a Standard Model Higgs boson, Monte Carlo simulations indicate an expected 500 Higgs events over a background of 200 continuum $b\bar{b}$ events (with $|\cos\theta|<0.8$), allowing a determination of the two-photon width to within 6%. If a 140 GeV Higgs were to be discovered, a machine with the parameters of Fig. 12b could be used. With 10 fb^{-1}, of integrated luminosity, we would expect near 1200 Higgs events above 70 continuum $b\bar{b}$ events, allowing a determination of the two-photon width to within 3%.

CONCLUSION

With the development of linear collider and laser technology, a new tool may be made available for particle physics research, namely a facility capable of high energy $\gamma\gamma$ collisions. Such a collider would provide a unique environment for the exploration of the spontaneous symmetry breaking sector of the Standard Model, providing a possible discovery mechanism for an intermediate mass Higgs boson, and certainly a means of studying such a particle once found.

REFERENCES

1. ALEPH Collaboration (D. Decamp et al.), preprint CERN-PPE-91-19 (1991)
   DELPHI Collaboration (P. Abreu et al.), preprint CERN-PPE-90-163 (1990)
   L3 Collaboration (B. Adeva et al.), Phys. Lett. **B257**, 450 (1991)

OPAL Collaboration (M. Z. Akrawy et al.), Phys. Lett. **B253**, 511 (1991)

2. J. F. Gunion et al., The Higgs Hunters Guide, Frontiers in Physics Series (Vol. 80), Redwood City, California (Addison Wesley,1990)

3. M. Schneegans, preprint LAPP-EXP-91-04 (1991)

   Z. Kunszt et al., preprint ETH-TH-91-17 (1991)

4. Mike Hildreth, private communication.

5. H. E. Haber and R. Hempfling, Phys. Rev. Lett. **66**, 1815 (1991)

   J. Ellis et al., Phys. Lett. **B257**, 83 (1991)

6. V. M. Budnev et al., Phys. Rep. **15C**, 181 (1975)

7. R. N. Cahn, preprint LBL-28672 (1990), in Proceedings of XVI International Symposium on Lepton and Photon Interactions, Stanford, California, 1989, edited by M. Riordan (World Scientific,1990), p. 60.

8. I. F. Ginzburg et al., Pis'ma Zh. Eksp. Teor. Fiz. **34**, 514 (1981); JETP Lett. **34**, 491 (1982)

9. C. Akerlof, SLC Workshop Notes, CN-39 (1981)

10. I. F. Ginzburg et al., Yad. Fiz. **38**, 372 (1983); Sov. J. Nucl. Phys. **38**, 222 (1983)

11. I. F. Ginzburg et al., Nucl. Inst. Meth. **205**, 47 (1983)

    I. F. Ginzburg et al., Nucl. Inst. Meth. **219**, 5 (1984)

12. J. C. Sens, preprint CERN-EP-88-99 (1988), in Proceedings of the VIIIth International Workshop on Photon-Photon Collisions, Jerusalem Hills, Israel, 1988, edited by U. Karshon (World Scientific, 1988)

13. V. I. Telnov, Nucl. Inst. Meth. **A294**, 72 (1990)

14. J. F. Gunion and H.E. Haber, preprint SCIPP 90/22 and UCD-90-25 (1990); presented at the 1990 DPF Summer Study on High Energy Physics, Snowmass, July 1990.

15. T. Barklow, preprint SLAC-PUB-5364 (1990); presented at the1990 DPF Summer Study on High Energy Physics, Snowmass, July 1990.

16. A. A. Sokolov et al., Quantum Electrodynamics, Moscow, USSR (Mir Publishers,1988)

17. T. Sjöstrand, Comp. Phys. Comm. **27**, 243 (1982)

    T. Sjöstrand, Comp. Phys. Comm. **39**, 3473 (1986)

    T. Sjöstrand and M. Bengtsson, Comp. Phys. Comm. **43**, 367 (1987)

18. SLD Design Report. SLAC-0273

    M. Breidenbach, preprint SLAC-PUB-3798

19. JADE Collaboration (W. Bartel et al.), Z. Phys. **C26**, 93 (1984)

20. C. Baltay et al., in SLAC-354, Proceedings of the SLD Physics Week, Kirkwood, CA, 1989, p. 495.

21. P. Mattig, preprint DESY-88-125 (1988)

22. C. Bortoletto et al., preprint UDINE-90-06-AA-REV (1991)

    T. Aziz, preprint TIFR-EHEP-91-2 (1991)

23. L. F. Landau, Dok. Akad. Nauk USSR **60**, 207 (1948)

    C. N. Yang, Phys. Rev. **77**, 242 (1950)

CHAIRMAN: D. Borden

Scientific Secretaries: A.V. Chizhov, P. Vikas

DISCUSSION

– *Sivaram:*

Have you considered the build up of the transverse polarization of the electrons in the beam as a result of the interaction of the magnetic moments with the guiding magnetic fields? How would this affect your parameters?

– *Borden:*

This is a technical detail of the machine physics with which I haven't concerned myself.

– *Sivaram:*

Could you clarify how the process $\gamma\gamma \to H$ compares with other processes of producing Higgs bosons in different energy ranges?

– *Borden:*

The $\gamma\gamma$ production mechanism is not competitive for Higgs masses below about 70-80 GeV, and above twice the Z mass hadron colliders do a much better job. It is only in the intermediate mass region of 80-180 GeV that $\gamma\gamma$ production is competitive.

– *Khoze:*

If I understood you correctly, your results connected with the suppression of the background from $\gamma\gamma \to b\bar{b}$ were strongly based on polarization arguments. As Prof. Dokshitzer explained in his talk, it is rather easy to emit a gluon from a quark, and gluon emission could spoil your polarization. It is especially important in the case of forward emission. Would you care to comment?

– *Borden:*

I admit that we have not considered radiative gluon emission. While the background from a $b\bar{b}g$ final state is important, especially near the beamline, in the central region it is not a dominant background. I agree with you, however, that it should not be ignored. This area is obviously an important one to include in a more detailed study of this process.

– *Barletta*:

This is an extended comment on the practicality of a high luminosity $\gamma\gamma$ and $e\gamma$ collider. If I take only present (or near term) technology, and try to determine if the required luminosities are possible, this is what I find. Beam spot sizes can be as small as 150 nm, and I assume 10^{10} particles per bunch. Using the DESY proposal (with 200 micro-bunches per pulse), a luminosity of $10^{33} \text{cm}^{-2}\text{s}^{-1}$ is possible at repetition rate of only 50 Hz! Present achieved emittances are near 10^{-6} mrad which implies a β^* of about 7 mm. All of this implies a laser with a 10 ps pulse length and laser jitter of about 1 ps. The laser power requirements are near 10^{18} W/cm^2 over a laser spotsize of about 10 μm. The only problem then is the pulse train energy. It is intriguing that such a collider may be possible in the not so distant future.

But is the laser really necessary? Perhaps beamstrahlung in a plasma lens can generate the needed photon flux, without the complications of a backscattered laser.

In any case, I conclude that a $\gamma\gamma$ or $e\gamma$ collider is an interesting near term possibility.

– *Borden*:

I cannot agree with you more.

– *Altarelli*:

While I think that $\gamma\gamma$ and $e\gamma$ machines would certainly be interesting, I wonder if the intermediate mass Higgs example is really a good one. In fact once one has an e^+e^--linear collider with $\sqrt{s} = 0.5$ TeV and L $= 10^{33} \text{cm}^{-2}\text{s}^{-1}$ the observation of the intermediate mass Higgs is not difficult in the e^+e^- mode.

– *Borden*:

I agree. Here in Erice in 1991 we are concerned with the origin of mass, and what is more intimately connected with the origin of mass than the Higgs boson? When I am not in Erice I am concerned with a number of other physics issues connected with a $\gamma\gamma$ or $e\gamma$ collider, including: $\gamma\gamma \rightarrow W^+W^-$; $\gamma\gamma \rightarrow$ hadrons; $\gamma\gamma \rightarrow H^+H^-$; $\gamma\gamma \rightarrow$ slepton pairs; $e\gamma \rightarrow e^*$; $e\gamma \rightarrow W\nu$; $e\gamma \rightarrow eZ$; and $e\gamma$ selectron-photino, just to name a few.

EXPERIMENTAL PHYSICS AT THE HIGHEST ENERGY (IN THIS CENTURY!)

John Peoples

Fermi National Accelerator Laboratory
P.O. Box 500
Batavia, Illinois 60510 USA

INTRODUCTION

During the last few days the students at the school have heard presentations on theory ranging from strings to the evolution of structure functions. There has been very little reported on how experiments are actually done. What I would like to do is discuss, without great detail, how experiments are done at a collider today. Thus the first part of my talk is entitled *Tevatron Today*. At the Fermilab Tevatron there are proton-antiproton collisions at $\sqrt{s}$ = 1800 GeV with a peak luminosity 2 x 10^{30} cm^{-2} sec^{-1}. The major detector that was built to use that luminosity (CDF) will be described, and then I will describe the Tevatron tomorrow. The report will be along the following outline:

Detection Strategies

Experimental Results from the 1988-89 Data

Prospects for 1992-94 and Beyond

Achieving the needed Luminosity

Summary

DETECTION STRATEGIES

The two major goals of the Fermilab collider program are to discover the top quark and to measure its properties, while a third goal is to measure the properties of the W boson. These goals are interrelated, since efficient detection of W bosons will be crucial to the discovery of the top. Let me begin this discussion by describing how W's and Z's are detected.

For the W at the Tevatron Collider, the process observed is
$\bar{p}p \to W^{\pm}$ + anything.

The fundamental interaction is at the quark level, and the remainder of the proton and antiproton scatter away more-or-less down the beam pipe. The average transverse momentum carried by the components other than the up quark and the anti-down quark is quite small. It is important to note that in addition to the fundamental interaction at the quark level there is also the $O(\alpha)$ strong process in which a hard gluon is emitted either by the up quark or the anti-down quark. In these events the W can have an appreciable transverse momentum. This will be important because there are a whole series of events of proton + antiproton → W + 1,2,3 and 4 jets but the jets have a significant amount of transverse momentum. The W decays as follows:

$$W^+ \to u\bar{d}, \; c\bar{s} \qquad \text{(quarks)}$$

$$\to e^+ \nu_e, \; \mu^+\nu_\mu, \; \tau^+\nu_\tau \qquad \text{(leptons)}$$

When the W decays into a neutrino and a charged lepton, these particles appear in the final state. When the W decays into a quark-antiquark pair the quarks do not appear in the final state, but instead two or more jets of hadrons appear. When there are two jets the direction of the hadron jets corresponds to the direction of the quarks. While the probability that a W will decay into a quark-antiquark pair is a factor of two or more greater than the probability to decay into a neutrino and a charged lepton, the quark-antiquark decay mode cannot be distinguished from hard scattering between the partons in the proton and antiproton.

At the Tevatron energies it is not possible to distinguish the decay of W into two jets. It is not possible to distinguish W decaying into a quark-antiquark pair from the QCD background. This is because the quark-antiquark pair forms two or more hard jets, but so does the QCD process of quark + antiquark → quark + antiquark + jets, or the QCD process in which the two initial state partons produce two outgoing partons with large transverse momentum. These have a much larger cross section than the W production cross section. Consequently, W decaying into quark-antiquark pairs cannot be distinguished from $\bar{p}p \to$ jets.

In practice, W's and Z's are detected via their leptonic decay modes, and we will discuss these in more detail. Firstly, the decay leptons (e, μ, ν) have to be identified. This is done by measuring the momenta of the charged particles in the vertex chambers (VTPC) and in the central track chamber (CTC), and comparing the momenta to the energy deposited in the calorimeter. The fact that e, μ and hadrons interact differently in matter allows the three types of particles to be separated. Electrons interact very strongly with high-Z materials and lose roughly half of their energy in one radiation length of matter. The process by which this is done is that the electrons emit photons through bremsstrahlung as they pass near the nucleus of a high-Z material; the photons then convert into electron-positron pairs in the high-Z material. The electrons and photons will be absorbed almost entirely after 15 to 20 radiation lengths of material. Lead is chosen because it absorbs electrons very rapidly but does not rapidly absorb hadrons such as pions, protons and neutrons since 20 radiation lengths of lead is only about 0.6 absorption lengths. Thus a charged pion will only lose on average a third or half of its energy in the lead. The calorimeter is constructed by having 20 radiation lengths of lead and scintillator followed by 4 absorption lengths of iron and scintillator. Thus an electron will lose almost all of its energy in the lead-scintillator component of the calorimeter and very little of its energy in the iron-scintillator part of the hadron calorimeter. Conversely, a pion will lose a relatively small fraction of energy in the lead-scintillator portion of the calorimeter and lose most of its energy in the iron-scintillator part of the calorimeter. The scintillator samples the amount of energy that is being lost per gram of material at the location of the scintillator. Thus the longitudinal structure of the shower can be used to determine whether the shower is caused by an electron or a pion. Typically electrons can be separated from pions at the level of 1 part in 300.

The calorimeters are calibrated by placing one of the calorimeter modules in a beam of known momentum. By using a Cerenkov counter one can separate electrons from pions in the beam and thus obtain the response of the calorimeter to both electrons and hadrons. During operation of the detector, particle momenta and corresponding calorimeter response are measured, allowing separation of electrons and pions. Figure 1 shows schematicaly how an electron appears in the CDF detector central calorimeter. Listed below are some of the relevant CDF detector properties.

(i) Momentum measurement in the central drift chamber

$$\frac{\delta p_t}{p_t} = 0.0011 p_t \qquad (|\eta| > 1.1)$$

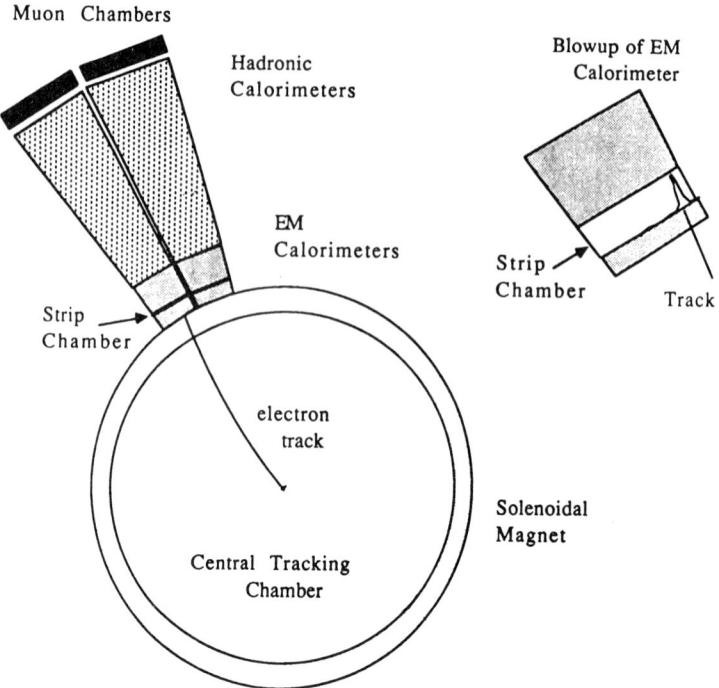

Fig. 1. Schematic view of an electron in the CDF central calorimeter.

(ii) Electron energy measurement in the central calorimeter

$$\frac{\delta E_t}{E_t} = \left[\left(\frac{0.135}{\sqrt{E_t}}\right)^2 + (0.02)^2\right]^{1/2} \qquad (|\eta| > 1.1)$$

(iii) Trigger thresholds

    1 "e" $E_t^e > 12$

                     $E_t^e > 7$ for 1/300 of events

    1 "μ" $P_t^\mu > 9$ GeV/c

    2 "μ" $P_t^\mu (1) > 3$ GeV/c

                     $P_t^\mu (2) > 3$ GeV/c

While very few of the electrons from W decay have a transverse energy E_T less than 12 GeV, the trigger is set at 12 GeV in order to detect the more copious source of electrons that come from semileptonic decays of B hadrons. This copious source of electrons allows CDF to carry out an extensive cross calibration of the electromagnetic calorimeter by comparing the track momenta (p) to sample energy deposited in the electromagnetic section of the calorimeter (E). The response of the 480 towers of the central calorimeter is adjusted to ~ 2% by making E/P constant across the calorimeter. The calibration of the momentum scale is verified using decays of J/Ψ and upsilon into two muons.

When measuring μ⁺μ⁻ pairs, CDF first measures the particles' momenta $\vec{p}^+, \vec{p}^-$ in the central drift chamber, and then identifies them as muons if they penetrate through the calorimeter with minimum energy loss. The invarient mass of the dimuon system is given by

$$m^2_{\mu\mu} = (E^+ + E^-)^2 - (\vec{p}^+ + \vec{p}^-)^2$$

$m_{\mu\mu}$ is the mass of a particle which decays into the two muons. The accuracy with which CDF has measured the masses of known particles which decay into two muons is illustrated in the following table.

Particle	Current Mass GeV	CDF Mass GeV
J/Ψ	3.0969 ± 0.0001	3.0963 ± 0.0005
Υ(1s)	9.4603 ± 0.0002	9.457 ± 0.005
Z⁰	91.175 ± 0.021	90.71 ± 0.45

We show in Figure 2 dimuon and dielectron mass plots from CDF demonstrating the production of Z⁰'s. Finally, the overall energy scale of the central electromagnetic calorimeter is adjusted by comparing the E/P distributions for a clean sample of W → e decays and a detailed simulation including internal and external radiation (as electrons pass through matter they radiate photons. These photons are approximately collinear with the electron and their energy is registered in the calorimeter). The absolute energy calibration is good to ~ 0.24%.

W± Detection

A single W can be detected by observing that the event has an electron or a muon that can clearly be separated from the hadronic background, and there is a substantial amount of missing transverse energy. The calorimeter resolution is such that the 20 to 50 GeV of missing transverse momentum that accompanies leptonic decay of a W can be easily separated from the fluctuations in the measurement of the transverse momentum that depends on the resolutions of the calorimeter and the fluctuation in transverse momenta that are carried away by the spectator particles.

The statistical errors are just a consequence of the limited size of the event sample. The systematic errors are to a large extent also dependent upon the size of the event sample. They are dominated by the calorimeter response, the fluctuations in the transverse momentum and longitudinal momentum that the spectator particles carry away as they go down the beam pipe and finally by the knowledge of the structure functions of the interacting quarks.

In the reaction $\bar{p}p \rightarrow W^{\pm} + X$, $W^{\pm} \rightarrow \mu^{\pm} + \nu$,

the neutrino transverse momentum, $\vec{p}_\nu^t$ is given by $-\left[\Sigma \vec{h}_i^t + \vec{p}_\mu^t\right] = \vec{E}_t$ where $\vec{h}_i^t$

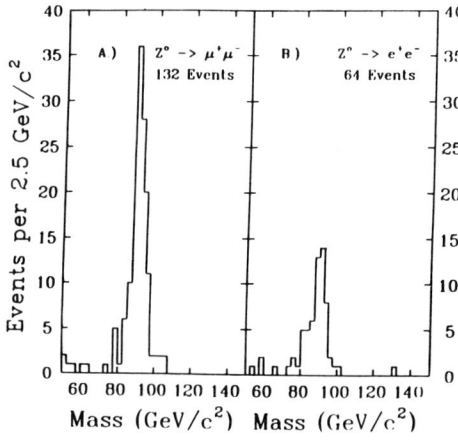

Fig. 2. Dimuon and dielectron mass plots.

is a hadron momentum, and the sum is taken over all hadrons. (A significant fraction of the longitudinal momentum is lost down the beam pipe).

Since $\Sigma E_i \hat{n}_i = \Sigma \vec{h}_i$ (where the sums are taken over the calorimeters), and $\Sigma \vec{h}_i$ (undetected) ≈ 0 (~ a few GeV), then the missing transverse energy $\vec{E}_t$ which is equal to $\vec{p}_\nu$, is given by $-\left[\Sigma E_i \hat{n}_i + \vec{p}_\mu\right]$

In CDF, the W selection criteria are

$$p_t^\ell > 20 \text{ GeV} \quad (\ell \text{ is a muon or electron})$$

$$E_t > 20 \text{ GeV}$$

$$S = \frac{|\vec{E}_t|}{\sqrt{\Sigma E_t^i}} > 2.4 \text{ GeV}^{1/2}$$

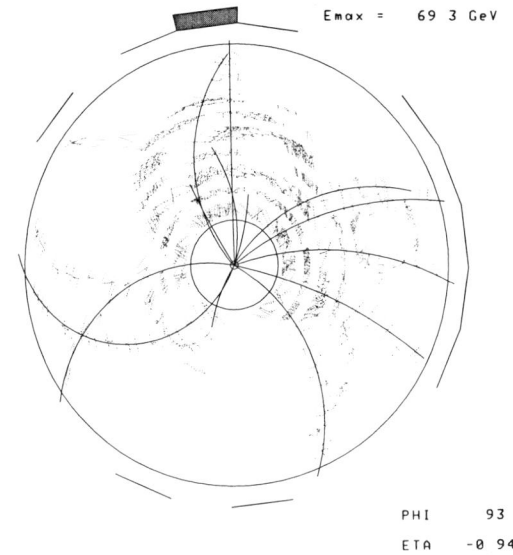

Fig. 3.   W → eν event.

It is observed, from minimum bias events, that the resolution in E_t is given by

$$\sigma(E_{x,y}) = (0.47 \pm 0.03) \left[\Sigma_t^i\right]^{1/2}$$

Figures 3 and 4 show a W → e + ν event.

EXPERIMENTAL RESULTS FROM THE 1988-89 DATA

We will report here on CDF's measurements of

(i) $W^\pm$ mass
(ii)   Production cross sections for $W^\pm$ and Z^o
(iii) Search for massive gauge bosons
(iv)   Production cross sections for b quarks
(v) $B - \bar{B}$ mixing
(vi)   Search for the top quark

(i) W Mass

Figure 5 shows the transverse mass distributions deduced from CDF's electron and muon data, described earlier, together with best fits for the W mass. After corrections, these give the most recent values.

$M_{W^{\pm}}$ (electrons) (GeV/c)^2 = 79.91 ±0.35 (stat) ±0.24 (sys) ±0.19 (scale).

$M_{W^{\pm}}$ (muons) (GeV/c)^2 = 79.90 ± 0.53 (stat) ± 0.32 (sys) ± 0.08 (scale). The above results are obtained from 1130 W → e and 592 W → μ "Superclean" samples.

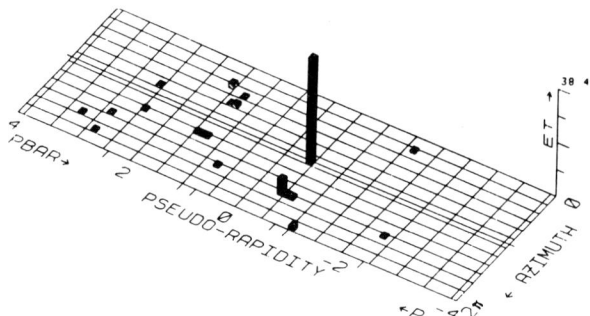

Fig. 4.   Lego plot of W → eν event.

The combined W mass from CDF is given by

$M_{W^{\pm}}$ = 79.91 ± 0.39 (GeV/c)^2

When combined with M_{Z^0} obtained from LEP experiments, CDF obtains

$$\sin^2\theta_w = 1 - \frac{M^2_{W^{\pm}}}{M^2_{Z^0}} = 0.2317 \pm 0.0075$$

It is worth noting that by increasing the event samples substantially, the errors on the weak mixing angle can be reduced. The

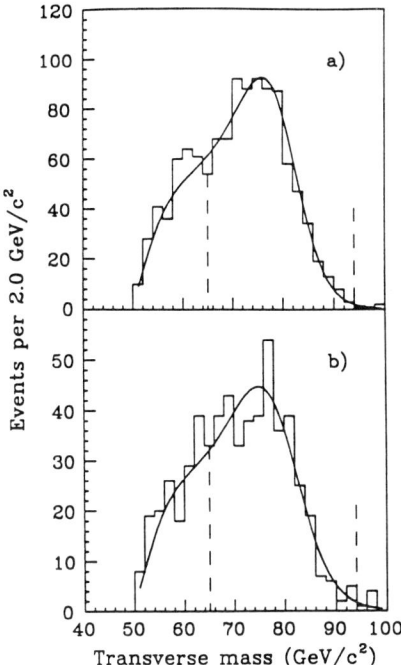

Fig. 5. a) The transverse mass distribution for all W → eν candidates; overlaid is the best fit to the data. Indicated with dashes is the range of transverse mass used in the fit. b) The transverse mass distribution for all W → μν candidates.

measurement of the ratio of the W to Z cross sections times the branching ratio into leptons affords a measurement of this W width.

Define R as the experimentally measured

$$R = \frac{\sigma(p\bar{p} \to W)}{\sigma(p\bar{p} \to Z)} \cdot \frac{BR(W \to e\nu)}{BR(Z \to ee)}$$

Then $$R = \frac{\sigma(p\bar{p} \to W)}{\sigma(p\bar{p} \to Z)} \cdot \frac{\Gamma(W \to e\nu)}{\Gamma(W)} \cdot \frac{\Gamma(W)}{\Gamma(Z \to ee)}$$

$$\Gamma(W) = \frac{1}{R} \cdot \frac{\sigma(p\bar{p} \to W)}{\sigma(p\bar{p} \to Z)} \cdot \Gamma(W \to e\nu) \cdot \frac{\Gamma(Z)}{\Gamma(Z \to ee)}$$
↑ ↑ ↑ ↑
Measured   Theory   Theory   Theory or LEP

A precise measurement of W width is important because it can be used to place a model-independent bound on the top quark mass.

331

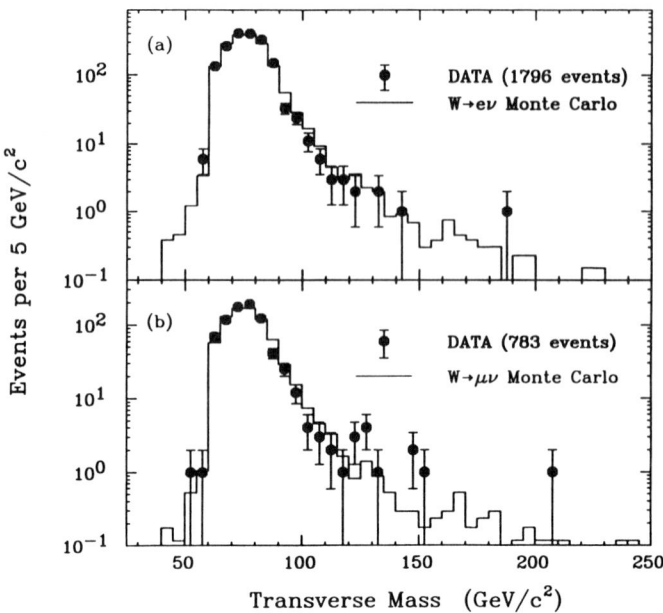

Fig. 6. a) The observed transverse mass distribution compared to a Monte Carlo prediction for $W \to e\nu$. b) As a), for $W \to \mu\nu$.

(ii) <u>Production cross sections for $W^\pm$ and Z^0</u>

CDF's data are based on 2664 events from an inclusive sample of $W^\pm \to e^\pm \nu$, and 243 events of $Z^0 \to e^+e^-$.

The results are

$$\sigma(W^\pm \to e^\pm \nu) = 2.19 \pm 0.04 \pm 0.2 \text{ nb}$$
$$\sigma(Z^0 \to e^+e^-) = 0.209 \pm 0.013 \pm 0.017 \text{ nb}$$

where the first error given is statistical and the second is systematic.

From the above,

$$\Gamma(W^\pm) = 2.12 \pm 0.20 \text{ GeV} \quad \text{(CDF)}$$

to be compared with the standard model value with 5 quarks of

$$\Gamma(W^\pm) = 2.07 \text{ GeV}$$

This result rules out decays of the $W^\pm$ of the form

$$W^\pm \to t+b \text{ or } W^\pm \to H^\pm + b$$

and this places a lower limit on the t and $H^\pm$ mass

(iii) <u>Search for massive gauge bosons</u>

Searches for massive gauge bosons are made by comparing the transverse mass distribution formed from a charged lepton and the missing E_t with Monte Carlo predictions for these quantities that are due to the decay of heavier W bosons, W'. Figure 6 shows two such plots, with Figure 7 illustrating the limits obtained for a massive gauge boson W'. Limits on Z' decays to e^+e^- is obtained from non-observation of peaks in the e^+e^- mass spectrum.

The 95% confidence limits obtained by CDF for vector boson masses, assuming standard model couplings, are

$M_{W'} > 520$ GeV/c^2

$M_{Z'} > 412$ GeV/c^2

(iv) <u>Production cross sections for b-quarks</u>

The production mechanism searched for is

$$\bar{p} + p \rightarrow \bar{b} + b + X$$

where $\bar{b}$ + b are observed as $\bar{B}_h + B_h + X'$. The primary production process is gluon fusion leading to a b quark and an anti-b quark. B hadrons containing b quarks are "observed."

We first discuss observation of B hadrons via the inclusive electron sample. The decay mode is

$$B \rightarrow \nu_e + e + Y$$

where Y usually contains a charmed particle.

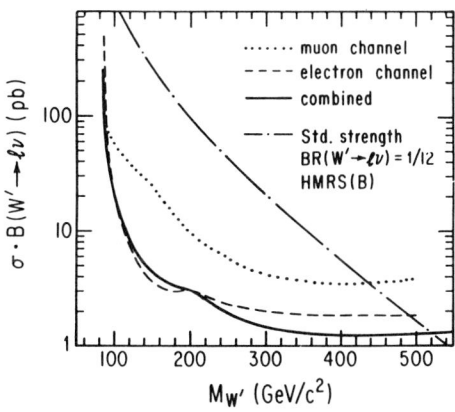

Fig. 7. Limits obtained for the mass of a massive gauge boson W'.

Figure 8 gives the inclusive electron P_t spectrum of 17K events obtained with the 12 GeV threshold triggers and a pre-scaled sample of the 7 GeV threshold triggers. Following a cut on the transverse mass and ee pair mass to remove W's and Z's, the lower spectrum of Figure 8 is obtained. There is a background in this data from non-electron events (15 ± 15%) and photon conversions (20 ± 5%).

If the "low" p_t electrons of Figure 8 are from semi-leptonic B decays, then D^0 decays should be observed through the following process.

$$B \rightarrow D^0 + e^- + \bar{\nu} + X \, , \, D^0 \rightarrow K^- + \pi^+$$

$$\overline{B} \rightarrow \overline{D^0} + e^+ + \nu + X \, , \, \overline{D}^0 \rightarrow K^+ + \pi^-$$

(Note that by D^0, we also include D^{0*}, etc.)

We see that the charge of K and electron are identical, and so the charge of the lepton tags that of the K.

The experimental method is as follows. A search is first made for a pair of oppositely charged particles that are nearby in rapidity with respect to the charged lepton. If Y denotes the

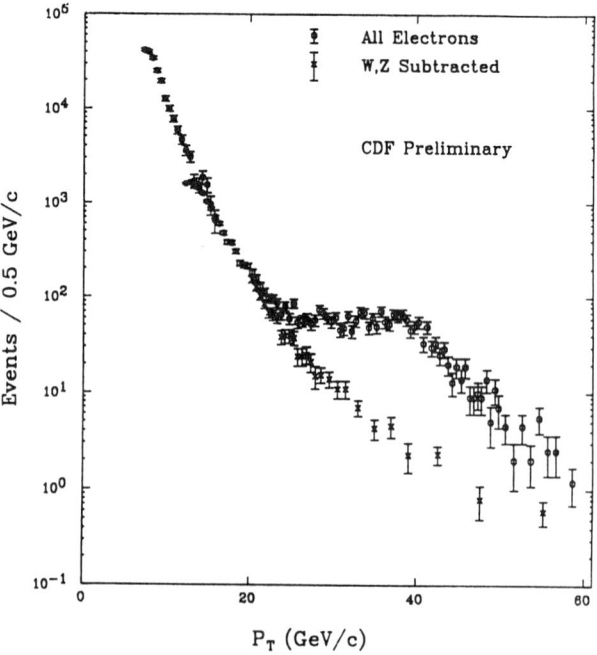

Fig. 8. Inclusive electron p_T spectrum.

pseudo rapidity of the particle and ϕ the azimuthal angle then the variable ΔR is defined as

$$\Delta R = \sqrt{\Delta y^2 + \Delta \phi^2} \text{ where}$$

$$\Delta y = |y_e - y_{K,\pi}|, \quad \Delta \phi = (\phi_e - \phi_{K,\pi})$$

The particle with same charge as the lepton is called the K and the other particle the π.

Figure 9 shows a Kπ mass plot for opposite charge mass combinations within $\Delta R = \sqrt{\Delta y^2 + \Delta \phi^2}$ of 0.6, and $P_K > 1.5$ GeV, where the K is taken as the charged particle with the same charge as the lepton. A D^0 peak is observed with 95 ± 17 events, showing that indeed semi-leptonic B decays are being observed. If the K is defined as the particle with the opposite charge from the lepton, no peak is seen.

A process that can result in the same charged correlation between electrons and kaons from D^0 decays is

$$\bar{p}p \to c\bar{c}X$$
$$c \to D^0 X'$$
$$\bar{c} \to e^- \bar{s} X''$$

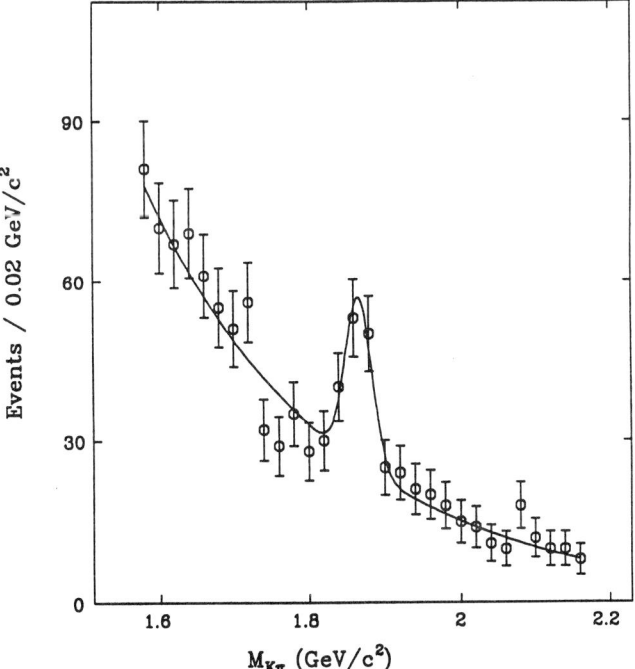

Fig. 9. K$^-\pi^+$ invariant mass distribution in the electron events.

This is especially significant when the $c\bar{c}$ pair is produced close together in phase space via the "gluon splitting" process $g \to c\bar{c}$

However, Monte Carlo calculations predict that $c \to e$ accounts for only approximately 15% of the observed prompt electron signal, assuming $\sigma(b\bar{b}) = \sigma(c\bar{c})$ at high P_t. This conclusion is supported by the observed relative rates of strange and anti-strange vector mesons accompanying the electron. (From $b \to \bar{e} \nu c, c \to s$ we expect strange particles; from $\bar{c} \to e \bar{\nu} \bar{s}$ we expect anti-strange particles). Distributions of the electron momentum component perpendicular to the jet axis are also consistent with a charm fraction of order 15%.

The results so far are (for $|y| < 1$)

P_t^e (GeV/c)	$p_t^{min}(b)$ (GeV/c)	$\sigma_b(P_t > P_t^{min})$ (nb)
10–15	15	1150 ± 450
15–20	23	210 ± 80
20–25	32	53 ± 21

CDF has also observed B decays directly, via the process

$$B^\pm \to J/\Psi \ K^\pm$$

Events are selected that have the dimuon pair with a mass equal to that of the J/Ψ. The invariant mass of the J/Ψ combined with any charged particle that is in a cone of $\Delta \leq 1$ is formed. The resulting invariant mass spectrum (Figure 10) shows a peak of 16 ± 6 events at the B mass.

From $J/\Psi K^\pm$ decays, the present result is

$$\sigma_b(p_t^b > 10, |y^b| < 1) = 8.2 \pm 2.9 \ \text{(stat)} \pm 3.3 \ \text{(sys)} \ \mu b$$

(v) **B–B̄ Mixing**

$B\bar{B}$ mixing was first observed by UA1 when they compared the number of events with like-sign dimuons to opposite sign dimuons. That result was subsequently confirmed and made more quantitative by the ARGUS and CLEO experiments. CDF has carried out a similar analysis by looking at μe events. In this case one compares μe events with the same sign to μe events with the opposite sign. μe events with the same sign can be

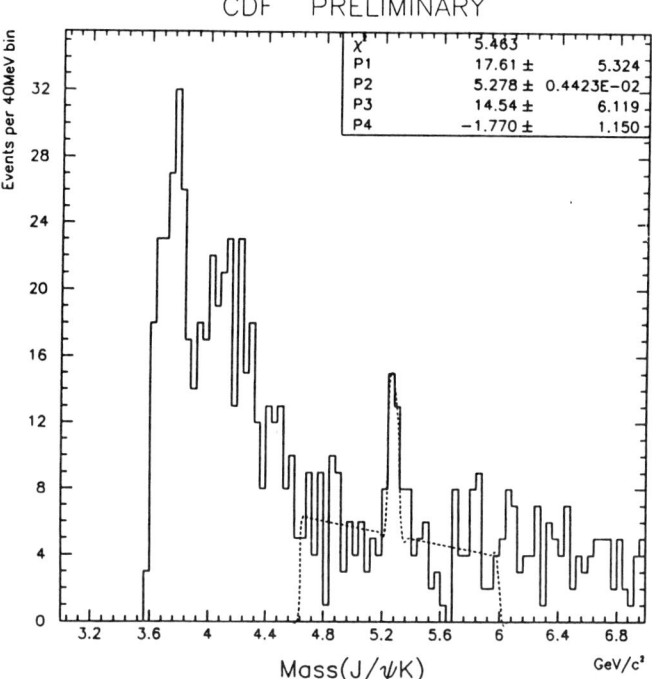

Fig. 10. J/ΨK invariant mass.

produced when a B^0 transforms itself into a $\bar{B}^0$ through the mixing process.

The normal $B^0\bar{B}^0$ production process is $\bar{p} + p \to \bar{B}_x^o + B_x^o + \ldots$

Since with mixing B_d^0 (B_s^0) can change into $\bar{B}_d^o$ ($\bar{B}_s^o$), one expects to observe

where
$$\bar{p} + p \to B_x^o + B_x^o + \ldots\ldots$$
$$B_x^o \to \mu^- + \nu + D_x + \ldots$$

and
$$B_x^o \to e^- + \nu + D_x + \ldots$$

The observation of like-sign μe pairs can be due to mixing by the above process, or by the sequences

$$B_x^o \to e^- + \nu + D_x \quad , \quad D_x \to h + \ldots..$$
$$\bar{B}_x^o \to h + \bar{D}_x \quad , \quad \bar{D}_x \to \mu^- + \nu + h$$

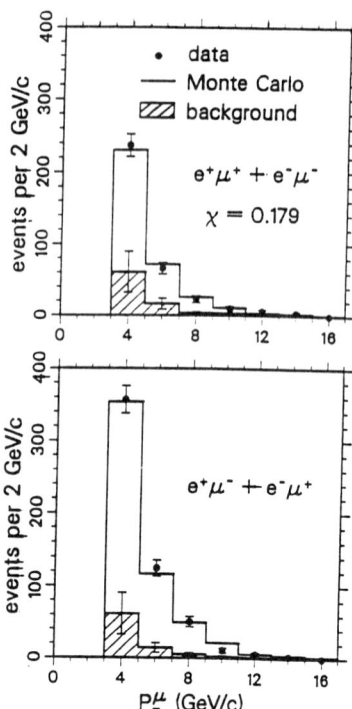

Fig. 11. Muon p_T spectra for the data, the Monte Carlo model with the observed mixing, and background in like-sign and opposite-sign $e\mu$ events.

and also by

$$B^o \to h + D, \quad D \to e^+ + \nu$$

Some distributions for opposite-sign and like-sign are displayed in Figure 11.

If $$R = \frac{N(\mu^+e^+) + N(\mu^-e^-)}{N(\mu^-e^+) + N(\mu^+e^-)}$$

it can be shown that, for no mixing, $R = 0.23 \pm 0.06$. However, the experimentally observed value is

$$R = \frac{256}{464} = 0.556 \pm 0.048 \text{ (stat)} {}^{+0.035}_{-0.042} \text{ (background)}$$

This gives a value of the mixing parameter χ for a mixture of B_d and B_s mesons

$$\chi = 0.179 \pm 0.027 \text{ (stat)} \pm 0.022 \text{ (syst)} \pm 0.032 \text{ (MC)}$$

combining this with a similar analysis in the ee channel, we get

$\chi = 0.176 \pm 0.031$ (stat + sys) ± 0.032 (MC)

This result for mixing is consistent with results obtained by UA1, ARGUS and CLEO. Since it is a combination of the mixing due to both B_d and B_s, it gives an opportunity to place limits on B_s mixing. This is shown in Figure 12. At the moment it is not possible to derive a useful limit on B_s mixing using the CDF data but the result shows the power of the method when eventually the number of B decays of this type will be substantially increased. Moreover it will be possible to separate the B_d decays from the B_s decays by means of a vertex detector, allowing a good measurement of B_s mixing.

(vi) <u>Search for the top quark</u>

The top quark has spin = 1/2 and electric charge = 2/3 e and weak isospin $I_3 = +1/2$. Because of its color properties, t is strongly coupled to gluons, and hence it will be produced in $t\bar{t}$ pairs by $\bar{p}p$ collisions.

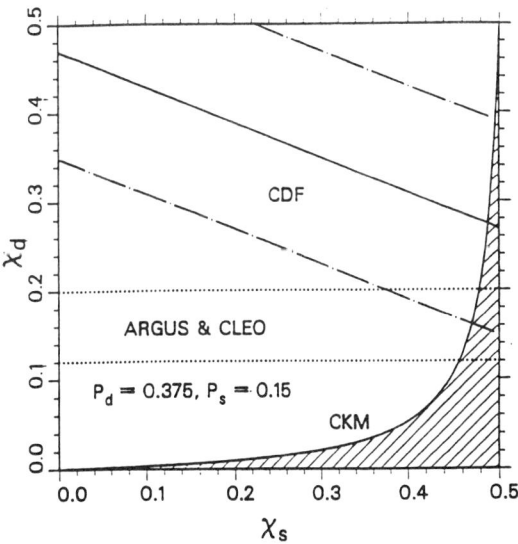

Fig. 12. The mixing probability of B_d^0 vs B_s^0, assuming that B_u, B_d, and B_s are produced in the ratio 0.375: 0.375: 0.15. The χ_d range is the ARGUS and CLEO combined result of 0.16 ± 0.04. The shaded region is allowed by the Standard Model. The bands represent $\pm 1\sigma$ uncertainty.

In the standard model, top quarks decay through the weak charged current t $\rightarrow$ b + (W$^+$)*, where the (W$^+$)* is real or virtual depending on the top mass. The (W$^+$)* then decays into leptons or quarks. A $t\bar{t}$ decay chain will look as follows:

$$p\bar{p} \rightarrow t\bar{t}$$
$$|\quad |_\rightarrow (W)b$$
$$|\qquad |_\rightarrow e\nu, \mu\nu, \tau\nu, u\bar{d}, c\bar{s}$$
$$|_\rightarrow (W)b$$
$$\quad |_\rightarrow e\nu, \mu\nu, \tau\nu, u\bar{d}, c\bar{s}$$

The quarks in the final state will materialize in the detector as hadronic jets. The fully hadronic final states cannot be distinguished from much more copious QCD processess

$$p\bar{p} \rightarrow \text{multi-jets}$$

Therefore only events with at least one electron or muon in the final state are useful in a search for top. The branching ratios are

W $\rightarrow$ eν,	W $\rightarrow$ q$\bar{q}$	12/81	(e + jets)
W $\rightarrow$ $\mu\nu$,	W $\rightarrow$ q$\bar{q}$	12/81	(μ + jets)
W $\rightarrow$ eν,	W $\rightarrow$ $\mu\nu$	2/81	(eμ)
W $\rightarrow$ eν,	W $\rightarrow$ eν	1/81	(ee)
W $\rightarrow$ $\mu\nu$,	W $\rightarrow$ $\mu\nu$	1/81	($\mu\mu$)

The most recent CDF results use the eμ, ee and $\mu\mu$ channels to establish a 95% C.L. upper limit on the $t\bar{t}$ production cross section. Using theoretical predictions for $\sigma(t\bar{t})$, this limit can be turned into a limit M_{top} > 86 GeV/c^2. While the signal/background ratio of a single lepton plus jets is not favorable, it can be improved by tagging the low energy b jet through its semileptonic decay into a muon. Combining the previous results with those of the complementary search in the e + jets and μ + jets channels, with a b $\rightarrow$ μ tag, improves this limit to 91 GeV/c^2 (see Figure 13).

PROSPECTS FOR 1991-94 AND BEYOND

Figure 14 shows the Tevatron layout and the two detectors, CDF and D0, which will take data in the 1992-93 Collider run. CDF will be upgraded

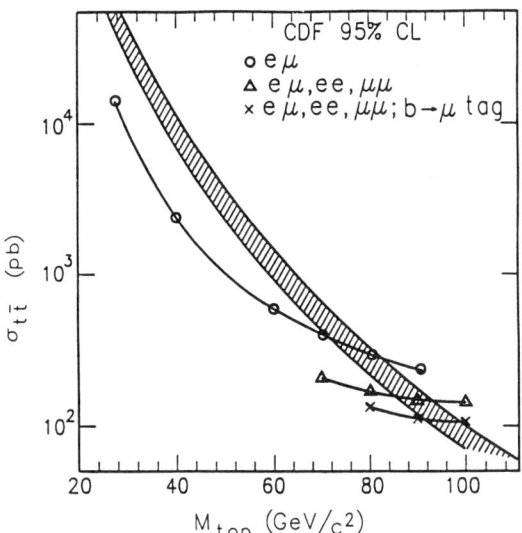

Fig. 13. 95% C.L. limits on $\sigma_{t\bar{t}}$ compared with a band of theoretical predictions; limits are shown from three different experimental techniques.

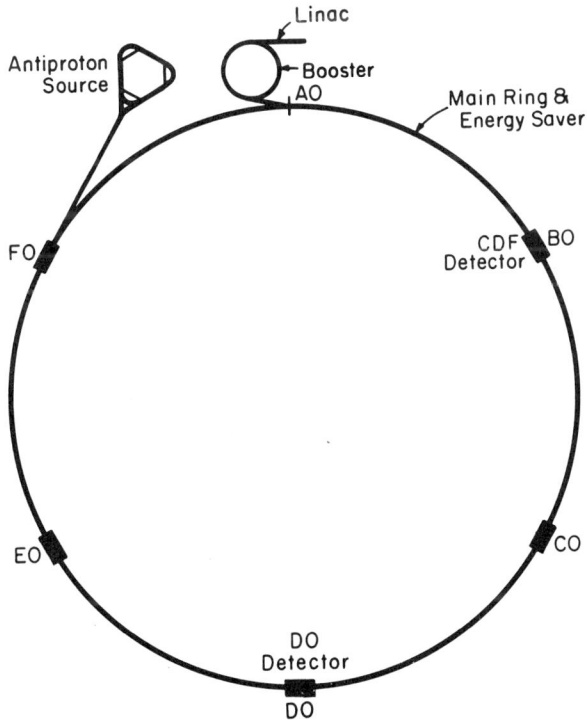

Fig. 14. Tevatron layout showing the locations of the CDF and D0 detectors.

from its current configuration with a silicon microvertex detector, increased muon coverage, and increased capability to handle peak luminosities of over 10^{31} cm^{-2} sec^{-1}. The new detector, D0, has fine grain calorimetry, and no solenoid magnetic field. It has full coverage for muons, with magnetized iron to give a muon momentum measurement.

In 1992, the Collider will have its performance upgraded, with separated orbits for p and $\bar{p}$ to reduce beam-beam interactions and thus increase luminosity. The typical peak luminosity should be 5x10^{30} in 1992 and 10^{31} cm^{-2} sec^{-1} in 1993. The goal will be to deliver 100 pb^{-1} to both CDF and D0. The prospects for improving the W measurements in the 1992-93 run are as follows.

	Current value GeV	Current error GeV	Error from 1992-93 data GeV
W mass	79.91	± 0.39	≈ ± 0.15
W width	2.19	± 0.20	≈ ± 0.05

100 pb^{-1} of integrated luminosity will allow a top quark discovery in the μe + jets channel if its mass is less than 150 GeV.

Beyond 1993, there will be major changes to CDF's calorimeter data acquisition system, and to its microvertex detector. For D0, upgrades will include electronics, triggers, data acquisition system, and tracking, including the introduction of a microvertex detector.

ACHIEVING THE NEEDED LUMINOSITY

The luminosity goals for 1993-93 are given below:

	Peak Luminosity	Integrated Luminosity
1992	5 x 10^{30} cm^{-1} sec^{-1}	25 pb-1
1993	10^{31} cm^{-1} sec^{-1}	75 pb^{-1}

In 1994-95 the number of p and $\bar{p}$ bunches in the Collider will be increased from the current 6 to 36. While this will not in itself increase the luminosity it will make it possible to increase the luminosity when it is possible to produce more $\bar{p}$'s. It will reduce the typical number of interactions per crossing from 1 to .16. Since this will reduce the time between bunches from the present 3500 ns down to 400 ns, there will need to be changes in the detector electronics.

Beyond 1996, to increase the luminosity further, the Main Ring (the present injector to the Tevatron) will be replaced by the Main Injector, a new 150 GeV accelerator in its own tunnel. This will increase the peak luminosity to 5×10^{31} cm^{-2} sec^{-1}. With improvements to both CDF and D0, this should allow integrated luminosities of 500 pb^{-1}, and a search for the top up to masses greater than 200 GeV, and possibly up to 250 GeV.

SUMMARY

I have reviewed the current status of the Tevatron and the CDF analysis, giving the present results for vector boson masses and widths, b quark production, and the lower limit on the top mass. Briefly given are the plans for upgrades to Fermilab's facilities and the expected physics improvements that will be obtained when they are implemented.

CHAIRMAN: J. Peoples

Scientific Secretaries: R. McNulty, X. Wu, P. Vikas, A. Hasan

DISCUSSION

– *Wadhwa:*

Once the top quark has been discovered and its mass measured, do you think that we will have enough sensitivity to give results on the Higgs mass?

– *Peoples:*

There are a few ideas in which the Higgs is some how related to the top mass. Most of them are for very heavy Higgs, in the 200 - 250 Gev range. One has to wait to see what comes out. Then one decides what to look for next.

– *Etzion (comment):*

The Higgs and the top masses enter in the radiative corrections for $\sin^2\theta_W$ measurement. As it is most sensitive to the top mass and much less sensitive to the Higgs mass, the measurement of the top mass would not make any serious change in the search for the Higgs mass.

– *Peoples:*

If the top mass is measured precisely and you really know $\sin^2\theta_W$ very well, then you get the Higgs mass to an order of magnitude with minimum Standard Model.

– *Wadhwa:*

Do you have any results on the CKM matrix elements?

– *Peoples:*

No. To measure any of the matrix elements you have to measure an absolute rate, or use branching ratios from another experiment. CDF is unable to measure any branching ratios. It is very difficult to do this in a hadron collider. In fact, even for an e^+e^- machine, it's very hard to do. The only branching ratios that have been measured are for D^0 and D^+.

– *Hasan:*

What are the mass limits on the SUSY particles as calculated by CDF?

– *Peoples:*

The CDF results that have been published so far come from the 1987 data sample of 20 pb^{-1}. This is a rather small sample compared to their 1988-89 data sample. With the 1987 data, the published mass limits are around 75 Gev, and they expect to reach 150 Gev with the 1988-89 data. With 500 pb^{-1} they expect to achieve a sensitivity of 200-220 Gev. The search for supersymmetric particles depends on the observation of missing transverse momentum.

– *Neubert:*

Is it possible to disentangle the contribution of B_d and B_s to the mixing parameter χ, using the measured value of χ_d obtained from ARGUS and CLEO? This would be important, since from the ratio X_d/X_s one can derive the ratio $|V_{ts}/V_{td}|^2$ of KM matrix elements. Knowledge of this ratio is essential in determining the unitarity triangle and in estimating the amount of CP violation in flavour-changing decays.

– *Peoples:*

The answer is yes. In the preprint on B-mixing, CDF has tried to do this. The problem is that the precision with which the χ's are measured is not good enough to draw a strong conclusion. The results suggest that the mixing is large. In the next run they will have the microvertex detector and they hope that they can make some separation of B_s and B_d decays.

– *Etzion:*

Why does the CDF not look at the τ channel, at least its leptonic decay channel?

– *Peoples:*

They are working on it.

– *Etzion:*

What is the idea behind building the D0 central tracking chamber without a magnetic field?

– *Peoples:*

In the previous talk Don Perkins talked about what committees did to experiments. In 1983, the CDF proposal existed and it was under construction and another very large group of people wanted to build a second detector to study the same physics. The committee which reviewed the proposal said that they should not have a magnetic field. So, it did not have one. The issue in these large detectors is whether you make them the same, or you make them different. So, D0 was built to have a substantially better calorimeter and far better muon detection than CDF. The idea was to shrink the tracking volume down so that pion decays to muons would be small. That was the perspective in 1983. Now, in 1991, D0 has proposed to put in a magnet in order to reconstruct momenta.

– Hernandez:

Can CDF improve the lower bound on the Higgs mass?

– Peoples:

I don't think that CDF or D0 can do better than LEP on the Higgs. The Higgs is very difficult to find at a hadron collider, especially the low energy Higgs. CDF has published a result on a search for a low mass Higgs but I am not familiar with it. I think it is for very light masses, a few hundred MeV, because they are looking for Higgs particles that decay inside the tracking volume.

– Hernandez:

Is there any observable that you can measure with higher precision than in any present or proposed experiment?

– Peoples:

Yes. The reason I talked about the mass of the W and its width is because I believe that they can be measured better at Tevatrons than anywhere else and that includes LEP II. The issue is going to be systematic errors at the Tevatron and the issue at LEP II is going to be whether they get enough integrated luminosity. The CDF and D0 believe that they can reduce the error on the mass of the W down to 35-50 MeV. They will require 500 pb^{-1} to do this. It will be an order of magnitude improvement on the current CDF numbers.

– McNulty:

You have an upper limit of 89 GeV on the mass of the top. It is known that you have a number of candidates of higher masses which are consistent with background processes. Would you care to comment on the nature and number of these candidates?

– Peoples:

Yes, there is one candidate - a muon-electron candidate. It would be unlikely to be purely background. I don't have a number in my head for the probability that the event is consistent with backgrounds, but I think that it is of order of 1 in 20. But, with one event it is really hard to figure out what mass it would have. That is the reason that the 95% confidence limit on the top mass is as low as 89 GeV. The event is consistent with a mass in the range of 100-150 GeV. I can't comment any more. I think the people on CDF just want to wait till they have at least 10 of such events before they claim to have seen the top.

– *Barletta:*

Would you comment on the capabilities to study CP violation in the B sector once the luminosity upgrade is completed?

– *Peoples:*

The cross-section for $B\overline{B}$ production is very large; the question is how to trigger on $B\overline{B}$ pairs. In the results I presented today on $B\overline{B}$ pairs, the B meson typically has 15-20 GeV, thus CDF was looking at a relatively small part of the cross-section. To do CP violation one will have to be sensitive to most of the cross-section. However, it is possible. I don't think that the existing detectors (CDF and D0) can do that without some improvements. My expectation is that if CDF can collect 100,000 straightforward B decays and prove that the microvertex detector can be made to work, then one can do it. It will be necessary to get a 2-D microvertex detector (the one CDF has is 1-D). Like Professor Zichichi, I am not particularly worried about the data acquisition. We can get zillions of microprocessors with 100 MIPs each. Already, the CDF data acquisition and processing system is larger than the Fermilab central computer division was last year.

– *Barletta:*

What is the limiting technology in the final luminosity upgrade?

– *Peoples:*

I think $\mathcal{L} = 10^{32}$ is an outside possibility, with the limiting factor being the $\bar{p}$ source. We have to make some major changes in the $\bar{p}$ source. The core cooling system will have to work at 8 20 GHz, and the pickup with that bandwidth is hard to design. But these things don't look impossible. I am not sure that the target system can be made to work. That's going to be the biggest problem. You have to put the protons on the target in sufficient abundance to get the antiprotons. It is not clear that the target is going to stay together when it is struck with the required beam intensity.

– *Barletta:*

Has there been any interest in allowing for p-p collisions at the highest luminosity?

– *Peoples:*

p-p collisions require two rings. That would probably be useful for doing b-physics, because the b's don't care whether the gluons come from p or $\bar{p}$. However

– *Sivaram:*

Has there been any estimate of the W magnetic moment from scattering experiments?

– *Peoples:*

The CDF collaboration and the D0 collaboration hope to get about 100 events each in the next run in which a W is produced in association with a hard photon. This gives some information on the magnetic moment. To make a good measurement these experiments will need about 500 inverse picobarns.

– *Sivaram:*

You showed a plot for the cross-section for $t\bar{t}$ state versus the mass of the top. Does this imply that beyond a particular mass, the possibility of observing $t\bar{t}$ state is negligible?

– *Peoples:*

I think this is merely the way the graph has been drawn. For top masses above 250 GeV, it will be very difficult to produce more than 10 or 20 observable decays.

– *Shabelski:*

There is a problem with charm production total cross-section, because the different data do not agree with each other. Have you any plans to measure the total cross-section and spectra of charmed hadrons?

– *Peoples:*

That would be feasible once the micro-vertex detector has been installed. At the moment CDF is able to identify D^0's with the standard trick of looking at the low energy pion from the D^* decay, by assuming that the two charged particles of opposite charge are a K and a π from the D^0 and then looking for a peak in the invariant mass spectrum. However this is a relatively poor measurement compared to what could be produced with a vertex detector.

– *Schulze:*

Did CDF also study multi-particle production, especially multiplicity distributions? Are they still a negative binomial at this energy?

there would be considerable expense involved in building another superconducting ring, so it's not a consideration in the immediate future.

– *Peoples:*

CDF has published some results from its 1987 data and the paper gives the multiplicity as a function of rapidity. The reason why the newer data has not been published yet, is that the threshold of the trigger for single jet events is enormous, and is thus relatively insensitive to any small momentum process. There was also an attempt to detect the particles produced in association with a diffractively scattered proton.

– *Ozdes:*

Is there any study on diffraction?

– *Peoples:*

There was an experiment (done with the CDF detector) that used the accelerator as a spectrometer to detect a diffractive p or p̄ emerging from the collision. The particles produced with a rapidity magnitude of less than 3 in association with the diffractive p or p̄ was detected in the CDF central detector. Data was taken but not analyzed. It would be interesting to see some low p_t phenomena because they have not been observed at these energies.

THE SSC PROJECT AND EXPERIMENTAL PROGRAM

Frederick J. Gilman
Superconducting Super Collider Laboratory
Dallas, Texas 75237

THE SSC PROJECT

The Superconducting Super Collider (SSC) will take a giant step beyond existing high energy physics facilities, providing research opportunities well into the 21st century. Not only will probing measurements be made of the validity of our current understanding of particle physics as summarized by the Standard Model, but the factor of 20 increase in center-of-mass energy over current accelerators will make possible broad explorations of what lies beyond.

A brief chronology of the SSC Project is given in Table 1. The project originated at the 1982 Snowmass Summer Study where it was seen as the next major step needed to address many of the most important open issues in high energy physics. From there the idea was developed and translated during the 1980s into the reality of the present SSC Laboratory in Texas and the SSC project baseline.

Table 1. SSC Chronology

1982	Concept originated at Snowmass Summer Study[1]
1983	Recommendation by DOE's High Energy Physics Advisory Panel to initiate the SSC project
1984	SSC Central Design Group formed
1986	SSC Conceptual Design Report[2] issued
1989	Selection of the Texas site
	Selection of URA as the Management Organization by DOE
	Establishment of the SSC Laboratory
1990	Site-Specific Conceptual Design Report[3]
	Cost/Schedule Baseline established[4]

In broad outline, the SSC project involves building a proton-proton collider with a center-of-mass energy of 40 TeV and a luminosity of $10^{33}/cm^2$ - sec. Fig. 1 shows the location of the project and its 87-km ring near Dallas, Texas. A schematic of the collider ring (not to scale) and the chain of injector accelerators is shown in Fig. 2.

Physics at the Highest Energy and Luminosity, Edited by A. Zichichi
Plenum Press, New York, 1992

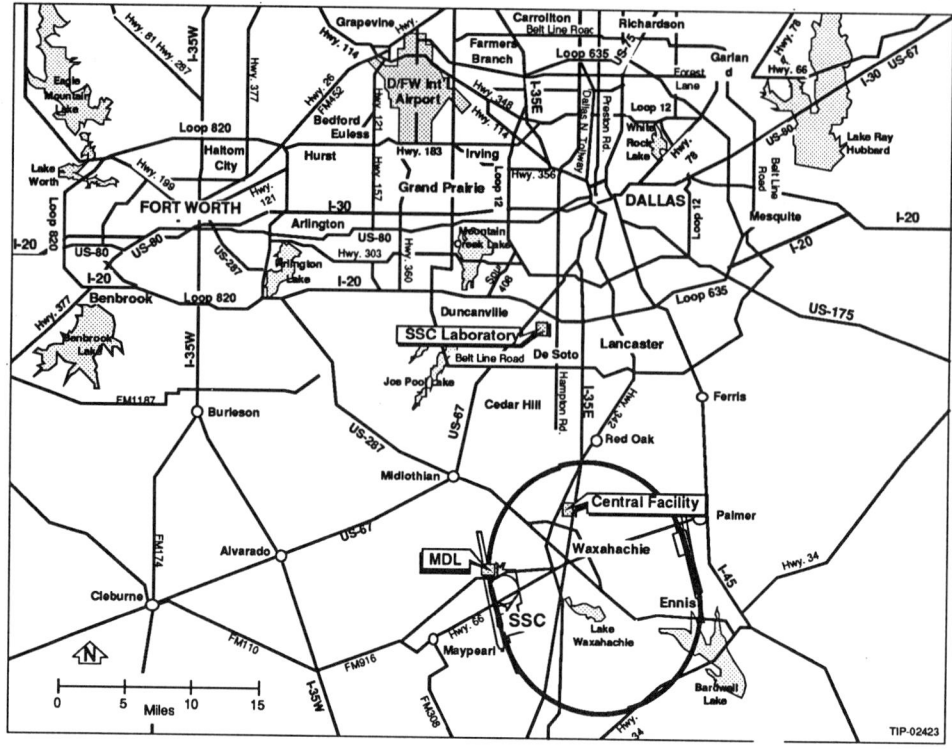

Fig. 1. Location of the SSC Project

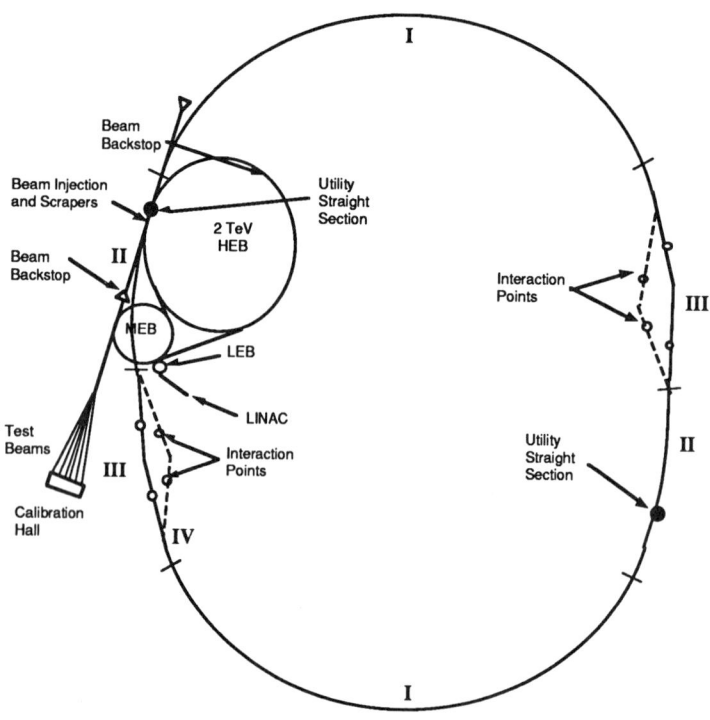

Fig. 2. Ring Schematic (not to scale)

Table 2. SSC Parameters

Energy	20 TeV
Particles/bunch (N)	0.75×10^{10}
Circumference	87,120 m
No. of bunches (B)	17,424
NB	1.3×10^{14}
$f$rotation	3.4 kHz
$f$collisions	60 MHz
S_b	5.0 m
$\varepsilon_N (\sigma)$	1 π mm-rad
β^*	1/2 m
σ^* (μm)	5
Luminosity ($\mathcal{L}$)	1×10^{33} cm^{-2} s^{-1}
$\mathcal{L}$/hit	1.6×10^{25} cm^{-2} s^{-1}
$\Delta\nu_{HO}$ (total)	0.003
$\Delta\nu_{LR}$ (total)	0.004
Power (kW) at 1.3×10^{14} (NB)	8.75/ring

Table 2 lists the principal parameters of the collider itself.

The ten-year schedule began in October 1989. Since then the SSC Project has moved from the design stage into start of construction. Temporary laboratory buildings were established in Dallas in 1989. Space was acquired in 1991 in the city of Waxahachie (see the Central Facility inside the ring) sufficient to house most of the SSCL project team by early 1992. Master planning has been initiated for all site activities under the direction of the Laboratory and the architect-engineering firm of Parsons-Brinckerhoff/Morrison-Knudsen, selected in 1990. The first buildings on the west side of the ring, the Magnet Development Laboratory and the Accelerator Systems String Test facility, will be finished in 1991. Nearby, a 60 by 30 foot elliptical shaft is being started that will permit access for a tunnel boring machine to excavate the first section of tunnel for the collider in 1992. Later, this same shaft will be used for installing magnets in the tunnel.

Progress on the technical components has been excellent. Tests of half a dozen model 50 mm aperture dipole magnets have yielded quenches at well above the design current without training. Full length superconducting dipoles will be tested at both Brookhaven and Fermilab later this year. A full string test of these magnets is to take place in Texas in 1992, and in-house development and test efforts on dipole, quadrupole, and correction magnets have been established. The vendors for the first 500 collider magnets—General Dynamics and Westinghouse—were selected toward the end of 1990, and initial delivery from their own plants is scheduled for 1994. Similarly, Babcock and Wilcox was chosen in 1991 to supply the first of the superconducting quadrupoles for the collider, and other major hardware procurements, such as that for cryogenic systems, have been initiated as well. In short, the construction of the SSC is under way.

THE PHYSICS OF THE SSC

The SSC will be able to explore a wide range of possible new phenomena that will be accessible due to the enormous increase in hadron collider energy that 40 TeV represents. More specifically, the energy and luminosity of the SSC were chosen with the aim of uncovering the nature of electroweak symmetry breaking by being fully capable of exploring the 1 TeV scale for collisions of the proton's constituents.

In the Standard Model, symmetry breaking is accomplished through a non-zero vacuum expectation value of the neutral Higgs field, giving masses to the W and Z bosons as well as to the quarks and leptons. Left as the telltale evidence of this mechanism is a single physical particle, the Higgs boson. This is but the simplest model and hardly the favorite of most theorists, who expect that there may be a number of scalar particles (sometimes still all subsumed under the name Higgs bosons) to be found (as, for example, in supersymmetry); or that electroweak symmetry breaking comes about with the existence of a whole new dynamical sector and the equivalent of the Higgs boson being a composite particle with many relatives. Such is the case for Technicolor theories. In any event, the Higgs boson of the Standard Model has become one of the benchmarks for both theory and experiment with regard to electroweak symmetry breaking, and it serves more generally to give one indication of the power of the SSC to explore new physics.

In very high energy proton-proton collisions such as those at the SSC, the Higgs boson is produced primarily in gluon-gluon and W–W collisions of the constituents of the protons. The resulting cross section is shown in Fig. 3. Note that the SSC has roughly an order of magnitude advantage over the LHC when M_H is many hundreds of GeV.

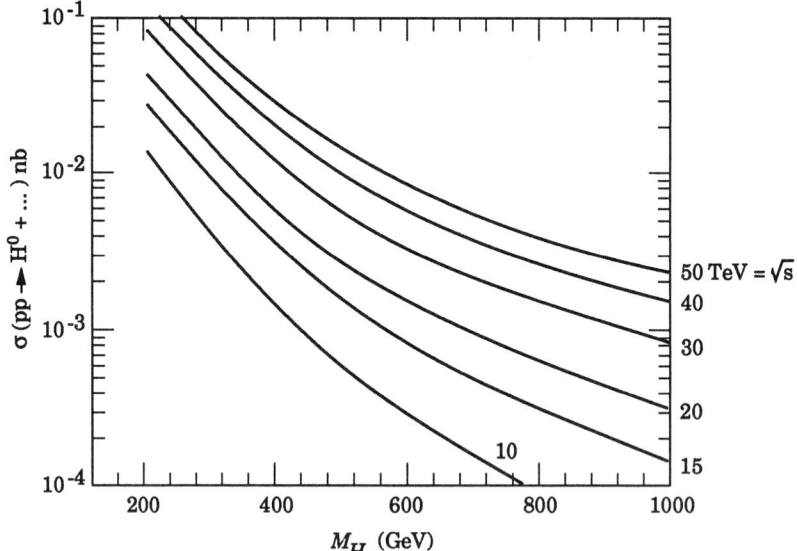

Fig. 3. Cross section for Higgs production in proton-proton collisions for various center-of-mass energies.

A few graphs illustrate the situation. The intermediate mass ($M_W < M_H < 2M_W$) range is well explored at the SSC. Figure 4 is representative of the possibilities at the SSC and shows in particular the capabilities of the proposed SDC detector[5] to see a Higgs boson in this range in one year at design luminosity using the decay into four charged leptons (originating in the decay into a real and virtual Z).

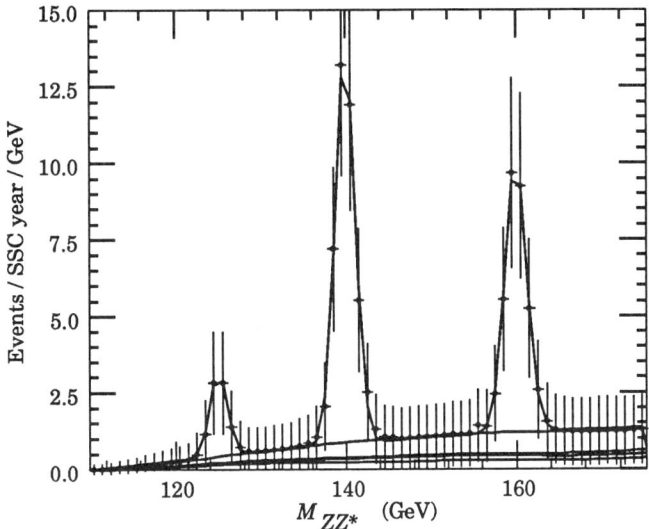

Fig. 4. Signals for H → $e^+e^-e^+e^-$, $\mu^+\mu^-e^+e^-$, and $\mu^+\mu^-\mu^+\mu^-$ using the SDC detector[5] and Higgs masses of 125, 140, and 160 GeV.

At the lower end of the intermediate mass range, the four lepton decay has a very small branching ratio that makes this mode too small to detect. Here the decay into γγ, although small, may be particularly relevant, and Fig. 5—from the Expression of Interest for the GEM detector[6]—shows the signal at the SSC in one year of running with different resolutions for the electromagnetic calorimeter.

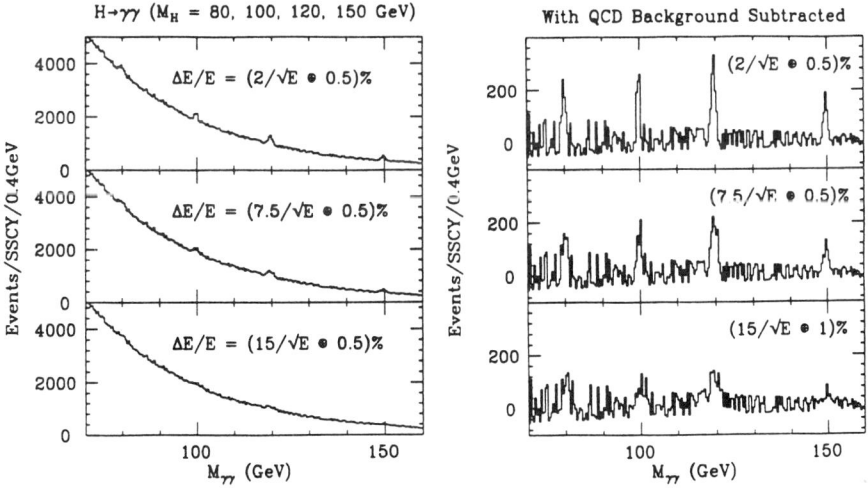

Fig. 5. Signals for H → γγ with three different calorimeter resolutions.[6]

For $M_H > 2M_W$, a high mass Higgs boson, the signals resulting from H → ZZ → four leptons are appreciable, as shown in Fig. 6. It should be possible to pursue the Higgs boson up to masses of about a TeV in several years of running at design luminosity, especially with the larger signal obtained by the use of other decays such as H → ZZ → $\ell^+\ell^-\nu\bar{\nu}$.

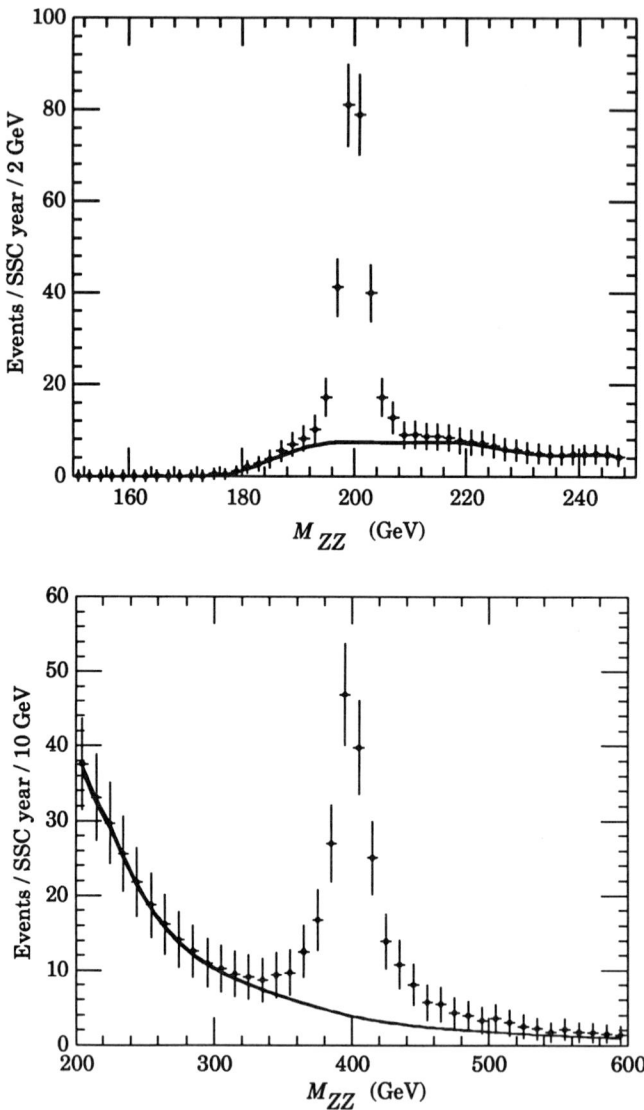

Fig. 6. Signals for H → ZZ → 4 charged leptons for a high mass Higgs in the SDC detector.[5]

This is only one small portion of the physics capabilities[7] of the SSC. Even within the Standard Model, the production of tens of millions of top quarks and trillions of bottom quarks per year makes detailed studies of their decays possible, including the very interesting questions associated with CP violation in B decays. Outside the Standard Model, it is possible to search for a larger gauge sector and the corresponding W' and Z' bosons up to a mass scale that is two orders of magnitude greater than that given by the ordinary W and Z masses. Other examples include the elucidation of proposed technicolor or supersymmetry models. When one turns to processes signalling a threshold for new physics in collisions of the quark constituents of the proton, a few events per year are generated above 20 TeV in the quark-quark center-of-mass for a cross section as small as tens of femtobarns.

Over the past decade or so the Standard Model has been tested repeatedly with ever increasing accuracy and over a broader range of energies. So far no violations have been found; indeed, many options have been closed or greatly limited. More than ever, we need to address the TeV scale experimentally in a direct way if the answers to the fundamental questions about what lies beyond the Standard Model are to be found.

THE SSC EXPERIMENTAL PROGRAM

Although there were many workshops, symposia, and conferences related to the physics and potential experiments at the SSC throughout the 1980s, in a formal sense the SSC Experimental Program began with the first meeting of the SSC Program Advisory Committee (PAC) in February 1990. Out of that meeting came a call for Expressions of Interest on any type of experiment that might be part of the initial experimental program. The initial fifteen Expressions of Interest were presented and considered at the PAC

Table 3. Expressions of Interest

Number	Title of EOI
SSC-EOI0001	Inclusive Spin-Spin Effects and Cross Sections Near 20 TeV
SSC-EOI0002	Low P_t Physics at the SSC
SSC-EOI0003	Solenoidal Detector Collaboration (SDC)
SSC-EOI0004	A Very Long Baseline Neutrino Oscillation Experiment
SSC-EOI0005	High Luminosity at the SSC
SSC-EOI0006	Electrons Muons Partons with Air Core Toroids (EMPACT)
SSC-EOI0007	10^{34}
SSC-EOI0008	Bottom Collider Detector (BCD)
SSC-EOI0009	A Compact Diamond-Based Detector for the SSC
SSC-EOI0010	L*
SSC-EOI0011	A Calorimeter-Based High-Rate Detector for the SSC (TEXAS)
SSC-EOI0012	(Proposal to be Submitted)
SSC-EOI0013	Internal Target Beauty Physics at the SSC
SSC-EOI0014	Super Fixed Target Beauty Facility
SSC-EOI0015	Search for Magnetic Monopoles
SSC-EOI0016	A High Resolution Detector
SSC-EOI0017	Relativistic Atomic Physics at the SSC
SSC-EOI0018	Electromagnetic and Muon Detector for the SSC
SSC-EOI0019	A Full-Acceptance Detector for SSC Physics at Low and Intermediate Mass Scales
SSC-EOI0020	An Expression of Interest To Construct A Major SSC Detector
SSC-EOI0021	Ideas for an SSC Open Geometry Forward Collider Detector to Measure CP-Violation in B-Decay

meetings in June and July 1990. Table 3 presents a list of the twenty-one Expressions of Interest submitted to the Laboratory up to now.

These potential experiments span the gamut from very large collider experiments that probe physics at the highest mass scales available to those that involve the chain of injector accelerators and deal with neutrino oscillations and even atomic physics.

Because of their scale and long lead-time, there was a need to concentrate on the major SSC detectors. These experiments are of unprecedented scale and complexity. In fact, each can well be considered a laboratory in itself, with many hundreds of collaborating physicists and hundreds of millions of dollars involved. Much of the scientific life of the SSC Laboratory will be carried on in these detector-laboratories. Following the July PAC meeting, there was a call for Letters of Intent for such detectors. A strong recommendation was made to the Laboratory to have two such detectors. One was to have as broad a range of capabilities as possible; the other to be competitive and complementary, emphasizing particularly robust muon detection and calorimetry.

Three such Letters of Intent were received: SDC[5], L*[8], and EMPACT/TEXAS[9] (formed from the combination of the groups that had put forward the EMPACT and TEXAS Expressions of Interest). These Letters of Intent became the focus of Laboratory attention in the months following their submission. In particular, they were the subjects of intensive debate and examination at PAC meetings in December 1990 and March 1991.

The first to be given the go-ahead to proceed toward the development of a full technical design report was SDC (see Fig. 7).

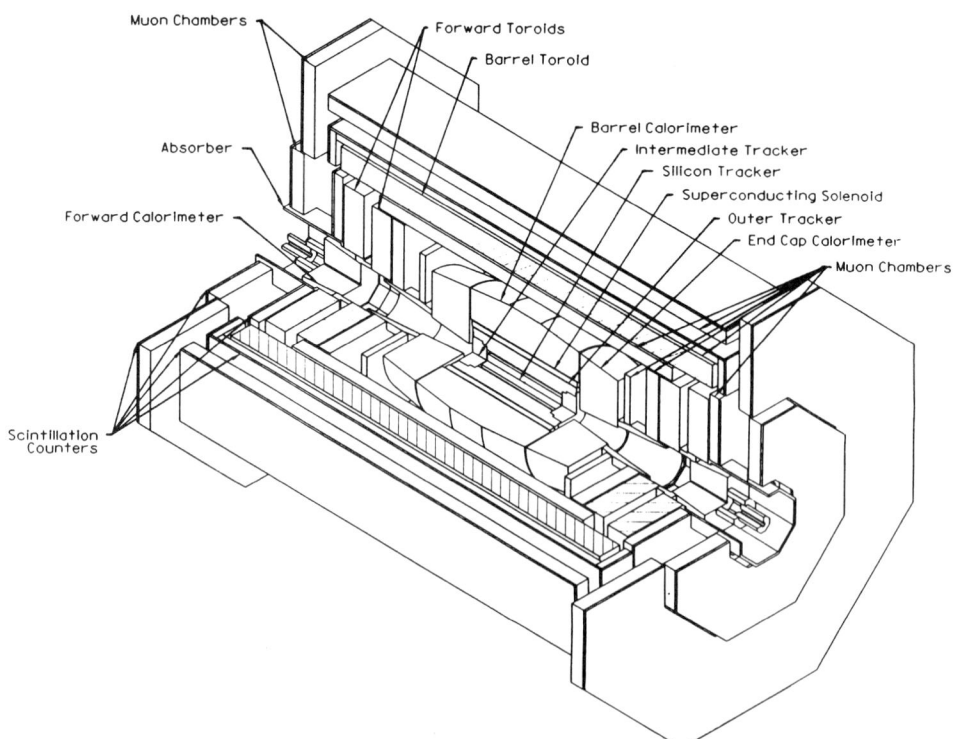

Fig. 7. The SDC detector concept presented in the Letter of Intent.[5]

It was felt that the SDC design goals emphasizing good central tracking, hermetic calorimetry, lepton energy measurement and identification, and high resolution vertex detection satisfied the broad range of capabilities desired for the first detector described in the call for the Letters of Intent. While important technical questions remained to be resolved at that time through further R&D, it was felt that the collaboration had the technical strength to address and resolve them, and then go on to build the detector.

EMPACT/TEXAS and L* were then the candidates for the detector complementary to SDC. After January 1991, attention focussed on L*, but by late spring unresolved questions pertaining to funding and collaboration management led the SSCL to conclude that L* should not be supported to proceed. Nevertheless, much excellent technical work had been done and it was possible to build on it. Following an open meeting in June at the Laboratory, a new international collaboration formed that was able to submit an Expression of Interest before the July meeting of the PAC. The detector concept, given the name GEM (for gammas, electrons, and muons) is shown in Fig. 8.

At its July 1991 meeting, the PAC reiterated the desirability of having two major detectors at startup of collider operations with complementary as well as overlapping capabilities. SDC is proceeding toward a full technical design report in April 1992. The Letter of Intent for the major detector complementary to SDC is due November 30, and the corresponding Technical Design Report in the fall of 1992. Both detectors are to be treated as having potentially equal importance and priority. Each will involve substantial in-kind contributions from non-U.S. participants and is currently estimated to entail a total cost (using "U.S. accounting methods") in the neighborhood of $500M.

At both the PAC meetings in July of 1990 and 1991, consideration was also given to smaller experiments, although the time pressure for them is considerably less than for the major detectors. Very useful Expressions of Interest continue to be received by the Laboratory in this regard. However, formal proposals are only envisaged in late 1993 or early 1994. In the meantime the Laboratory will try to encourage and focus efforts for these smaller

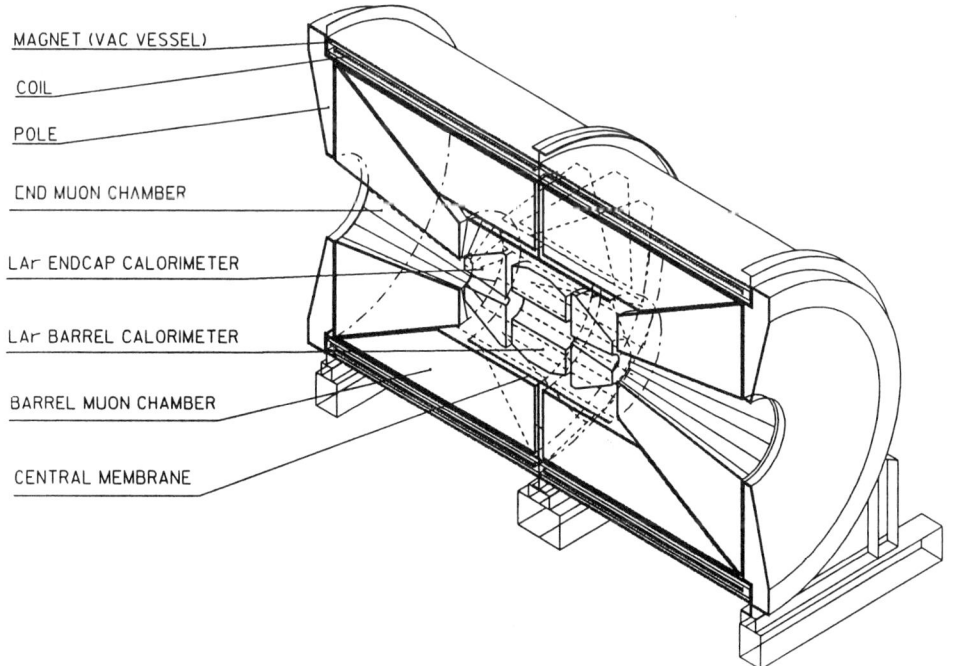

Fig. 8. The GEM detector concept as given in the Letter of Intent[10]

experiments. One particular area where this is the case is that of B-physics. The SSC has special capabilities here, beyond and complementary to what is hoped to come from possible electron-positron machines in the same time frame. With an eye toward the progress that is being made in both fixed target, hadron collider, and electron-positron experiments, a workshop will be held in late 1992 at the SSCL to examine in depth the best strategy for B-physics at the SSC—whether collider or fixed target, or both.

ACKNOWLEDGEMENT

The SSC Laboratory is operated by the Universities Research Association, Inc., for the U.S. Department of Energy under Contract No. DE-AC35-89ER40486.

REFERENCES

1. Proceedings of the 1982 DPF Summer Study on Elementary Particle Physics and Future Facilities, June 28–July 16, 1982, Snowmass, Colorado, edited by R. Donaldson, R. Gustafson, F. Paige (Fermilab, Batavia, 1982).
2. SSC Conceptual Design Report (Central Design Group, Berkeley, 1984) SSCL-SR-2020.
3. Site-Specific Conceptual Design of the Superconducting Super Collider, edited by J. R. Sanford and D. M. Matthews (SSC Laboratory, Dallas, July, 1990) SSCL-SR-1056.
4. Report on the Superconducting Super Collider Cost and Schedule Baseline, DOE/ER-0468P (U.S. Department of Energy, Washington, January, 1991).
5. Letter of Intent by the Solenoidal Detector Collaboration, November, 1990.
6. An Expression of Interest to Construct a Major SSC Detector, July, 1991.
7. See, for example, the Report of the Ad-Hoc Committee on SSC Physics, SSC-250 Rev. (SSC Laboratory, Dallas, December, 1990) and references therein.
8. Letter of Intent to the Superconducting Super Collider Laboratory by the L* Collaboration, November, 1990.
9. EMPACT/TEXAS Letter of Intent for the Superconducting Super Collider, November, 1990.
10. GEM Letter of Intent, November, 1991.

CHAIRMAN: F.J. Gilman

*Scientific Secretaries: E. Eskut, N. Khalatyan, A. Kuzucu,
R. Malik, M. Neubert*

DISCUSSION

– *Borden:*

I heard much discussion at the time Waxahachie was chosen as the SSC site about the problem of fire-ants. I understand these ants have a voracious appetite for electrical equipment and wonder if they pose any serious problems for the SSC. If so, what solutions have been considered?

– *Gilman:*

They are not a serious problem; what problem that actually existed has been solved.

– *Hasan:*

What is the reason to put the maximum energy of the SSC at 40 TeV? Is it some technological limit or is it some kind of threshold that will prove or disprove a theory?

– *Gilman:*

The energy and luminosity of the SSC were chosen so that one could be rather confident of fully exploring the nature of electroweak symmetry breaking, independent of the mechanism.

– *Junk:*

1) All accelerators, at some point or another, have been upgraded. Are there any plans for possible upgrades, and have steps been taken so as not to preclude such upgrades by committing to a fixed design?
2) A related question: what is an estimate of the physics lifetime for the SSC experiments?

– *Gilman:*

1) Our initial goal is to get to a luminosity of $10^{33}/cm^2$ sec for 40 TeV pp collisions, and the design of the machine is made to assure that goal. In the SSC design care is being taken to allow future upgrades, and in particular to allow for greater luminosities by an order of magnitude or more.
2) It is difficult to answer in advance, but if recent history at other accelerators is a guide, one would guess that the two large experiments will stay 10 years

or longer, with various improvements and upgrades along the way. Smaller experiments would mostly finish sooner and be replaced by other experiments that focus on particular physics issues of the time.

– Hernandez:

1) Would it not be more profitable, considering the SSC tunnel is not done yet, to try to go to higher energies in order to reduce the overlapping region that the LHC and SSC will be able to explore?
2) Is there anything that the SSC will definitely do better than the LHC in its high luminosity option?

– Gilman:

1) First, the overall parameters of the SSC are now fixed. As I indicated earlier, this was done on the basis of being able to fully address the nature of electroweak symmetry breaking. Raising the energy of the SSC still higher would in any case not decrease the physics overlap with LHC, but rather would simply allow additional possibilities at the SSC.
2) Second, the SSC has an advantage over the LHC in that there is a larger cross-section for most phenomena at higher energy. For example, near 1 TeV in mass, the cross-section for Higgs boson production is roughly an order of magnitude larger at the SSC. In third instance, an experiment operating at an order of magnitude less luminosity and with far less demands on detector technology at the SSC is competitive with an experiment with similar goals at LHC. If the technology detector is available, the possibility always exists of operating the SSC at roughly the same luminosity as LHC and having more events because of the increased cross-section.

– Sivaram:

1) You said that detection of some rare decays is possible only in SSC. Could you specify these and the type of information one could get from these decays?
2) Also, what constraints could be put on alternative mechanisms of generating mass like dynamical symmetry breaking?

– Gilman:

1) Rare B meson decays, like $B \to \mu\mu$, which in the Standard Model occur below 10^{-10} in branching ratio, can, by definition, only be seen if at least 10^{10} B mesons are produced. This is beyond the capability (in one year of running) of the electron-positron machines now envisaged, but a small fraction of the B's produced in one year at the SSC. Observing such rare decays typically gives information on the (virtual) heavy particles involved in the loop diagrams responsible for the decay, e.g., the mass and couplings of the top quark, or con hadronic decay parameters, e.g. f_B.

2) There are many options. As one example, technicolor theories have been discussed widely, where there would be a technirho meson in the 1 to 2 TeV mass range that decays to two weak gauge bosons. Such possibilities have been simulated by Monte Carlo programs and could be seen at the SSC.

– *Peoples (comment):*

I would like to comment that so far an efficient vertex detection has only been implemented in fixed target experiments. I see no reason that e^+e^- machines would be of any advantage in this respect as compared to hadron colliders.

– *Neubert:*

You mentioned the impressive amount of B → $\mu\mu$ pairs produced at hadron colliders. What really is the capability for reconstructing exclusive rare B decays (CP-violating processes, penguin-induced transitions, etc.) at the SSC?

– *Gilman:*

Up to now, we have only just begun to learn to exploit the possibilities for doing B physics at hadron colliders (or in hadron fixed target experiments, for that matter). At CDF, one has been able to reconstruct quasi two-body decays of B mesons where there is a clear signature. Whether the full capabilities at hadron colliders can be attained has neither been proven nor disproven. The key issue will be using vertex information (to help tag the B decay) at the trigger level. We will have to learn from the experiments at CERN and Fermilab in the next few years to see what event reconstruction efficiency is achievable and how much vertex information in the trigger is feasible.

– *Gallo:*

In spite of the fact that there are hundreds of particles in the central tracking chamber, what kind of test of QCD do you think can be performed at the SSC?

– *Gilman:*

Although there are hundreds of hadronic tracks, it is possible to identify jets, particularly at high transverse momentum. This allows one to test QCD at very high mass scales (above 10 TeV in jet-jet mass). There are many other interesting aspects of QCD that can be tested. Up to now they have not been as well studied as they could be, with more attention given to electroweak symmetry breaking, additional gauge bosons, exotic phenomena, etc.

– *Khalatyan:*

What are the prospects for fixed-target experiments at the SSC?

– *Gilman:*

Fixed target experiments would fall within the "smaller experiments" area that I discussed in my lecture. There is a range of possible options including:
1) The use of particles produced in the interaction regions, such as gammas, neutrinos,...
2) A gas-jet or droplet experiment intercepting one 20 TeV beam,
3) An extracted beam of perhaps 10^7 protons per second that is produced in a way that does not interfere with collider operation. These are all being discussed, and it would not be surprising if more than one of them was used in an initial experiment.

– *Barletta:*

If you can increase the luminosity to 10^{34}, there will be many events per crossing. Why not use the extra power to increase the energy rather than luminosity?

– *Gilman:*

It is always good to have both options, but aside from technical feasibility, which one is chosen depends on the physics process. If one is only dealing with detection of muons, turning up the luminosity significantly may be the best option. On the other hand, if one wants at least partially reconstructed hadronic events then you may want to keep the luminosity down at 10^{33}, or even 10^{32} in some cases.

– *Wang:*

Why has Dallas been chosen to build the SSC, and not FNAL or BNL? Would not this have reduced the total cost?

– *Gilman:*

There was a governmental decision to have an open competition as to the site of the SSC. Many states submitted possible locations and Texas won the competition with what was the best site overall.

– *Syed:*

What is the possibility of having more than one event/bunch crossing? How do you plan to do the analysis of missing energy and decays of long-lived particles in cases where the vertices of multiple interactions are mixed?

– *Gilman:*

At a luminosity of $10^{33}/cm^2$ sec, there are in fact 1.6 events/crossing, on average, at the SSC. However, the most likely events are peripheral collisions (for example, elastic or quasi elastic scattering) with a small number of particles produced that primarily go along the beam direction. In the case of multiple interactions per crossing, one therefore often has a high transverse momentum event that is triggered on plus other events that do not add much, if anything, to the particles at high transverse momentum. The difficulties at higher luminosity then depend on the detector (its granularity, and correspondingly, the occupancy of particular elements) and the physics one is after. Working with only high transverse momentum muons, for example, is much easier in this environment than trying to trigger upon, and then do detailed track reconstruction, of the products of the decay of a long-lived particle.

– *Ozdes:*

How many Higgs bosons do you expect to see at SSC energies?

– *Gilman:*

At 40 TeV, there are of order 20,000 Higgs bosons with a mass of 800 GeV produced per year. Note though that if one only looks for the Higgs decaying to ZZ and then the Z's decaying to charged leptons, the combination of branching ratios results in tens of decays of this sort per year.

– *Duff:*

This morning you discussed the experimental physics programme at the SSC. Could you say a few words about the theoretical physics programme?

– *Gilman:*

We are planning to have a Theory Department numbering 20 to 30 persons. A number of theory visitors have come to the SSC in the past few years. We are looking to make the first permanent appointments within the next year, as well as to continue an active visitors' program.

– *Kaur:*

The SSC project and RHIC were sanctioned almost at the same time. Could the SSC ring not be used for accelerating heavy ions as well?

– *Gilman:*

As you indicated, the two projects were given independent go-aheads. RHIC is devoted to heavy ion collisions and is the next step in this area of physics. The SSC is devoted to doing pp collisions with a full, frontier experimental programme that stretches for years. We do not foresee exercising an option of doing ep or heavy ion collisions as part of the initial SSC experimental programme.

MAXIMIZING THE LUMINOSITY OF ELOISATRON, A HADRON SUPERCOLLIDER AT 100 TeV PER BEAM

William A. Barletta

Department of Physics
University of California Los Angeles
and
Lawrence Livermore National Laboratory
Livermore, CA 94610

INTRODUCTION

In general as one raises the energy of a collider one must simultaneously increase the luminosity to compensate for decreasing cross sections. Applying the presently available accelerator technology embodied in the designs of the LHC (8 TeV per beam at 10^{34} cm^{-2} s^{-1}) and the SSC (20 TeV per beam at 10^{33} cm^{-2} s^{-1}) to the ELOISATRON, a proton collider operating at 100 TeV per beam yields a collider design with a luminosity of 10^{34} cm^{-2} s^{-1}. To extend the physics reach of supercolliders to the maximum possible extent will require designing the collider to achieve luminosities > 10^{35} cm^{-2} s^{-1}.

This paper presents a general context for assessing the performance of hadron supercolliders in general and ELOISATRON in particular. It begins with an illustration of machine trends and with definitions of key collider characteristics. A brief description of design strategies to accommodate limiting beam physics and limiting technologies constitutes the next section. A simple spreadsheet based computer code incorporating these considerations allows one to perform self-consistent parameter searches to yield design characteristics of ELOISATRON (ELN) with maximum possible luminosity. To underscore the point that near term technology is applicable to the limits of the high energy frontier, the paper concludes with a sketch of the characteristics of the Ultimate ELOISATRON (UELN), a hypothetical collider of higher energy than the ELOISATRON.

COLLIDER TRENDS IN ENERGY AND LUMINOSITY

The search for understanding the nature of mass and the dynamics underlying the physical universe have led particle

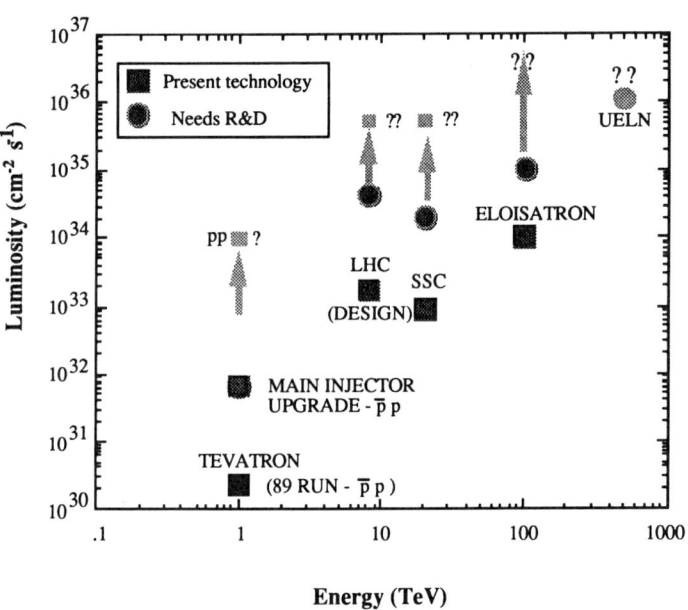

Figure 1. The luminosity goals of present and future hadron colliders

physicists to seek to build colliders with ever higher beam energies and luminosities. The performance trends in present and future hadron colliders are illustrated in figure 1. The limits on collider performance are determined both by beam physics and by available technology. This paper offers a framework for exploring the systematics and in particular the energy dependence of limiting beam physics and limiting technologies.

Luminosity

For bunches of equal population, N, and equal sizes at the interaction point colliding at a frequency f_{coll}, the luminosity is given by the well known expression

$$L = \frac{N_1 N_2 f_{coll}}{4 \pi \sigma_x \sigma_y (1 + q^2)^{1/2}} \quad (1a)$$

where σ_x and σ_y are the Gaussian horizontal and vertical radii and where q accounts for the luminosity loss due to a crossing angle, α in the vertical plane;

$$q = \frac{\alpha \sigma_z}{2 \sigma_y}. \quad (1b)$$

Writing the beam radius in terms of the normalized ε_n, the relativistic factor, γ, and the value of the β-function at the interaction point, β^*, and ignoring the effects of a non-zero crossing angle, one has

$$L = \frac{N^2 c \gamma}{4 \pi \varepsilon_n \beta^* S_B} = \frac{1}{e\, r_p} \left(\frac{N\, r_p}{4 \pi \varepsilon_n}\right)\left(\frac{\gamma I}{\beta^*}\right) \equiv \frac{1}{e\, r_p} \xi\left(\frac{\gamma I}{\beta^*}\right). \quad (2)$$

where I is the average beam current and r_p is the classical radius of the proton. The quantity ξ is the linear (head-on) tune shift produced by the beam-beam interaction. Eq.(2) displays a natural linear growth of the luminosity with beam energy. The "pain" associated with increasing the luminosity faster than the natural linear growth of luminosity with energy derives from the necessity to increase the beam current simultaneously with increasing γ. How should one choose ε_n, β^*, S_B, and N as a function of the beam energy to maximize the luminosity? What constrains the choices?

DESIGN STRATEGIES

A recent look at maximizing the luminosity of the SSC and the LHC (Snowmass, 1990) suggested the following approaches to collider design:

    1) increasing the charge per bunch,
    2) increasing the number of bunches,
    3) increasing the crossing angle to allow more rapid bunch separation thereby reducing parasitic crossings,

4) tilting the bunch with respect to the direction of motion at the interaction point ("crab-crossing"),
5) minimizing the β function at the interaction point.

These strategies together with their attendant physics and technology issues are exactly those being pursued in design studies of very high luminosity B and ϕ factories at SLAC, CERN, Cornell, KEK, UCLA, and Laboratorio Nazionale Frascati (LNF). One of the major technical difficulties in raising the luminosity in proposed flavor factories from the level achieved in CESR (2×10^{32} cm^{-2}s^{-1}) to that required to explore the nature of CP violation is handling the intense synchrotron radiation that is generated by the multi-ampere beam currents.

Upon close examination of the challenge of reaching the highest possible luminosity both in ELOISATRON and in lower energy hadron supercolliders, one finds that the physical phenomenon that underlies nearly all design difficulties in the range from 10 to 100 TeV (and beyond) is the emission of synchrotron radiation by the protons. Even the practical difficulties of controlling the consequences of synchrotron radiation are similar to those for electron rings if one assumes (as is generally the case) that the vacuum pipe in the hadron supercollider must operate at cryogenic temperatures. If, however, the operating temperature of the new, high T_c superconductors can be pushed to $\approx$ 300 °K, very high electrical conductivity need not imply cryogenic temperatures. In that case even at 100 TeV the synchrotron radiation load from the high current proton beam will easily be within the range routinely handled in existing electron storage rings. In that case the ultimate energy per beam would not be limited to 100 TeV by synchrotron radiation; instead, a PeV collider could be considered.

Accelerator physicists generally expect that the design approaches listed above should achieve the desired performance goals. Nonetheless, although an ELOISATRON with a luminosity of 10^{34} cm^{-2} s^{-1} may be quite conventional with respect to its constituent technologies, achieving the highest possible luminosity at 100 TeV (>10^{36} cm^{-2} s^{-1}) will push both beam characteristics and accelerator technologies considerably beyond present practice. Nonetheless, the required levels of performance are not beyond reasonable extrapolations of state-of-the-art accelerator technologies.

Here, the skeptic might object that regardless of its design, a collider of conventional technologies with an energy five times higher energy than that of the SSC would cost roughly five times as much as the SSC, that such a cost is unacceptably high, and hence that conventional designs are not economically practical. Fortunately, in the context of an appropriate partnership between scientific laboratories and industry that maximizes the use of existing technical and manufacturing infrastructure, this crude economic argument does not have force. The costs of the SSC project are strongly related to the style of its execution and its construction which required the established of a new, large-scale technical infrastructure.

This choice has certainly received substantial criticism in technical circles.

From a technological point of view the cost of any hadron supercollider will depend strongly on the ultimate luminosity for which it is designed. Economic considerations notwithstanding, to achieve the highest possible luminosities (>10^{36} cm^{-2} s^{-1}), existing technologies must be pushed into new regimes (e.g., by finding practical, high T_c superconductors suitable for magnet windings). In that case the design strategies must receive detailed experimental exploration for one to arrive at level of confidence commensurate with the cost of a 100 TeV collider with a luminosity in the range of 10^{36} cm^{-2} s^{-1}.

The beam dynamics limiting luminosity are embodied in the maximum tune shift that can be achieved during collisions. In existing hadron colliders such as the Tevatron and the Sp$\bar{\text{p}}$S where radiation effects are negligible, the total tune shift from all of the multiple interaction points is found to be 0.024. As it is unclear whether such a large tune shift can be realized from a single interaction point, the Snowmass study took 0.01 as a limiting value. In the ELOISATRON operating at ≈100 TeV, radiation damping will become an important determinant of the radial distribution of the beam. In the absence of dilution effects to the collisions and feedback control of instabilities, this distribution would resemble that seen in electron storage rings. Then the maximum tune shift in a collider with a single interaction point may no longer be limited to 0.01, but might approach the value of 0.03 (or even 0.06) that has been achieved in high energy electron-positron colliders (PEP at SLAC). A more pessimistic scenario is that fast dilution plus radiation damping might actual broaden the beam distribution vis á vis that in electron rings. Then the tune shift may actually be lowered.

One of the best ways to study the limits of stability in any system is to make frequent excursions into the unstable operating regime. Unfortunately the consequent loss of beam would impermissible in existing storage rings that must also provide high integrated luminosity to users. To develop the requisite experimental database one might instead consider using a low energy e^+-e^- collider operating at ≈100 MeV per beam with high currents and with several bunches. With a flexible lattice such a storage ring would permit both high and low emittance tunes. With the installation of radio frequency deflection cavities ("crab" cavities), the sensitivity of crab-crossing to synchro-betatron resonances can be tested. These features are similar to those of DAΦNE, the φ factory under construction at the Laboratorio Nazionale Frascati.

The radiation damping characteristics of e^+-e^- colliders and hadron supercollider are shown in Fig. 2. The number encircled is the number of damping times per luminosity lifetime at 10^{34} cm^{-2}s^{-1} luminosity.

Designing for maximum luminosity yields a collider with a very large number bunches that implies beam crossing rates of the order of 0.1 − 1 GHz with several tens of collisions per crossing.Determining whether detectors and data acquisition and processing systems can be designed to accommodate the enormous rates implied by extremely high luminosities needs a serious evaluation by a group of detector specialists working with accelerator physicists familiar with hadron and e^+-e^- colliders.

The implications of trying to maximize luminosity of ELN in a self-consistent design can be explored most easily using a computer code to perform parameter searches. In the absence of experimental evidence of higher values of the maximum head-on tune shift per interaction point, one should take this value to be 0.01 as used in the Snowmass study. With all the cautions stated above, on the basis of such a parameter exploration it

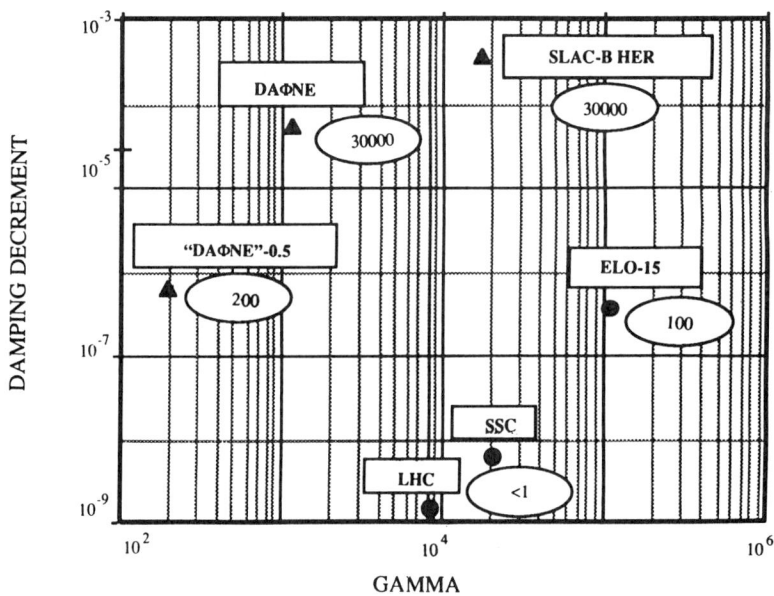

Figure 2. Radiation damping characteristics of colliders

seems practical to construct a 100 TeV per beam collider with a luminosity $>10^{34}$ cm^{-2} s^{-1} using the same technologies being realized for the LHC. At energies below its top value such an ELOISATRON would have an energy vs. luminosity dependence as shown in Figure 3.

Reducing the allowable radiation to 2 W/m would reduce the luminosity at 100 TeV by a factor of two. In contrast, raising the value to 20 W/m yields 10^{35} cm^{-2}s^{-1}. Parameters for both of these cases are given in Tables 1 and 2 respectively. In neither case does the use of crab-crossing seem to offer much increase in the design luminosity.

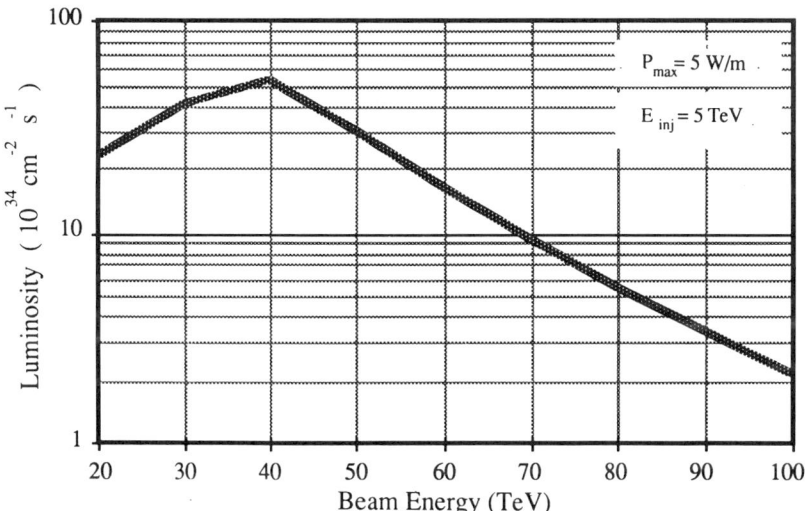

Figure 3. L versus energy for an ELOSIATRON with 5 W/m of radiation allowed on the walls of the vacuum chamber.

Table 1. Parameter set for ELOISATRON at 10^{34} cm^{-2}s^{-1} with present technology. Output is from scaling computer code, ELOISASCALE

Ring charcteristics		Interaction points	
* Max Energy (TeV)	100	* Number of IPs	2
* B dipole - max (T)	10	* Beta* (m)	0.5
* Dipole fraction	0.765	* Crossing angle (mr)	0.25
To - rev time (s)	9.13E-04	* L* - long range (m)	170
Circum (km)	273.78	* Crab cross - Y / N	y
		Beta* scale (m)	0.98
Beam characteristics		Sigma-z (cm)	6.0
* Energy (TeV)	100	l* to Q1 - (m)	44.7
* Norm emit@E (mm-mrad)	0.03	Σ-IP (μm)	0.5
* Bunch space buckets	10	Δnu-HO / IP	1.2E-02
gamma	1.1E+05	Δnu-LR	0.0E+00
Bunch space (m)	6.59	Δnu-tot	0.024
Number of bunches	4.2E+04	R (lum correct)	1.00
* NBunch (nC) < 0 for input	0.0	Δt crossing (ns)	22.0
Bunch population	2.9E+09	Interact / crossing	33.6
Current (A)	0.02	T-lum/t-damp-perp	6.9
		Luminosity half-life (hr)	7.5
Injection / Energetics		Luminosity / IP	1.1E+34
* Injection Energy (TeV)	5		
Fill time (hr)	2.0	**Instabilities**	
Stored Energy (GJ)	2.04	* R-pipe (cm)	2
D(E,Q1) MGy/yr	53.1	* Operating temp (°K)	20
Hadrons in Q1 (W/kg)	5.3	* Injection temp (°K)	4
Q1 survival (months)	23	Resistive wall -turns	210.5
Debris (kW per side)	20.0	RW at injection - turns	16.9
		μwave Z/n @ inj (ohms)	165.4
Synchrotron radiation			
* Max P(W/m) on walls	2	**Vacuum**	
Uo (GeV/turn)	2.3E-02	* Op. pressure (nTorr)	1
Damping decrement	2.3E-07	* Desorb coeff.	0.001
E- Damping time (s)	3.9E+03	E-crit (eV)	1.0E+04
Power (W)	5.2E+05	N-gamma (s-1 m-1)	3.6E+15
Power density (W/m)	1.9	Req. pumping (L/s/m)	107.7
Parasitic heat per beam		**RF systems**	
Resitive wall (kW)	0.0	* rf -Frequency (Mhz)	455
RF-HOMs (kW)	0.1	N cavities/ring	293
P to Compressors (MW)	39.5	Max. P to klystrons (MW)	3.4

Table 2. Parameter set generated with the computer code, ELOISASCALE for the ELOISATRON at 10^{35} cm^{-2}s^{-1}.

Ring charcteristics		Interaction points	
* Max Energy (TeV)	100	* Number of IPs	2
* B dipole - max (T)	10	* Beta* (m)	0.5
* Dipole fraction	0.765	* Crossing angle (mr)	0.15
To - rev time (s)	9.13E-04	* L* - long range (m)	170
Circum (km)	**273.78**	* Crab cross - Y / N	n
		Beta* scale (m)	0.98
Beam characteristics		Sigma-z (cm)	6.0
* Energy (TeV)	100	Σ-IP (μm)	1.7
* Norm emit@E (mm-mrad)	0.32	Δnu-HO / IP	1.0E-02
* Bunch space buckets	9	Δnu-LR	1.6E-04
gamma	1.1E+05	Δnu-tot	**0.021**
Bunch space (m)	5.93	R (lum correct)	0.97
Number of bunches	**4.6E+04**	Δt crossing (ns)	**19.8**
* NBunch (nC) < 0 for input	0.0	Interact / crossing	**271.2**
Bunch population	**2.8E+10**	T-lum/t-damp-perp	8.1
Current (A)	**0.23**	Luminosity half-life (hr)	**8.7**
		Luminosity / IP	**1.0E+35**
Injection / Energetics			
* Injection Energy (TeV)	8	**Instabilities**	
Fill time (hr)	**2.0**	* R-pipe (cm)	3.5
Stored Energy (GJ)	**21.38**	* Operating temp (°K)	60
D(E,Q1) MGy/yr	**476.0**	* Injection temp (°K)	20
Hadrons in Q1 (W/kg)	47.6	Resistive wall -turns	67.7
Debris (kW per side)	**179.1**	RW at injection - turns	**8.6**
		μwave Z/n @ inj (ohms)	**28.0**
Synchrotron radiation			
* Max P(W/m) on walls	20	**Vacuum**	
Uo (GeV/turn)	2.3E-02	* Op. pressure (nTorr)	5
Damping decrement	2.3E-07	* Desorb coeff.	0.001
E- Damping time (s)	3.9E+03	E-crit (eV)	1.0E+04
Power (W)	5.5E+06	N-gamma (s-1 m-1)	3.8E+16
Power density (W/m)	**20.0**	Req. pumping (L/s/m)	**225.9**
Parasitic heat per beam		**RF systems**	
Resitive wall (kW)	4.5	* rf -Frequency (Mhz)	455
RF-HOMs (kW)	10.3	N cavities/ring	293
P to Compressors (MW)	**119.0**	Max. P to klystrons (MW)	**34.2**

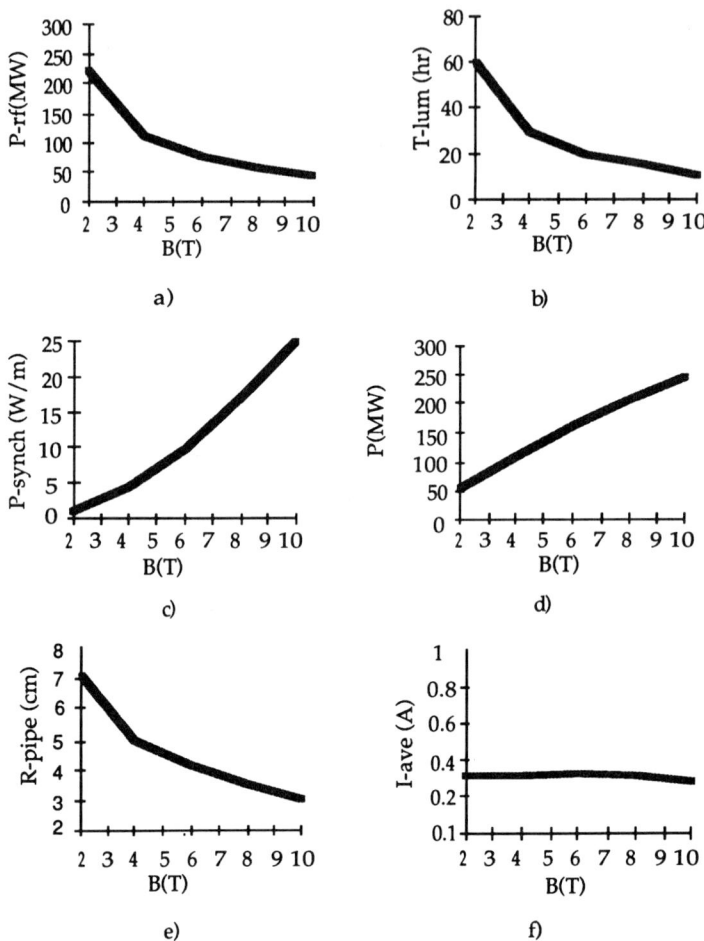

Figure 4. Variation of ELN characteristics with dipole field for ELN, a 100 TeV per beam proton supercollider, with a luminosity of 10^{35} cm^{-2}s^{-1}. Panel a) RF-power v. B_{dipole}, Panel b) Luminosity lifetime v. B_{dipole}, Panel c) Thermal load on vacuum chamber walls v. B, Panel d) Total operating power from mains v. B_{dipole}, Panel e) Vertical aperture to control

The design code also allows one to search for an optimum value of the dipole field. An examination of the variation of various characteristics of ELN with B_{dipole} for operation at 10^{35} cm^{-2} s^{-1} is shown in Fig. 4. It is difficult to draw any final conclusions about the choice of B_{dipole} with an adequate model of the variation in magnet cost with strength, length and aperture.

resistive wall modes v. B_{dipole}, and Panel f) Average current v. B_{dipole}

If new materials allow one to increase the radiation load to 100 W/m, a 100 TeV/beam, the ELOISATRON could provide a luminosity of $\approx 10^{36}$ cm^{-2}s^{-1}, especially when one operates the collider at intermediate energies (Figure 5). Realizing this latter possibility will depend on the existence of a broadly based experimental research effort prior to the final machine design. Another approach to allowing very large radiation loads it to search for magnet designs that would allow the radiation to be deposited on warm surfaces. In either case and most importantly, the possibility of achieving top luminosity ($> 10^{36}$ cm^{-2} s^{-1}) must be incorporated <u>ab initio</u> into the design, not considered as an after thought.

With respect to the longest term future, an examination of the systematics of collider design suggests the ultimate potential of conventional storage ring technology in the exploration of the energy frontier of elementary particle physics. If the vacuum chamber of the proton storage ring can be operated at or above room temperatures, then one should be able to construct a hadron collider with a center of mass energy of ≈ 1 PeV and a luminosity exceeding 10^{36} cm^{-2} s^{-1} at its peak energy. This machine, with a circumference twenty times larger than the SSC (see Table 3) may well be the ultimate

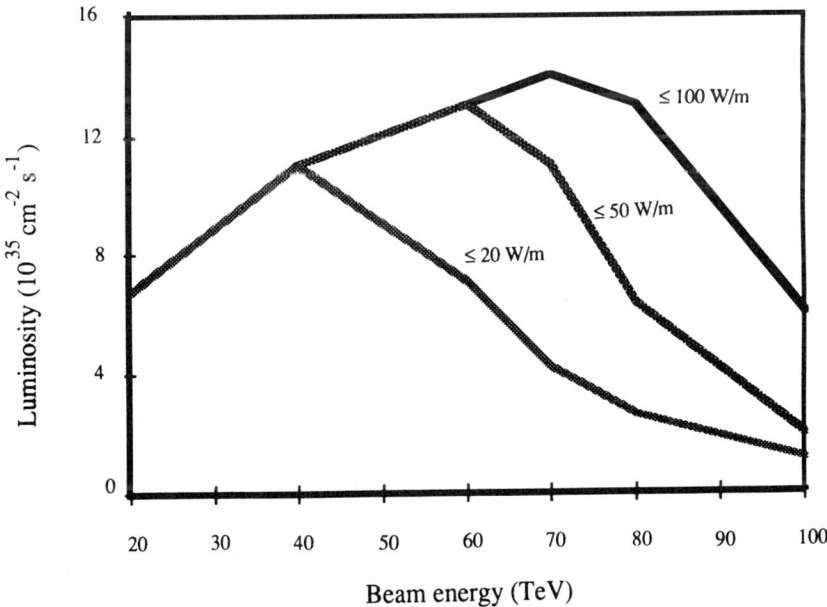

Figure 5. Luminosity limits for ELOISATRON if means of handling large radiation loads are developed

hadron supercollider, i.e., the Ultimate ELOISATRON (UELN). This exercise in exploring the limits of conventional technologies also suggests the regime in which linear colliders might be applicable to protons.

Table 3. Characteristics of the Ultimate ELOISATRON

Center of mass energy	1 PeV
Circumference	1600 km
B_{dipole}	20 T
Beam energy	500 TeV
Beam current	200 mA
Mains power	2 GW
$\langle P_{synch} \rangle$	10 kW/m
Interaction regions (IR)	2
Limiting technology	Magnet survival at IR
Tune shift	0.01 per IR
Luminosity	$\approx 10^{36}$ cm^{-2} s^{-1}

If the length of each linac of a linear UELN is not too much longer than the radius implied by Table 3, then a linac must be able to achieve gradients $\gg$ 1 GeV/m. The luminosity is directly related to the average power in the beams by

$$L\ (10^{33} \text{cm}^{-2}\text{s}^{-1}) = \frac{D\ H_D}{30} \left(\frac{1 \text{ mm}}{\sigma_z}\right) \left(\frac{P_{beam}}{1 \text{ MW}}\right) \qquad (3)$$

where H_D is the luminosity degradation due to the pinch effect and where D is the disruption parameter that measures the pinch;

$$D = \frac{r_p N_B \sigma_z}{\gamma \sigma_{x,y}^2} = r_p N_B \left(\frac{\sigma_z}{\beta^* \varepsilon_n}\right). \qquad (4)$$

The disruption is related to the tune shift in the ring by

$$D(\text{ring}) = 4\pi \Delta v_{HO} \left(\frac{\sigma_z}{\beta^*}\right). \qquad (5)$$

For $D < 2$, the value of $H_D \approx 1$. At 500 TeV β^* is limited to be ≈ 2 n the normalized emittance, ε_n, will be difficult to make less than 10^{-7} m rad. For a 100 GHz accelerating field, the bunch length $\sigma_z \leq 10^{-6}$ m. Thus the quantity in parentheses in Eq.(5)

in Eq.(5) is of order 1 m^{-1}. Even where it possible to generate bunches of 100 nC with such low emittances and to preserve the emittance in the presence of the extremely large wake fields in the linear accelerator, it would be difficult for $r_p N_B$ to exceed 10^{-6} m. Hence the disruption in a proton linear collider will be exceedingly small. Therefore, from (3) one sees that achieving a luminosity of 10^{33} cm^{-2} s^{-1} will require an average power of 30 GW per beam. As D decreases and P_{beam} increases with increasing energy, the practicality of a linear proton collider actually diminishes at even higher energies (above 500 TeV).Therefore, the ultimate hadron supercollider should be a synchrotron.

CONCLUSIONS

On the basis of a systematic parameter search it appears possible to build an ELOISATRON operating at 100 TeV per beam with a luminosity exceeding 10^{34} cm^{-2}s^{-1} by using presently available technology such as that being incorporated in the designs of SSC and LHC. Such a supercollider would have the physics reach of a 10 TeV e$^+$-e$^-$ linear collider, for which no reasonable design concept now exists. Assuming moderate advances in the state of accelerator technology during its design cycle the ELOISATRON could be expected to operate at luminosities ≈10^{35} cm^{-2}s^{-1} at 100 TeV/beam. Even higher luminosities would obtain at lower energies. With advanced technologies but based upon conventional approaches, a collider, UELN, with an energy five times that of ELOISATRON appears be possible.

ACKNOWLEDGEMENTS

The author wishes to thank for conversations Robert Siemann (SLAC); Robert Palmer, and Alex Chao have also been extremely helpful in framing the analytical approach. The author's participation in Snowmass 1990 also provided a broad source of the physics embodied in the ELOISACSALE design program. This work was partially performed under the auspices of the Lawrence Livermore Laboratory for the U. S. Dept. of Energy under contract W-7405-eng-48.

CHAIRMAN: W. Barletta

Scientific Secretaries: A.V. Chizov, M.T. Dova, E. Etzion, A.A. Syed

DISCUSSION

– *Etzion:*

What are the difficulties in focusing protons compared to focusing electrons? (lowering β^*)?

– *Barletta:*

The difference is only due to the momentum scale differences. I can see two main problems:

a) lowering β^* requires simultaneously reducing the pulse length. Doing so decreases the ability of the detector to distinguish between so many events actually occuring together, especially when one waits for undetermined new physics events;

b) large variations in β complicates the way the machine avoids beam losses, which can damage the beam pipe, the detector and, worse, can be dangerous to people in the area.

– *Barletta:*

The number of events per beam crossing that one can distinguish depends on the length of the bunch. The machine can produce a bunch length ranging from 5 to 50 cm long. Say that the number of events per beam crossing is about 20 over a 20 cm bunch length. If the vertex chamber can resolve the vertex to 0.5 cm, then probably the experimentalists will not care. This point needs to be discussed between the detector designer and the machine designers. Regarding the radiation damage from beam losses, as the ring will be very long, we will need to have several abort systems around it which send signals when they sense abnormal beam loss so that the beam (or the abnormal bunch) is kicked out and not deposited in the detector.

– *Khoze:*

The maximal luminosity for the Eloisatron was claimed to be $\sim 10^{36}$ cm^{-2}sec^{-1}. What is the ultimate luminosity one can expect for the e$^+$e$^-$ colliding beams in TeV energy region?

– *Barletta:*

The maximum luminosity attainable in electron positron colliders depends very strongly on the amount of electrical power that is consumed. It also depends on how much the beamstrahlung[†] can be suppressed. These two factors may seem unrelated, but in fact they are closely coupled. In the 1-2 TeV range the luminosities that are expected (and needed) are less than $10^{34} cm^{-2} sec^{-1}$. At 10 TeV one needs to reach the luminosity of 10^{36}, but the best estimates that I know of are 10^{35}. The cost of the Eloisatron with 200 TeV is equivalent to a linear collider with 15 TeV per beam. The Eloisatron should be able to get luminosity of $10^{37} cm^{-2} sec^{-1}$ if we can build it with room temperature walls. In contrast, I am pessimistic about meeting all the technical challenges involved in building an electron positron machine of 10 TeV per beam.

– Khoze:

Did I understand you correctly that the maximal luminosity for the B factories will not exceed $10^{34} cm^{-2} sec^{-1}$? It does not sound extremely optimistic for CP violation business.

– Barletta:

If we use standard design practices we can reach the luminosity of about $10^{34} cm^{-2} sec^{-1}$. Using the "tricks" that we are planning for the ϕ factory at UCLA (a quasi-isochronous), we can obtain a luminosity of about $10^{35} cm^{-2} sec^{-1}$ in a B factory. In the UCLA design with strong radiation damping the growth time of instabilities is longer than the damping time you may be able to get above 10^{35}, but we still have to prove that such beams are really stable.

– Khoze:

You have mentioned the importance of the Touchek effect. What are the typical values of transverse momenta of particles inside the beams?

– Barletta:

For Eloisatron the normalized emittance is between .1 to 1 mm mrad. The geometrical emittance is the normalized value divided by γ. This value implies a beam size in a collider of the order of 100 microns. The mean angle is just the emittance divided by the values and will be very small, about 1 micro radian.

– Zichichi:

Why are you pessimistic about the 10 TeV electron-positron collider?

– Barletta:

I am not convinced about the ability of having a luminosity of $10^{35} cm^{-2} sec^{-1}$ to 10^{36} because of the large energy spread induced by the pinch effect and because

---

† Beamstrahlung is the emission of radiation by the particle in the collective field of the other beam.

quantum fluctuations from beamstrahlung destroy the beam quality during the final focus. This effect could be suppressed if you put a plasma of metal density inside the detector but that will give you a "dirty" collision. Finally appropriate rf-power sources and accelerator structures do not exist.

– *Zichichi:*

It is important to emphasize that it is still impossible to construct such kind of machines.

– *Levi:*

I have heard that you want to put magnets inside the detector; can you please tell me where they will be, how big, and of which material?

– *Barletta:*

For Eloisatron the first quadrupole is in a distance of 45 m as compared to the length of 30 m for the planned detector, which means the quadrupole is well outside the detector. You may get rid of part of the hadronic debris if you allow a dipole field in the detector. I have mentioned it only as an idea that one has to look at.

– *Qureshi:*

How can crab crossing enhance luminosity?

– *Barletta:*

Crab crossing is a means of controlling the tune shift effects. The accelerator tune shift has two contributions: one from the head-on collisions and the second from the so-called parasitic crossing. To eliminate the effect of parasitic crossings you need 1) to force the beams to cross at a large angle, 2) to tilt the beams with respect to the trajectory so that they collide head-on and 3) then to change the bunch orientation again to be parallel to the beam trajectory. It is an important possibility that one needs to consider. However, my intuition is that it is more useful in electron-positron colliders than in hadron colliders.

– *Sivaram:*

1) Do the synchrotron and beamstrahlung losses scale the same way with energy, i.e. γ^4? Could you clarify about the 10 kW/m radiation losses?

2) Your unit of storage energy was in tons. Can you explain this unit?

– *Barletta:*

The synchrotron and the beamstrahlung both have the same scaling, γ^4. The 10 kW/m is not an upper limit. There have been designs for room temperature electron storage rings with a factor of two higher than that. The problem of the proton machines is that the emittance is extremely tiny, so that the 10 kW/m will be deposited along an extremely thin strip.

The standard units are 4.2 MJ per kg of TNT.

– *Buzuloiu:*

We have had today a series of beautiful lectures and I shall mention Professor Zichichi's talk first. So we have learned that it is worth while to build powerful accelerators even for their products. I mean even if no other ultimate law of physics comes out.

One supposes that like electron has been transformed in electronics over a century, the gluons will become gluonics.

Then we were delighted with the lecture of that grand unification is not far, and soon will be understandable by everyone like the Newton laws. Professor Duff's lecture was dangerously clear and left the impression that everyone should be able to start thinking about his ideas.

And finally, the two lectures supported the idea of the Eloisatron project: firstly from theoretical interest into domain of energy and secondly from the practical point of view, the feasibility study of Professor Barletta.

Let me make a remark with respect to this last one. It is true that firstly one has to know how to build the machine, secondly how to build detectors and only after that one should think about how to take the data. It is also true that electronics will evolve tremendously in the next five years. But the challange of high energy physics could be quite difficult even if the technology evolves as fast as it is evolving.

Until now high energy physics could take the signal processing technology just like a thirsty man takes the water from a small river in the countryside. But there comes a moment when you arrive at the seaside and even with all the water in the sea you will remain thirsty, so it is not only a question of technology development in electronics. One of today's revolutions is in the architectures and here everyone could have his needs and HEP has very special needs. It may be very likely that the God of telecommunications or of seismics will not have the right architecture for physics.

I would like to add a word of caution here: if you have sugar and salt and you mix them, in principle you can separate them again, but only in principle. Nobody will be able to separate 700 events mixed in a huge detector. I shall not enter into details but one can imagine that two detector structures of approximately the same sensitivity could be very different in their signal processing needs. The data processing people will be faced with impossible problems just because the system was thought as a detector plus a data processing and not as a whole.

– *Barletta:*

Until now hadron colliders have been designed with minimal attention to the detectors that will have to extract the physics. This continues to be true for SSC and LHC. But I agree with Professor Buzuloiu that this practice cannot continue for an Eloisatron at the highest possible luminosity. Such a collider must be designed from the interaction points outward. In my study of high luminosity designs I found that I was left with two free parameters. I shall make the following comment: the first degree of freedom completely unconstrained is the bunch length. I will allow the experimentalist to choose the bunch length parameter. I have an added degree of freedom if I take advantage of the fact that at 100 TeV one really can load the beam, damp the beam and bring the beam into collision. I can have even added damping wigglers to control the emittance as the interactions proceed. Thus the second degree of freedom to choose can be the bunch separation. There are also some other little details that constrain flexibility which I have not looked at yet. But I invite the detector designers to jump into the machine design right now so that as a unit the detector and the collider have compatible architectures.

– *Junk:*

It seems as if you experience a lot of the "pain" involved in achieving high luminosity in the injector because space charge causes tune spread and chromaticity, and the energy in the injector is lower than in the main ring. What keeps the size of the injector rings down? Why not build a bigger injector and install wigglers to keep emittance down, if necessary?

– *Barletta:*

The booster rings are actually very big; the linac is 1.7 GeV much higher than that for SSC, the low energy booster went up to 12 GeV, and may be pushed higher. It would be nice to stay with a reasonably small linac. My scheme was just to show the practicality of building an injection chain. It is certainly not the optimal one, just an acceptable one.

– *Khalatyan:*

What is the value of the interaction (crossing) angle? Does the vacuum tube in the collider experiments have the shape similar to the cone? Does the detector have any hole in the forward direction? Do you have estimation how many tracks are lost in this hole?

– *Barletta:*

The size of the crossing angle I worked with was around 100-200 μrad. An angle of 250 μRad does not reduce the luminosity by more than 2%. I chose that angle to keep the length over which the long range tune shift was operative of the order of 200 m as it was designed for the SSC. As that parameter was an area where designers have made careful calculations, I have decided to take that dimension as a constraint in order to calculate the angle effect on the luminosity reduction.

The reference material that I used to do the scaling calculation of the radiation within the magnet was based on results which the SSC design group believed were valid for a pseudo-rapidity less than 6, (that corresponds to 3 mRad), so the angle I have chosen for the beam crossing is small compared to the range over which they believed that their model was relatively accurate. In angular range of 100-200 μrad the details of the hadronic shower are not really well known.

– Niaz:

What will happen if one of the dipole fails?

– Barletta:

Presumably you mean that the magnet quenches. If you had a quench of a magnet and if B went to zero instantaneously, two things that can happen: 1) The beam is now misdirected, so it is going to hit the wall some place. Fortunately the bending beam angle is very small, but unfortunately the beam pipe is very small too. Therefore you have to be able to detect small resistivity changes in the magnets, and try to abort the beams into several special dumps, so that you will not have a very large number of bunches lost into the same area. The second effect is that if the thickness of the high conductivity coating on the wall is well chosen, the very large forces due to eddy currents can push the chamber walls apart. Fortunately the failure mode as described cannot occur; the field must take many revolution times to collapse because of the large inductance of the dipole. Hence there is time to act and avoid catastrophic beam loss into a place not designed for such loss. That consideration has been taken into account in both SSC and LHC designs of the vacuum chambers.

NEW DETECTORS FOR SUPERCOLLIDERS: LAA

A. Zichichi
CERN, Geneva, Switzerland

The LAA Project has been presented in previous courses, ["The LAA Project - Second Year of Activity", in *The Challenging Questions*, Plenum Press, New York, p. 221-286, (1990) and "The Main Achievements of the LAA Project", in *Physics up to 200 TeV*, Plenum Press, New York, p. 327-393, (1991)] together with the most significant results. The list of new achievements in technological R&D is unfortunately always very boring, especially to the reader. On the other hand, a work in technology needs a lot of details to be of interest. Once these are given there is no need to repeat the background information. Lack of time and interest in keeping the attention of the reader has led to the decision to publish the most interesting part of the game: i.e. the Discussion.

CHAIRMAN: A. Zichichi

Scientific Secretaries: D. Borden, M.T. Dova, A. Hasan,

K. Qureshi, A.A. Syed

DISCUSSION

– *Kaur:*

Out of the proposed future colliders like LHC, SSC and ELN, the energy at ELN is the highest. There will be a jump from some 40 TeV at SSC to 200 TeV at ELN. In addition to studying the rare phenomena and looking for new physics, physicists will also be interested in doing conventional physics. Is it justified to merely extrapolate the physics results in 40 TeV to 200 GeV? There may be exotic phenomena and discontinuities in the intermediate unexplored energy range!

– *Zichichi:*

I would be the happiest man on the planet if you were right, because we are looking for such discontinuities and the best that we can do is to extrapolate the physics from 40 TeV to 200 TeV. The Monte Carlo after this extrapolation, we hope, will sufficiently describe the backgrounds at 200 TeV. What we are looking for are processes like Weak Interactions becoming strong, as discussed by Roberto Peccei, and for whatever will be unique at these higher energies. There are two possible scenarios, the pessimistic scenario where nothing new will be discovered at LHC and SSC, in which case we must begin to plan now for what lies beyond at still higher energies. That involves resolving the technical problems in building the 200 TeV machine, the ELOISATRON, for which we must begin now. If we wait for the LHC/SSC programmes to be completed before we start planning, then we will not be able to convince people that high energy physics is not dead and we will not be able to raise the funds necessary for building ELOISATRON. In the optimistic scenario, some new physics will be discovered at LHC/SSC. We will then need at least ten years to prepare a higher energy machine to explore these new areas.

We are the conservators of the legacy of Galileo Galilei, this means that we must present to the general public the importance of our work, because no one else will speak for us. Without fighting in the cultural area of our society to get a strong support for our work, our field will surely perish.

– *Hernandez:*

If nature finally chooses to follow the pessimistic picture and there is no new physics found in the LHC and/or SSC in the form of new particle or rare processes, but in the form of small deviations in cross-sections, would not the cleaner e^+e^-

colliders do better, in high precision measurements, than the 200 TeV pp colliders even if they cannot reach the very high energy range?

– Zichichi:

I have been doing both high precision measurements, like the muon (g-2) and physics with high energies, and I am convinced that high energy is more important than low energy high precision measurements. But I am not against high precision experiments. However if I have to choose, I will choose high energy experiments. As explained in my lecture (e^+e^-) machines are out of competition in the multi-TeV range.

– Etzion:

As I understand, you were talking about the multidrift tube detector with its very good precision of 60 μm as your vertex detector. I wonder why you do not plan to use this as a tracking chamber and put inside a silicon microstrip detector as a vertex detector, which can give a better resolution of about 3 μm to detect secondary vertices near the interaction point?

– Zichichi:

My personal view is that the first component of the tracking system is still an open question. The Multidrift Tube Module technology is not able to disentangle the tracks in high multiplicity hadronic events. Silicon is not a good candidate: it is not sufficiently radiation resistant. More research is needed to find a suitable candidate.

– Giannotti:

1) Scientists working on the ELOISATRON project have found a new organic scintillator (PMP). Why with this type of material it is possible to make scintillating fibres of 15 μm instead of 30 μm as with scintillator materials of previous generations? What are the technical characteristics that have made it possible?

2) Do you think that the only way of looking for new physics phenomena is to go up in energy?

– Zichichi:

1) Standard scintillators do not allow the propagation of light in such a narrow fibre, whereas PMP does. Before (PMP) was used for scintillating fibre the maximum diameter was 1000 μm, not 30 μm. So the improvement is from 1000 μm to 15 μm. Among the invited scientists there is one of my collaborators who is responsible for this work: Dr. H. Leutz. I would like to ask Dr. H. Leutz to give some details on this very interesting work.

– *Leutz:*

In a normal scintillator system there are two components, a scintillator and a wavelength shifter. Normally the scintillator emits light at about 340 nm, which is in the opaque region of the core material (polystyrene). The wavelength must then be shifted to about 425 nm, after which it can be transported through the fibre. This is accomplished by reabsorbing scintillation light in a second compound doped into the scintillator (the wavelength shifter) which reemits the light at 425 nm. The reabsorption length is about 200-300 μm; this long reabsorption length allows cross-talk between fibres. PMP requires no wavelength shifter because it emits at 420 nm, so there is no cross talk. Also, in PMP the absorption and emission bands are well separated so that reabsorption of scintillation light is not a problem. This is illustrated in the following graph.

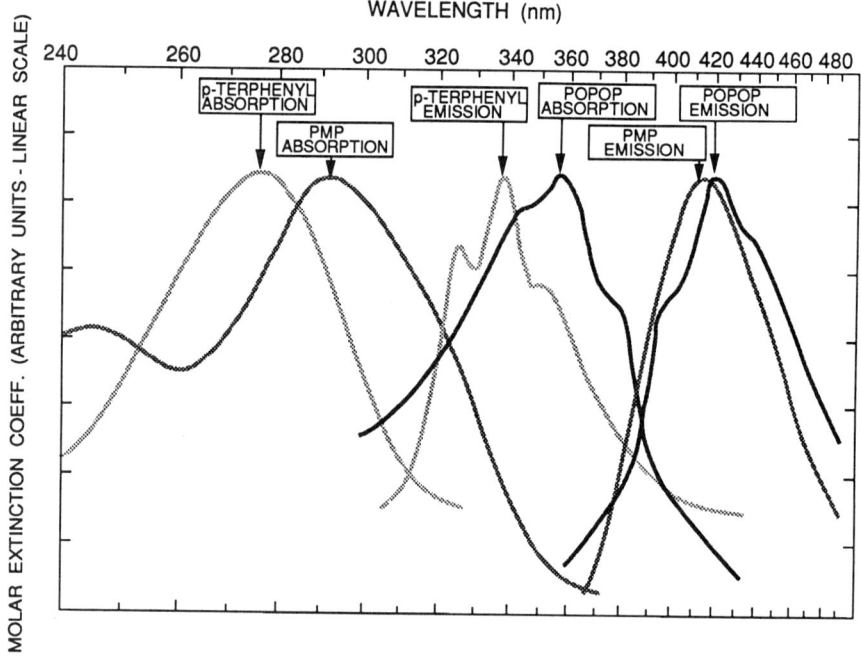

– *Zichichi:*

The answer to your question No. 2 is yes.

– *Gourdin:*

Let us suppose that tomorrow you receive in the mail the money to build the ELOISATRON, when we will see the first collision?

– *Zichichi:*

In ten years.

– *Gourdin:*

And these ten years are sufficient to prepare the whole of the experiment?

– *Zichichi:*

Since the inception of the LAA project, great progress has been made. I have no doubts that ten years will be more than sufficient.

– *Gourdin:*

What are the political considerations of building the ELOISATRON?

– *Zichichi:*

We have to distinguish between scientific and political concerns. Problems may arise when a scientist begins to think like a politician or vice versa. I have always made it clear that our responsibility as scientists is to make good proposals. Their realisation depends very much on the support that a project can get from "illuminated" political leaders.

– *Xexeo:*

Are you worried at all about the data analysis problems at the ELOISATRON?

– *Zichichi:*

The data acquisition and analysis technology is an important component but I am not worried about it at the present time, as these are the last components which need to be implemented, and we have time to think about them. First we have to solve all problems to build the Machine. Then all problems to build the Detector. Then comes the problem of DAQ (Data Acquisition and Analysis). One of the basic problems for DAQ is to know what we want to look for. The trigger and the data acquisition electronics must be tuned accordingly. We will learn, I hope, a lot of new physics in the following years. It would be too early to decide now.

– *Sanchez:*

What does the acronym "ELOISATRON" stand for?

– *Zichichi:*

E̲URASIATIC L̲O̲NG I̲NTERSECTING S̲TORAGE A̲CCELERATOR.

– *Passalacqua:*

Would there be problems with the beam dump and beam losses?

– *Zichichi:*

No.

– *Passalacqua:*

What about the inner tracking detector problems?

– *Zichichi:*

We need to do more research to find the best option. Perhaps scintillating fibres will be proved to be the best choice. We are also working to see if GaAs semiconductor strips will win out. We are fully engaged on this problem.

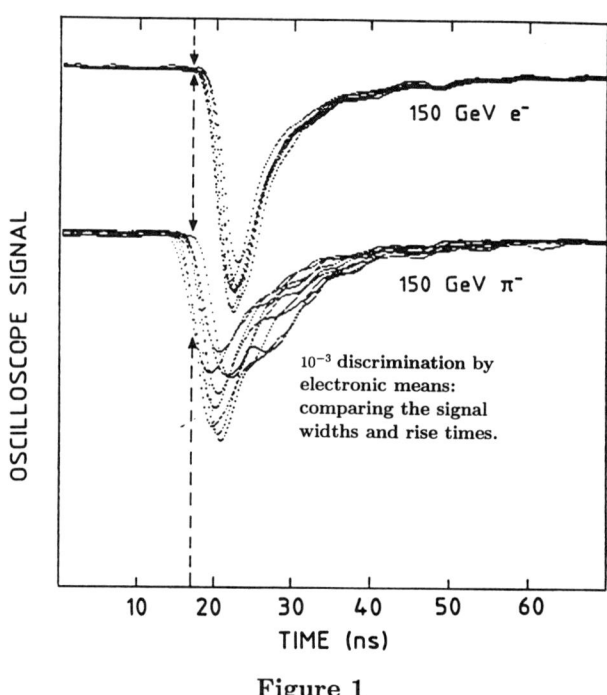

Figure 1

– *Gougas:*

In your presentation of the Spaghetti calorimeter you showed good electron pion separation at 150 GeV. Does the discriminatory capability extend to lower energies as well?

– *Zichichi:*

Yes. Let me repeat the crucial new point: electrons and pions can be separated within 10 ns, via the fast electronic study of the electrons and pions showers. This is illustrated in the following Figures 1 and 2.

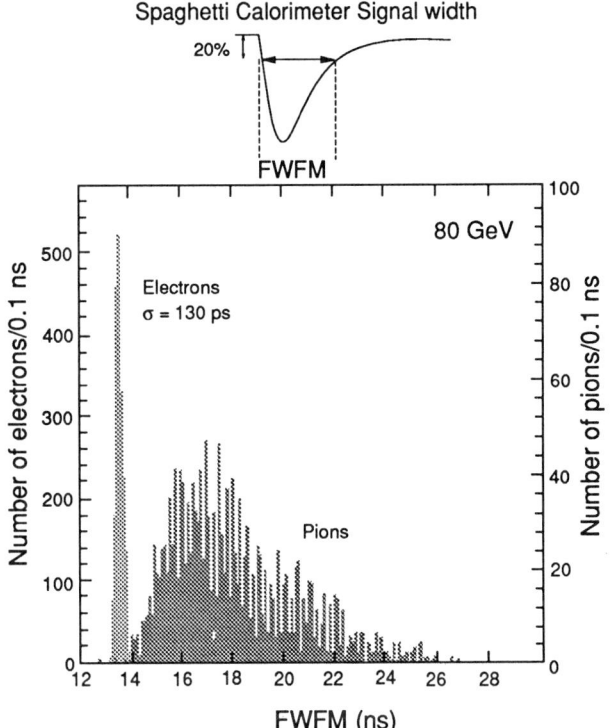

Figure 2

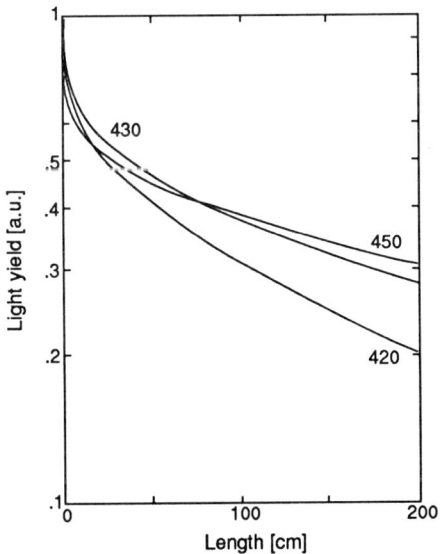

Figure 3

– *Gougas:*

What is the attenuation length of a 15 μm scintillating fibre?

– *Zichichi:*

The attenuation length of a 15 μm fibre is 25 cm. It can be 1 or 2 metres in the thicker fibres, which will have fewer internal reflections per unit length. As shown in Figure 3 the best case is, at present, PMP 450.

– *Khalatyan:*

What calorimeter do you envisage for the ELOISATRON detector?

– *Zichichi:*

The central calorimeter will be a Spaghetti Calorimeter.

– *Khalatyan:*

Is it radiation resistant?

– *Zichichi:*

We have tested it to 1 Mrad. I should mention that for the forward calorimeter we are also considering other options.

– *Khalatyan:*

What about the development of fast electronics?

– *Zichichi:*

We have a fast electronics group working to identify and solve these problems.

– *Papageorgiu:*

Is there any way we could ever test any of the fundamental principles upon which QFT are based, or do we have to take them as Kantian a priori, namely locality, causality and unitarity? In particular there have been recently models which can accommodate the SM without the Higgs-mechanism by introducing non-locality.

– *Zichichi:*

We are always hoping for unexpected results. The non-locality is one of them. But, whatever the prediction of the theory may be, the answer will always come from our ability to detect electrons, photons, pions, muons, missing energy, etc. It is then up to us to analyse these results and to draw conclusions about the nature of the Universe, including possible predictions of QFT.

– *Sivaram (comment):*

Since most of the speakers have spoken about the Planck scale, the ultimate unification scale, and we must think big (!) in terms of energy, if one extrapolates from figures for ELOISATRON, the length scale of an accelerator for Planck energy would be $\sim$ 1 kilo parsec $\sim$ 3000 light years! Of course the cosmic microwave background would limit the maximum energy to $\sim 3 \times 10^{12}$ GeV or so, since inverse Compton scattering would prevent attaining higher energies (a theoretical limit). No cosmic rays with higher energy have been seen! Providing indirect proof of the microwave background.

– *Zichichi:*

I am glad you want to build a kilo parsec accelerator but I am bound to the Earth.

– *Junk:*

How will you measure the luminosity at ELOISATRON as you will need a process of a known cross section at 200 TeV which you don't have?

– *Zichichi:*

This is a very interesting question. Fortunately it is a low priority item. If we could build a 10^{40} luminosity machine, we will not bother to measure the luminosity.

– *Syed:*

M.C. detector simulation already at LEP experiments is taking CPU time of the order of 300 sec/event. The simulation time grows with the event energy and detector size, and the event multiplicity, which are all increased by many orders of magnitude at ELOISATRON. What have you thought about that?

– *Zichichi:*

I am not worried about this. We have an excellent Monte Carlo group and this is one of the last components that we need to implement after building the machine and constructing the detector. I am very optimistic that by that time we will have computers fast enough to take care of this problem.

– *Ellis:*

Could you explain in a bit more detail as to what went into your calculations in your approach to the super symmetric grand unification scheme?

– *Zichichi:*

There are two important points in our supersymmetric approach to grand unification that I have presented. First is the measurement of α_s. We have taken

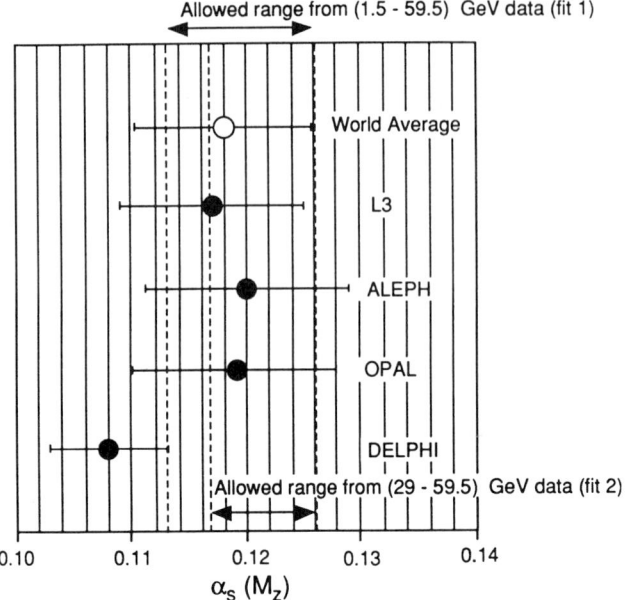

Figure 4

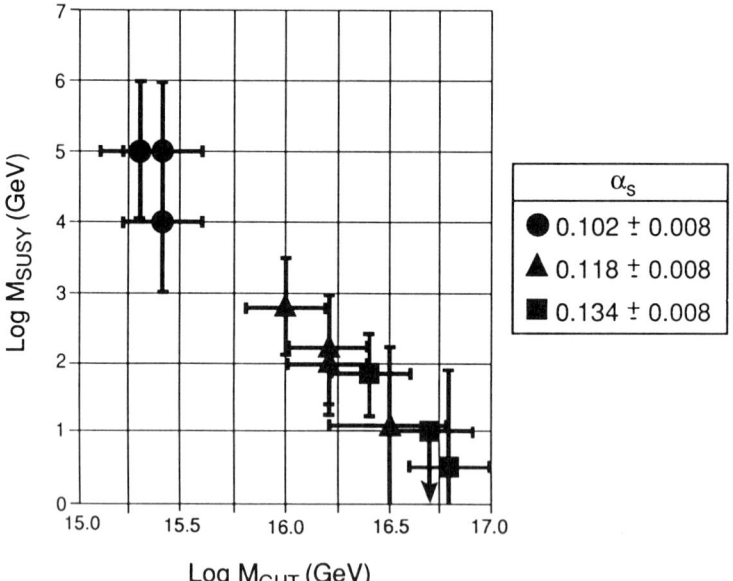

Figure 5

what we think is the correct world average. Here is the figure showing all data including the LEP results, Figure 4. Notice that the data before LEP are grouped into two classes: all energies (including the low part) (1.5 ÷ 59.5) and the high energy (29.0 ÷ 59.5). This grouping is shown because the data refer to very different experimental sources. We wanted to separate the low energy component from the data starting at 29 GeV. Nevertheless the overlap is significant. The LEP data are shown directly.

The second important point is the solutions of the coupled equations for $\alpha_1\alpha_2\alpha_3$. We have not limited our study to only one solution. We have investigated them all: they are three. In addition we have worked out a "numerical" solution, which should be the best. Finally we have taken $\pm 2\sigma$ as confidence level for the world average of $\alpha_s(M_{Z^0})$. When we take the world average for $\alpha_s(M_{Z^0})$, and add $\pm 2\sigma$ to the central value we have 3 values of $\alpha_s(M_{Z^0})$. But there are four possible solutions to the 2-loop coupled equations for $\alpha_1\alpha_2\alpha_3$. Putting all these possibilities together, we get, for M_{SUSY} and M_{GUT} the results shown in Figure 5. For M_{SUSY} we introduce, as everybody does, an "ad hoc" value as if all sparticles had the same mass. More details can be found in our paper (F. Anselmo, L. Cifarelli, A. Petermann and A. Zichichi, preprint CERN/PPE/91-123, 15 July 1991 and Il Nuovo Cimento **104A** (1991) 1817).

To draw conclusions outside the data shown in Figure 5 is not justified, on the basis of our present knowledge. In recent times claims on M_{SUSY} have been published. They are based on a single value of α_s, on the use of only one solution (not the numerical) of the coupled equations for $\alpha_1\alpha_2\alpha_3$ and on $\pm 1\sigma$. The use of the $\pm 1\sigma$ to make predictions has become the new fashion, as if the well-known laws of statistics had changed. This is a School and I would like to emphasize how incorrect is the case of $\pm 1\sigma$. The minimum range of safety is the 95% confidence level, i.e. $\pm 2\sigma$.

CLOSING CEREMONY

The closing ceremony took place on Sunday 21st July 1991. The School was especially honoured this year to have Madame Dirac in attendance at the Closing Session, at which she presented the Dirac scholarship to Ulf Danielsson.

The Prizes and Scholarships were awarded as specified below.

PRIZES AND SCHOLARSHIPS

- Prize for **Best Student** awarded to:

Ulf DANIELSSON, Princeton University, USA.

Matthias NEUBERT, Universität Heidelberg, Germany.

Ira ROTHSTEIN, University of Maryland, USA.

- Prize for **Best Scientific Secretary** awarded to:

Pilar HERNANDEZ, CERN, Geneva, Switzerland.

- <u>Twelve **Scholarships**</u> were open for competition among the participants. They were awarded as follows:

- **J.S. BELL** Scholarship to:

Ira ROTHSTEIN, University of Maryland, USA.

- **Patrick M.S. BLACKETT** Scholarship to:

Matthias NEUBERT, Universität Heidelberg, Germany.

- **James CHADWICK** Scholarship to:

Pilar HERNANDEZ, CERN, Geneva, Switzerland.

- **Amos DE-SHALIT** Scholarship to:

Erez ETZION, Tel Aviv University, Israel.

- **Paul A.M. DIRAC** Scholarship to:

Ulf DANIELSSON, Princeton University, USA.

- **Isidor I. RABI** Scholarship to:

Doug BORDEN, Stanford Linear Accelerator Center, USA.

- **Robert HOFSTADTER** Scholarship to:

Thomas JUNK, Stanford Linear Accelerator Center, USA.

- **Andreij D. SAKHAROV** Scholarship to:

Mark POTTERS, Princeton University, USA.

- **Jun John SAKURAI** Scholarship to:

Zbigniew PLUCIENNIK, Inst. of Theoretical Phys., Warsaw, Poland.

- **Gunnar KÄLLEN** Scholarship to:

Anwarl HASAN, World Laboratory and CERN, Geneva, Switzerland.

- **André LAGARRIGUE** Scholarship to:

Ujjwal VIKAS, World Laboratory and CERN, Geneva, Switzerland.

- **Giulio RACAH** Scholarship to:

Manjjt KAUR, World Laboratory and CERN, Geneva, Switzerland.

- One **EPS** Scholarship was awarded to:

Jan CZYZEWSKI, Inst. of Physics, Jagellonian Univ., Krakow, Poland.

- Two **WFS** Scholarships were awarded to:

Alexei V. CHIZHOV, Laboratory of Theoretical Physics, Dubna, USSR.
Ahpisit UNGKITCHANUKIT, Phys. Dept., Faculty of Scie., Bangkok, Thailand.

The following participants gave their collaboration in the Scientific Secretarial work:

Doug BORDEN	*Rukhsana MALIK*
Alexei V. CHIZHOV	*Ronan McNULTY*
Jan CZYZEWSKI	*Nars E. MOULAI*
Ulf DANIELSSON	*Matthias NEUBERT*
Maria Teresa DOVA	*Muhammad A. NIAZ*
Eda ESKUT	*Zbigniev PLUCIENNIK*
Erez ETZION	*Marc POTTERS*
Anwarul HASAN	*Khalid QURESHI*
Pilar HERNANDEZ	*Ira ROTHSTEIN*
Mark JOHNSON	*Hans-Josef SCHULZE*
Manjjt KAUR	*Ali A. SYED*
Norair KHALATIAN	*Patribha VIKAS*
Ayse KUZUCU	*Ujjwal VIKAS*
Stephen LAU	*Maneesh WADHWA*
Bengt A. LINDHOLM	*Xin WU*

Professor John Peoples, Director of Fermilab, Madame Margit Dirac, Professor Heinrich Leutz, Deputy Director of LAA, Professor Antonino Zichichi and Dr. Ulf Danielsson.

PARTICIPANTS

Gianluca ALIMONTI
Dipartimento di Fisica
Università di Milano
Via Celoria, 16
I-20133 MILANO, Italy

Guido ALTARELLI
C E R N
TH Division
CH-1211 GENEVA 23, Switzerland

William A. BARLETTA
Physics Department
UCLA/LLNL
East Avenue P.O. Box 808
LIVERMORE, CA 94550, USA

Knut Steinar BJORKEVOLL
Physics Department
University of Bergen
Allegaten, 55
N-5007 BERGEN, Norway

Doug BORDEN
S L A C, Bin 43
P.O. Box 4349
STANFORD, CA 94309, USA

Stefan BORNHOLDT
D E S Y
Theory Group
Notkestrasse, 85
D-2000 HAMBURG 52, Germany

Paolo BRANCHINI
Dipartimento di Fisica
Università di Roma II
Via Orazio Raimondo
I-00173 ROMA, Italy

Andrea BRIGNOLE	Dipartimento di Fisica Università di Padova Via Marzolo, 8 I-35100 PADOVA, Italy
David BROWN	C E R N PPE Division CH-1211 GENEVA 23, Switzerland
Riccardo BRUGNERA	Dipartimento di Fisica Università di Padova Via Marzolo, 8 I-35100 PADOVA, Italy
Vasile BUZULOIU	World Laboratory and C E R N CH-1211 GENEVA 23, Switzerland
Alexei V. CHIZHOV	Laboratory of Theoretical Physics Joint Inst. for Nuclear Research DUBNA, 141980 USSR
Luisa CIFARELLI	Dipartimento di Scienze Fisiche Università di Napoli Mostra d'Oltremare - Pad. 19 I-80125 NAPOLI, Italy
Jan CZYZEWSKI	Institute of Particle Physics Jagellonian University Reymonta, 4 PL-KRAKOW, Poland
Ulf DANIELSSON	Physics Department Princeton University Jadwin Hall - P.O. Box 708 PRINCETON, NJ 08544, USA
Luigi DEL DEBBIO	INFN- Sezione di Pisa Via Livornese 582/A San Piero a Grado I-56010 PISA, Italy

Yuri DOKSHITZER	Leningrad Inst. of Nuclear Physics Department of Theoretical Physics 188350 GATCHINA, USSR
Maria Teresa DOVA	World Laboratory and C E R N CH-1211 GENEVA 23, Switzerland
Michael DUFF	Center for Theoretical Physics Physics Department Texas A&M University COLLEGE STATION, TX 77843, USA
Huy Danh DUONG	Southampton University Physics Department Highfield SOUTHAMPTON, SO9 5NH, UK
John ELLIS	C E R N TH Division CH-1211 GENEVA 23, Switzerland
Eda ESKUT	World Laboratory and C E R N CH-1211 GENEVA 23, Switzerland
Erez ETZION	School of Physics and Astronomy Tel Aviv University 69978 TEL AVIV, Israel
Sergio FERRARA	Physics Department University of California 405 Hilgard Avenue LOS ANGELES, CA 90024, USA and C E R N TH Division CH-1211 GENEVA 23, Switzerland
Emanuele FIANDRINI	Dipartimento di Fisica Università di Perugia Via Pascoli I-06100 PERUGIA, Italy

Jeffrey R. FORSHAW   Physics Department
                                 University of Manchester
                                 MANCHESTER M13 9PL, UK

Elisabetta GALLO   Dipartimento di Fisica
                                 Università di Firenze
                                 Largo E. Fermi, 2
                                 I-50125 FIRENZE, Italy

Paola GIANOTTI   I N F N
                                 Sezione di Torino
                                 Via P. Giuria, 1
                                 I-10125 TORINO, Italy

Fred J. GILMAN   S S C Laboratory
                                 2550 Beckleymeade Avenue
                                 DALLAS, TX 75237, USA

Ferenc GLUCK   Central Res. Inst. for Physics
                                 P.O. Box 49
                                 H-1525 BUDAPEST, Hungary

Ariel GOOBAR   Physics Department
                                 University of Stockholm
                                 Vanadisvagen 9
                                 S-11346 STOCKHOLM, Sweden

Andreas GOUGAS   C E R N
                                 PPE Division
                                 CH-1211 GENEVA 23, Switzerland

Michel GOURDIN   Dept. de Physique
                                 Université P. et M. Curie
                                 4 Place Jussieu
                                 F-75252 PARIS, France

Thierry GRANIER   DSM / DPhPE / SEPh
                                 CEN - Saclay
                                 F-91191 GIF-sur-YVETTE, France

Reinhold J. GUTH	Werner-Heisenberg Inst. für Physik Max-Planck-Institut für Physik und Astrophysik Fohringer Ring 6 D-8000 MUNCHEN 40, Germany
Anwarul HASAN	World Laboratory and C E R N CH-1211 GENEVA 23, Switzerland
Pilar HERNANDEZ	C E R N TH Division CH-1211 GENEVA 23, Switzerland
Vidyut JAIN	Werner-Heisenberg Inst. für Physik Max-Planck-Institut für Physik und Astrophysik Fohringer Ring 6 D-8000 MUNCHEN 40, Germany
Mark JOHNSON	Physics Department University of North Carolina CB 3255 CHAPEL HILL, NC 27399, USA
Thomas JUNK	SLAC - Bin 96 2575 Sand Hill Road MENLO PARK, CA 94309, USA
William A. KAUFMAN	Randall Laboratory of Physics University of Michigan ANN ARBOR, MI 48109, USA
Manjjt KAUR	World Laboratory and C E R N CH-1211 GENEVA 23, Switzerland
Norair KHALATYAN	Yerevan Physics Institute Alikhanian Brothers St. 2 YEREVAN, 375036 Armenia, USSR
Valery KHOZE	Leningrad Inst. of Nuclear Physics Department of Theoretical Physics 188350 GATCHINA, USSR

Bruce James KING	Fermi National Accelerator Lab. P.O. Box 500 BATAVIA, IL 60510, USA
Ralph KRETSCHMER	Fachbereich Physik Universitat Siegen Adolf-Reichwein-Strasse 2 D-5900 SIEGEN, Germany
Ayse KUZUCU	World Laboratory and C E R N CH-1211 GENEVA 23, Switzerland
Stephen LAU	Physics Department University of North Carolina CB 3255 - Phillips Hall CHAPEL HILL, NC 27399, USA
Heinrich LEUTZ	C E R N PPE Division - LAA CH-1211 GENEVA 23, Switzerland
Giuseppe LEVI	INFN-Sezione di Perugia Via Pascoli I-06100 PERUGIA, Italy
Bengt-Åke LINDHOLM	Physics Department University of Stockholm Vanadisvagen. 9 S-11346 STOCKHOLM, Sweden
Eligio LISI	Dipartimento di Fisica Università di Bari Via Amendola, 173 I-70100 BARI, Italy
Sergio LUPIA	Dipartimento di Fisica Teorica Università di Torino Via P. Giuria, 1 I-10125 TORINO, Italy

Carmen MAIDANTCHIK	World Laboratory and C E R N CH-1211 GENEVA 23, Switzerland
Rukhsana MALIK	World Laboratory and C E R N CH-1211 GENEVA 23, Switzerland
Ronan McNULTY	Physics Department University of Liverpool Oliver Lodge Lab., Oxford Street LIVERPOOL L69 EBX, UK
Mario MIRAGLIUOLO	Dipartimento di Scienze Fisiche Università degli Studi di Napoli Mostra d'Oltremare - Pad. 19 I-80125 NAPOLI, Italy
Nasr-Eddine MOULAI	World Laboratory and C E R N CH-1211 GENEVA 23, Switzerland
Matthias NEUBERT	Institut für Theoretische Physik Universität Heidelberg Philosophenweg 16 D-6900 HEIDELBERG, Germany
Muhammad Arif NIAZ	World Laboratory and C E R N CH-1211 GENEVA 23, Switzerland
Lev B. OKUN	I T E P B. Cheremushkinskaya ul. 89 117259 MOSCOW, USSR
Orlando OLIVEIRA	Departamento de Fisica Fac. de Ciências e Tec. Univ. de Coimbra P-3000 COIMBRA, Portugal
Nazife OZDES	World Laboratory and C E R N CH-1211 GENEVA 23, Switzerland
Elena PAPAGEORGIU	Rutherford Appleton Laboratory High Energy Physics Theory Chilton DIDCOT, OX11 0RA, UK

Youngchul PARK	Physics Department University of California SANTA BARBARA, CA 93106, USA
Stefano PASSAGGIO	Dipartimento di Fisica Università di Genova and INFN-Sezione di Ge Via Dodecaneso, 33 I-16146 GENOVA, Italy
Luca PASSALACQUA	Lab. Naz. di Frascati-(LNF) Ist. Naz. di Fisica Nucleare-(INFN) Via E. Fermi, 40 I-00044 FRASCATI, Italy
Massimo PASSERA	INFN-Sezione di Torino Via P. Giuria, 1 I-10125 TORINO, Italy
Roberto PECCEI	Physics Department University of California 405 Hilgard Avenue LOS ANGELES, CA 90024, USA
John PEOPLES	F N A L P.O. Box 500 - Wilson Road BATAVIA, IL 60501, USA
Donald H. PERKINS	Nuclear Physics Laboratory University of Oxford Keble Road OXFORD, OX1 3RH, UK
Zbigniew PLUCIENNIK	Institute of Theoretical Physics Warsaw University Hoza, 69 PL-00-681 WARSAW, Poland
Marc POTTERS	Princeton University Physics Department P.O. Box 708 PRINCETON, NJ 08544, USA
Khalid QURESHI	World Laboratory and C E R N CH-1211 GENEVA 23, Switzerland

Renzo RAGAZZON
Dipartimento di Fisica Teorica
Università degli Studi di Trieste
Strada Costiera, 11
Miramare
I-34014 TRIESTE, Italy

Patrick ROBERTS
Institute of Theoretical Physic
Elementary Particle Physics
S-41296 GOTEBORG, Sweden

Ira ROTHSTEIN
Dept. of Physics and Astronomy
University of Maryland
COLLEGE PARK, MD 20742, USA

Norma SANCHEZ
Observatoire de Paris
Section d'Astrophysique
5, Place Jules Janssen
F-95125 MEUDON, France

Pietro SANTORELLI
Dipartimento di Scienze Fisiche
Università degli Studi di Napoli
Mostra d'Oltremare - Pad. 19
I-80125 NAPOLI, Italy

Hans-Josef SCHULZE
Inst. für Theoretische Physik
Universitat Heidelberg
Philosophenweg 19
D-6900 HEIDELBERG, Germany

Michael SEYMOUR
Physics Department
Cambridge University
CAMBRIDGE CB3 OHE, UK

Yuly SHABELSKI
Leningrad Inst. of Nuclear Physics
Department of Theoretical Physics
188350 GATCHINA, USSR

Chidambaram SIVARAM
Dept. of Physics and Astrophysics
Indian Astrophysical Institute
560034 BANGALORE, India

Vasilios SPANOS	Institute of Nuclear Physics NRPS "Demokritos" GR-15310 ATHENS, Greece
Aly Aamer SYED	World Laboratory and C E R N CH-1211 GENEVA 23, Switzerland
Tommaso TABARELLI	Dipartimento di Fisica Università degli Studi di Milano Via Celoria, 16 I-20133 MILANO, Italy
William TRISCHUK	C E R N PPE Division CH-1211 GENEVA 23, Switzerland
Ahpisit UNGKITCHANUKIT	Physics Department Faculty of Sciences Chulalonghorn University BANGKOK 10330, Thailand
Pratibha VIKAS	World Laboratory and C E R N CH-1211 GENEVA 23, Switzerland
Ujjwal VIKAS	World Laboratory and C E R N CH-1211 GENEVA 23, Switzerland
Maneesh WADHWA	World Laboratory and C E R N CH-1211 GENEVA 23, Switzerland
Zhao Min WANG	World Laboratory and C E R N CH-1211 GENEVA 23, Switzerland
Daniel WESELKA	Institut für Hochenergiephysik Osterreichischen Akademie Nikolsdorfer Gasse 18 A-1050 WIEN, Austria
Mark WEXLER	Physics Department Princeton University P.O. Box 708 PRINCETON, NJ 08544, USA

Xin WU   Dépt. de Physique Théorique
                            Université de Genève
                            24 Quai Ernest-Ansermet
                            CH-1211 GENEVA , Switzerland

Geraldo XEXEO   World Laboratory and C E R N
                            CH-1211 GENEVA 23, Switzerland

Qinghao YE   World Laboratory and C E R N
                            CH-1211 GENEVA 23, Switzerland

Quiping ZHANG   Physics Department
                            University of Stockholm
                            Vanadisvagen, 9
                            S-11346 STOCKHOLM, Sweden

INDEX

Abelian QED, 14
  vector potential A^m, 171
Acceleration of a particle
  defined, 3, 12
  force other than gravitation, 12
Adiabaticity, 284-285, 292-294
Adler Bell Jackiw anomaly, 96
Antiparticles, mass, 13
Atiyah Singer theorem, 99

B^*, lowering, 380
B + L violations, 112, 113
  amplitude, Sphaleron, 116-118
  high energy, 120
  neutrino activation, 125
  pp collisons, 123
  size, 115-117
  valley approach, 121, 128
  weak interactions (coupling
    theories), 89, 97-99, 102
B production, Tevatron, 246
Barden's identity, ø-vacua, 96
Baryons
  numbers
    high temperatures, 123
    universe, 85
BB mixing, 336-339
Beamstrahlung
  compared with synchrotron, 382-383
  defined, 381
  plasma lens, 322
Beryllium, neutrinos, 295, 302
Bessel function, modified, 31
Beta decay
  barrier penetration factor, 251
  beta spectrometer, 259
  electron spectrum, 251
  Kurie plot, 252-254, 257, 269, 277
  see also Neutrino physics
Betti numbers, 197, 198
Bianchi identity, 136, 173, 175, 178, 180
  Yang-Mills action, 184
Big Bang nucleosynthesis, neutrino physics, 270-271
Bjorken process, 248, 303
Black p-branes, collapse, 202-203
Bogonol'nyi bound, 181
Boltzmann factor, 123, 124

Borel summability, $\lambda\phi^4$, 90, 114, 129
Born term form factor, 27
Boron neutrinos, 295
Bosons
  copious, 90
  Goldstone boson, 273
  mass, one-loop order, 60
  massive gauge, 332-333
  space-time coordinates, 66
  W boson
    detection, 323-329
    results, 330-331
  see also Higgs bosons
Bottomonium decays, 59
Bound systems, 7-8
Brane-scan, 172, 192-193
  superthree-brane, 189
Bremsstrahlung, QCD, 46
Bronshtein, theory of the world as a whole, 21

Calibi-Yau approach, Lagrangians, 204
Calibi-Yau compactifications, 67
  deformations, 164
  Euler characteristic, 85
  N=Z field theories, 131, 133
  smooth, massive states, 154
  string theory, 162
  vortex formation, 162-163
Calibi-Yau threefold, 146, 150
CDF see Fermilab collider
Cherenkov detector, 279
Chern-Simons multiplet, 155
Chern-Simons terms, Yang-Mills action, 184-185
Chlorine experiment
  capture rate, 280-281, 282
  results, 294
Clebsch-Gordon coefficients, Yukawa couplings, 164
CLEO experiment, 336, 339
Closed universe, expansion, 8, 15
Compton cross section and kinematics, Standard Model, 305-307
Compton length, 17
Conformal and modular invariance, string theory, 66
Constant magnetic field, spin precession, 293-289

415

Coupling theories see Weak interactions
CP violation, 249
 B sector, 347
Crab crossings, 369, 382
Cryptons
 dark matter, 88
 masses, 86
Cube of theories, 21-22

Dark matter, cryptons, 88
Decoupling theorem, 242
Dedekind function, 74, 143, 144
Deformation of complex structure, and target-space duality symmetry, superstring vacua, 137-145
Dilaton vacuum expectation value, 198
Dirac equation
 Euclidean space, 97, 98
 unitary transformation, 13
Dirac magnetic moment, 280, 281
Dirac neutrinos, 269-270, 272
Dirac quantization rule, 170, 174, 186
Distorted Gaussian, 32, 33
DO central tracking chamber, 345-346
Double logarithmic approximation, 28-29
Duff-Lu string, 181
Durham algorithm, jet cross sections, 25-26

E, mass of a system of free particles, 6-7
Effective form factors, jet cross sections, 29-30
Effective string unification scale, Standard Model, 73-76
Eigenvalue spectrum problem, 97-98
Electrons, and pions, 392
Electroweak theory
 computations, 111
 damping mechanisms, 114-131
 LEP tests, 245
 m_t effects, 242-243
 precision tests, 211-221
 and unitarity, 119
 ø-Vacuum, 94-104
 weak interactions (coupling theories), 89, 97-99, 102
 see also Precision electroweak data
Eloisatron hadron supercollider, 367-379
 catastrophic beam loss, 380, 385
 discussion, 380-385
 luminosity
  beam dynamics, 371
  design strategies, 367-379
  goals, 369-370
  parameter sets, 374, 375
  top luminosity (ultimate ELN), 377
 trends, 347

magnet quenches, 385
 particles, transverse momenta, 381
 ultimate ELN
  characteristics, 377-378
  maximum attainable, 381
Energy
 binding, 7-8
 definitions, 6-7
Energy-momentum tensor, 4
Epsilons
 defined, 229
 Standard Model, 235
Equivalence principle, 12
Euclidean action, 98
Euclidean space
 Dirac equation, 97, 98
 $\lambda ø^4$, 105-106
Euler density, 6-D, 198
Euler number, Betti numbers, 197
Expressions of Interest (EOI), SSC project, 357

Fayet-Iliopulos term, 73
Fermi scale, 9
Fermilab collider
 detection strategies, 323-329
 discussion, 344-349
 experimental results, 329
  BB mixing, 336-339
  massive gauge bosons, 332-333
  production cross sections, 332, 333-336
  search for top quark, 339-340
  W boson mass, 330-331
 goals, 323
  luminosity goals, 342-343
 prospects, 340-342
Fermions
 charge, Standard Model, 97
 supersymmetric, D 66
 unitary transformation, 13
 violation, 90
 Yukawa coupling, 86
Five-branes
 discussion, 192-207
 elementary, 176-178
 heterotic, 193, 203, 204
 outstanding problems, 196
 perturbation, 184
 quantization, 192
 solitonic, 178-181
 and string mass spectra, 200
 string/five-brane duality, 182-185
Flavour(s), 49
 heavy, LEP, 235-238
 quark-lepton universality, 228-229, 230, 234
Fourier transform
 classical field, 91
 Higgs mass, 101, 102
 KRT formula, 109
 W field, 102
Frenkel-Kac mechanism, 202
Galileo Galilei, historical notes, 4-6, 24

Gallium experiment, expected
        result, 298
Gauge couplings
    constants, 9, 11
    heterotic string
        compactification, 151
    higher loop corrections, 154-156
Gauge invariance
    first coined, 19
    standard model, 19
Gaugino, mass term, 205
Gauss-Bonnet term, gravity, 162
Gaussian, distorted, 32, 33
Gaussian spectrum, 33
GEM detector, 355, 359
Generating functionals (GF), 27-28
Germanium crystal detector, 265
Gluons
    compared with photons, 19
    emission, radiative, 321
    form factor, 28
    jet, DL Master Equation, 28
    massless, 18
    radiation, 48
    relic, gluonic strings, 17
    scattering, 46
Goldstone boson, 273
Goldstone-Hoppe result, 200
Grand Universal Theory (GUT)
    flipped supersymmetric SU(5)
        GUT[4], 68-70
    models, 68-72
    one-loop and two-loop levels, 58
    simplest model, 50-51
    SU(3) x SU(2) x U(1), 67-70
    and supersymmetric GUTs, Standard
        Model, 50-53
    testing models, 51-53
    see also Standard Model;
        Supersymmetry
Gravitational field, space-time, 24
Gravitational properties of
        particles, 3-4, 12
Gravitino, mass term, 205
Graviton, massive, 14
Green-Schwarz mechanism, 153, 155,
        156
    anomaly-cancellation term, 184
    superstring p=1, 192
Green's functions, 91, 94, 96
    Minkowski space, 99
    point-like amplitudes, 101
    standard electroweak theory, 99

Hadron colliders see Eloisatron
        hadron supercollider
Hadrons
    charged, 37
    hadronization models, 246-247
    light (pions), 36
    LPHD hypothesis, 34, 42
    mass, 8, 38
    massive
        MLLAs, 38
        truncated QCD cascades, 31
    technihadrons, 10
Higgs boson, 10-11
    CP-even, CP-odd, 82
    gamma collisions, 310
    gamma production mechanism, 321
    heavy, 218
    intermediate mass, search, 313-
        318, 322
    light, 218
    mass, 54, 210
        discussion, 344
        effective potential formalism,
            83
        Fourier transform, 101, 102
        light compared with heavy, 60
        radiative corrections, 80-81
    mixing parameters, supersymmetric
        GUT, 62
    neutral, 309-310
    particle, Kobayashi-Maskawa
        mixing angles, 54
    photon-photon collider, 303-319
        discussion, 321-322
    production in SSC project, 354
    Standard Model, 210, 303-319
    in supersymmetric models,
        Standard Model, 47
    symmetry breaking, 354
Higgs Breit-Wigner, 310
Higgs doublets, 70
Higgs fields, 68
    black p-branes, 203
    residue, 101
    term X^{mnp}, 174
Higgs scalars, 10-11
Higgs sector, exotic nondoublet,
        248
Higgs vacuum, 55
Hodge dual operation, 173
Hubble constant, 8
Hubble expansion, 8

IBEC spectrum, 266-267
Inertia, and momentum, 3
Instanton-antiinstanton pairs, 117
Instantons
    high energy, 126
    semiclassical, 115
    solution, 179
    strong, 124
    Yang-Mills equations of motion,
        98

Jacobian factor, 100
JADE algorithm, 25, 26, 42
Jauch plot, 265
Jet cross sections
    calculation, 28
    Durham algorithm, 25-26
    effective form factors, 29-30
    JADE algorithm, 25, 26, 42
Jet rates
    double logarithmic approximation,
        28-29
    generating functionals, 27-28
Jet-jet invariant mass resolution,
        312
Jets, neural networks, 48

Kähler class deformations, 132, 135
Kähler manifolds, 133, 134
Kähler potential
 equation, 145
 modified S-Kähler, 156
 scalar fields, 164
Kaluza-Klein analysis, 136
Kaon spectra, 31
Khoze-Ringwald calculation, 128
Klein-Fock-Gordon equation, 19
Klein-Gordon operators, 110
Kobayashi-Maskawa matrix, 249
Kobayashi-Maskawa mixing angles, Higgs particle, 54
Kosterlitz-Thouless transition, 162
KRT formula
 functional integral, 126
 valley approach, 108-114
Kurie plot, beta decay, 252-254, 257, 269, 277

Lagrangians
 Calibi-Yau approach, 204
 low energy effective, 195-196
 one-loop, 204
$\lambda \phi^4$
 amplitudes, 107, 108, 114
 approximation, 107
 Borel summability, 90, 114, 129
 compared with electroweak theory, 105
 Euclidean space, 105-106
 high order estimates, 90, 120
 N-loop amplitude, 127
 tree graph amplitudes, 105
Landau poles, 244
Laser photons, Compton backscattering, 304-307
Lattice gauge theory, 127
Lead, in calorimeter construction, 325
LEP
 heavy flavours, 235-238
 m^z results, 209-214
Leptonic decay, 324
Leptons, mass, 9
Letters of Intent, SSC project, 358
Lightcone formulation, 206
Lightcone gauge, 193, 194, 200
Liouville model, 85, 87
Lipatov, large order perturbation theory, 104-108
Lorentz invariance, 192
Lorentz scalar, 17
LPHD hypothesis, 34, 42
Luminosity distribution, Standard Model, 307-309
Lund strings, 46

Magnetic field, sun, 290-291
Majorana neutrinos, 269-270, 272
'Majoron', 273
Mass
 discussion, 12-24
 and energy, 7
 Majorana, 10
 nature of
  and scalars, 10-11
  XVI-XVII centuries, 4-6
  XVIII-XIX centuries, 6
 negative, 13
 Newton's definition, 5
 of a particle, defined, 2
 patterns, elementary particles, 9-10
 relativistic, 2
 system of free particles, 6-7
Mass-shell definition, 59, 61
Maxwell equations, 6
 in D=4, 173
Mellin transformed spectrum, 32
Meson decays, B mesons, 362
Minkowski space
 asymptotic, 15
 external, 16, 24
 Green's functions, 99
MLLA effects
 defined, 34
 parton cascades, 34-38
MLLA equations, 27
MLLA evolution equations, 32
MLLA form factors, 27
Momentum, time derivative, 2
Monopoles, elementary, solitonic, 174-175
MSW solution, Weinberg-Salam mode, 281, 284
Multidrift tube detector, 389-390
Muon-electron, 346
Muons
 dimuon system, 327, 328
 spectra, 338

Nambu-Goto action, 201
Neural networks, jets, 48
Neutrino: antineutrino scattering, 229
Neutrino physics, 251-273
 17 keV neutrino, 254-262
  recent experiments, 262-268
 Big Bang nucleosynthesis, 270-271
 discussion, 268-269, 275-278
 kinematics of beta decay, 251-254
 supernova neutrinos, 271-273
Neutrino-nucleus deep inelastic scattering, 230
Neutrinos
 activation, B + L violations, 125
 beryllium, 295, 302
 gallium, 298
 heavy, 237
 mass, 10
 solar, wave length, 14
 solar neutrino problem, 279
  case of varying magnetic field, 290-291
  discussion, 301-302
  expected Gallium result, 298
  fitting the data, 294-297
  four by four case, 292-294
  Kamiokande signal, 279, 282, 301
  SAGE experiment, 298

Neutrinos (cont.)
  solar neutrino problem (cont.)
    spin precession in constant
      magnetic field, 283-289
    standard solar model (SSM),
      279
    time-varying, 279
  supernovae, 302
Newton, historical notes, 5
Newtonian constant, 8
Newtonian potential, 4
Nitrogen neutrinos, 295
Noether charge, string, 178, 180
Non-Abelian gauge theories, 94
NSNP see Neutrinos, solar neutrino
  problem, time varying
Nucleons, antinucleons, 271

p-branes, 170
  black, 202
  super, 202
Parasitic crossings, 369, 382
Particle, free, energy, defined, 13
Partons
  cascades
    MLLA effects, 34-38
    predictions, 41, 43
  multiplicity, 31
  production, 324
Pauli matrix combinations, 99
PDEs, non-linear, 199
PEP/PETRA, 237
Perturbation
  QCD, pomeron, 124
  radiative corrections, 212
  SUSY-breaking, 207
Perturbation theory, 66, 90
Photons
  compared with gluons, 19
  mass, 9
  photon-photon collider, Higgs
    boson, 303-319
Picard-Fuchs equations, 131, 133,
  145-151
Pions, 36, 41
Planck scale
  QFT, 17
  superunification, 11
PMP, scintillators, 389
Poincare scalar, 17
Polarizations, measurements, 241-
  242
Pomeron, high e scattering, 124
pp collisions, 45-46, 347
  B + L violations, 123
Precision electroweak data, 209-238
  discussion I and II, 241-250
  model independent analysis of
    data, 221-235
  see also Electroweak theory
Proton colliders, see also
    Eloisatron hadron
    supercollider; SSC project
Proton lifetime, non-SUSY, Standard
    Model, 81
Pseudoparticles, 25

$Q_\emptyset$, 32
Quantum chromodynamics (QCD)
  bremsstrahlung, 46
  MLLA effects, 34
  phenomenology
    discussion I, 41-44
    discussion II, 45-48
    jet rates and truncated parton
      cascades, 25-40
  sum rules, 8
Quantum mechanics
  consistency, 193
  and relativity, 63-64
Quark-lepton universality, 227-229,
  230, 234
Quarks
  and gluons, confined in hadrons,
    8
  heavy, 45, 46-47
    mass, 47
  mass, Fermi scale, 9
  top, 9
    Fermilab collider program,
      323, 339-340
    LEP analysis, 242
    mass, 18, 209-210
  Yukawa coupling, 18

Reissner-Nordstrom limit, 188
  black hole solution, 202
Relativity
  general, 64
  and quantum mechanics, 63-64
  relativistic particle, weak G
    field, 4
  special, 64
Riemann tensor, scalar fields, 164
Ringwald-Espinosa formula, 90, 99,
  100, 117

s-model sources, Yang-Mills action,
  185-188
Saddle points, 103, 104, 107, 112
  approximations, 126
SAGE experiment, solar neutrino
  problem, 298
Sargent rule, 270
Scalar fields, Riemann tensor, 164
Scalars, nature of mass, 10-11
Scintillators, PMP, 389
SDC detector, 354
Semiclassical formula
  n-leg amplitude, 109
  quantum corrections, 104
  representative correction, 109
SLD detector, Fast Monte Carlo
  Simulation, 312, 314
SNP see Neutrinos, solar neutrino
  problem
Solar neutrino problem
  Kamiokande signal, 279, 282, 301
  standard solar model (SSM), 279
Solitons
  five-branes, 198-199
  solution, 188
Space-time
  curved, duality, 162

Space-time (cont.)
  D, 170-173
  gravitational field, 24
  singularities, 207
  string compactification, 162
Sparticle mass, 80
Sphaleron, B + L amplitude, 116-118
Spin precession, constant magnetic field, 293-289
Spontaneous symmetry breaking sector see Standard Model
SSC Project
  chronology, 351
  discussion, 361-365
  experimental program, 357-360
  Expressions of Interest (EOI), 357
  fixed-target experiments, 364
  Letters of Intent, 358
  location, 352
  luminosity, upgrades, 361, 364
  parameters, 353
  physics, 354-357
  prospects, 353
  ring schematic, 352
Standard electroweak theory, weak interactions (coupling theories), 96
Standard Model
  Coleman-Weinberg limit, 79
  Compton cross section and kinematics, 305-307
  discussion I, 79-83
  discussion II, 83-88
    effective string unification scale, 73-76
    fermion charge, 97
    flipped supersymmetric SU(5) GUT, 68-70
    derived from string [4], 70-72
  gauge invariance, 19
  GUTs and supersymmetric GUTs, 50-53
  Higgs bosons, 210, 303-319
    supersymmetric models, 47
  light stop scenario, 81
  linear combinations of $\epsilon1, \epsilon2, \epsilon3$, 222-227
  luminosity distribution, 307-309
  mass problem, 49, 53-56
  model independent analysis of data, 221-235
  non-SUSY, 80
    proton lifetime, 81
  precision tests, LEP, 209-211
  predictions ZSHAPE, 216
  problems, 49-50
  string model-building, 63-68
  superstring 'prediction' for m_t, 72-73
  supersymmetric SU(5) GUT, 58-63
  symmetry breaking, 354
  t quark, 53
  WW-scattering, 20
  Yukawa couplings, 54, 57
  see also GUT

Stanford Linear Accelerator Center, 312
Stefan's Law, 271
String theory
  2-D world sheet, loop expansion, 197
  Calibi-Yau, 162
  compactification, space-time, 162
  conformal and modular invariance, 66
  d=2 sigma model, 196
  discrete or continuous spectrum, 195
  duality symmetry, 202
  elementary, 176-178
  and fivebranes, 169-189
    discussion I, 192-194
    discussion II, 204-207
  heterotic, left-moving and right-moving sectors, 67
  loop diagrams, 166
  mass spectra, and five-branes, 200
  model-building, Standard Model, 63-68
  Noether charge, 178, 180
  quantization, 194
  scales, 165
  solitonic, 178-181, 188
  String Unification scale, 88
  Theory of Everything, 50, 64-68
  see also Superstring theory
String/five-brane duality, 182-185
Sudakov form factor exponent, 29
Sun, magnetic field, 290-291
Supercolliders
  acronym defined, 391
  calorimeter, 394
  data analysis, 391
  new detectors, 387
    discussion, 388-397
  safety, 397
  time constraints, 390
  see also ELN; SSC project
Superconducting Super Collider see SSC project
Supermembranes, 170
Supernova neutrinos, neutrino physics, 271-273, 302
Superstring theory
  D=10, 87
  type IIA and B, 199-200
  'prediction' for m_t, Standard Model, 72-73
  see also String theory
Superstring vacua, 131-157
  deformation of complex structure and target-space duality symmetry, 137-145
  discussion, 162-167
  moduli dependence of gauge couplings, 152-161
  Picard-Fuchs equations, 131, 133, 145-151
  special geometry, 132-136
  target-space duality anomaly cancellation, 152-161

Supersymmetric GUT
  breaking
    gauge group, 86
    scale, 79-83
  CDF, mass limits, 344-349
  Higgs mixing parameters, 62
Superthree-brane, brane-scan, 189

Target-space duality anomaly
    cancellation, superstring
    vacua, 152-161
Technicolour models, 248
Tevatron
  B production, 246
  layout, 340-343
  m_t effects, 243
Thallium, Standard Model
    discrepancy, 81
Theories, cube of, 21-22
Theory of Everything, 17, 21, 22
  string theory, 50, 64-68
Thomson limit, 59
't-Hooft tunnelling, 100-101
  B + L violation, 120
  damping factor, 103
  estimate, 117
't-Hooft-Polyakov monopole, 203
Three-branes, 201
TOE see Theory of Everything
Toroidal compactifications, moduli
    space, 142
Transition magnetic moment, 280
Tritium, decay experiments, 254-257
Truncated QCD cascades, massive
    hadrons, 31

Unification, SU(3) x SU(2) x U(1),
    49, 50
Unitarity, and electroweak theory,
    119
Universe
  baryon number, 85
  closed, 8, 14, 15, 16
  massless, 15
  total energy, 15

ø-Vacuum, electroweak theory, 94-
    104
Vacuum, energy density, 8
Valley approach, 127
  B + L violations, 121
  KRT formula, 108-114
Valley method, B + L violations,
    128
Velocity, defined, 2, 3
Vortex formation, Calibi-Yau, 162-
    163

W bosons
  detection, 323-331
  mass, experimental results,
      Fermilab collider, 330-331
W field, 100, 101
  bond function, 103
  Fourier transform, 102
  polarization sum, 102
W mass, 111, 112

  bond function, 117-118
  multi-W exchange, 112
Weak interactions (coupling
    theories), 89-121
  B + L current, 97-99
  B + L violation, 89, 102
  discussion I, 123-125
  discussion II, 125-129
  phase space growth, 90-94
  semiclassical approximations,
      90-94
  standard electroweak theory, 96
Weil-Petersson metric, 137
Weinberg-Salam mode, MSW solution,
    281, 284
Weyl invariance, 201
Winding number solutions, 124
World sheet
  bosons and fermions, 203
  Euler number, 197
World as a whole, Bronshtein
    theory, 21
World-line, world-sheet, 64, 65-66
WW
  fusion, 303
  initial state branching, 116
WW-scattering, 20

Yang-Mills action
  Chern-Simons terms, 184-185
  D=10 supergravity, 179, 180
  internal symmetry, 204
  quadratic, 169, 183-184, 185, 188
  quartic, 169, 183-184, 186-188
  s-model sources, 185-188
  supersymmetric SU(N) model, 200
Yang-Mills equation of motion
  instantons, 98
  solution, 185, 187
Yang-Mills theory (N=1), 154
Yukawa couplings
  $a^{(3)}$ coefficient, 151
  automorphic tensor, 164
  Clebsch-Gordon coefficients, 164
  constant, 72, 73, 80
  quark, 18
  Standard Model, 54, 57
  tensor, Calibi-Yau, 165
Yukawa sector, CP violation, 86

421